Children with Specific Language Impairment

Language, Speech, and Communication

Children with Specific Language Impairment

Laurence B. Leonard

A Bradford Book
The MIT Press
Cambridge, Massachusetts
London, England

First MIT Press paperback edition, 2000

© 1998 Massachusetts Institute of Technology

This book was set in Palatino on the Monotype "Prism Plus" PostScript Imagesetter by Asco Typesetters, Hong Kong and was printed and bound in the United States of America.

Library of Congress Cataloging-in-Publication Data

Leonard, Laurence B.
  Children with specific language impairment / Laurence B. Leonard.
     p.   cm. — (Language, speech, and communication)
  "A Bradford book."
  Includes bibliographical references and index.
  ISBN 0-262-12206-5 (hardcover : alk. paper), 0-262-62136-3 (pb)
  1. Language disorders in children. I. Title. II. Series.
RJ496.L35L46   1997
618.92'855—dc21   96-37594
                                                              CIP

Contents

Preface

During at least two periods in the 1990s, the media devoted considerable print space and air time to a group of individuals with serious but seemingly isolated deficits in language ability. The common thread running through the cases described in the media might not have been obvious; the precise focus of the news accounts varied from one story to the next, and the affected individuals were not given the same clinical label. But to professionals working in this area, it was if the 1990s had been declared the decade of "specific language impairment." This term refers to a significant impairment in spoken language ability when there is no obvious accompanying condition such as mental retardation, neurological damage, or hearing impairment. Furthermore, the language problems in such cases were present from the beginning; they did not materialize at age two or three as the sequelae of some illness or psychological trauma. All of the individuals in these news accounts reportedly fit this pattern.

This volume is designed to provide the kind of information about specific language impairment that news accounts probably can't provide. In these pages, the history of the study of specific language impairment is covered, and a great many facts (at least as we know them) about the linguistic and nonlinguistic characteristics of this type of disorder are presented. Clinical issues are discussed, and the hypotheses that have been advanced to explain the nature of specific language impairment are reviewed and evaluated. It was not possible to include everything, but the coverage is surely representative, and then some—from cerebral morphology to grammatical morphology.

The emphasis is on children. Language deficits are obvious during childhood, and children with specific language impairment have received the greatest attention. However, adults are included as well. Many of the adults discussed in this volume were diagnosed as exhibiting specific language impairment as youngsters; others are parents of children with specific language impairment whose own language abilities became the object of study.

The answers to the big questions—What is the cause of specific language impairment? How can it be prevented or remedied?—are not to be found here, unfortunately. However, in reviewing much of the relevant evidence and highlighting potentially important directions for future inquiry, I hope that I have created a volume that moves the study of specific language impairment forward.

I have tried to organize the volume so that it is useful to readers with quite different interests. For most readers, the usual cover-to-cover reading will be the most informative. These will be investigators, practitioners, and graduate students with direct interest in language disorders from a research and/or clinical perspective. However, some readers will have interests that lie primarily in what specific language impairment has to say about language organization and development in general. For them, parts I, V, and VI will be of greatest importance, with parts II, III, and IV serving as a reference when evidence especially pertinent to one theory or another needs to be consulted. Because the chapters of parts II, III, and IV cover a great deal of evidence, each of these chapters concludes with a section that highlights important issues arising from the evidence just reviewed.

I first wrote a preface around 25 years ago, during an overnight train trip from London to Edinburgh. The fact that I didn't have a book-length manuscript to go with it didn't concern me. The rest of the book would come, I figured (though maybe after graduate school); in the meantime, the setting was too inspiring to pass up. I never used that preface, but its main theme holds as true today as when I wrote it: anything that I have accomplished would not have been possible without the support—professional as well as personal—of my wife, Jeanette Leonard. There are probably words that could adequately express my gratitude, but it will be a long time before I'm a skilled enough writer to find them.

I would also like to acknowledge and thank the many graduate students (and the smaller number of postdoctoral fellows) with whom I've had the privilege of working on matters of research. These individuals (a list of whom would be cause for envy) taught me a great deal during our time together, and some of these insights are reflected in subtle if not obvious ways in this volume.

I thank as well my nonstudent collaborators, whose contributions have always made me look good (or at least better). High on this list are my collaborators on studies of languages other than English. Without their expertise, my sense of specific language impairment would be shallow and narrow.

If this book contains a respectable level of argumentation, it is due in part to my culture. The people who study childhood language disorders (including specific language impairment) are the best of critics, quick to take issue with an unsupported or illogical point, but cordial and constructive in conveying the criticism. I feel fortunate to be working in this area. Another stroke of luck was finding myself at Purdue University. It would be hard to find a place more conducive to research and scholarship.

Appreciation is extended to the funding agencies, especially the National Institute on Deafness and Other Communication Disorders (NIDCD), for their support of the research on specific language impairment that is reviewed here. Although this area has always seen some excellent studies, the overall quality of research has risen dramatically since NIDCD arrived on the scene.

Several people had a very tangibile role in the preparation of this volume. Bernard Grela was a master at creating graphs and other figures to specification. His handiwork is seen in all figures but 7.3 and 9.1. For the latter two, I thank Warren Garrett. Erika Gerber did a superb job of moving a very large number of references around a very large manuscript, and Wanda Posto was a major help in getting many separate pieces of text properly placed in a single document.

I thank Amy Brand, senior editor at The MIT Press, for her support and patience during the preparation of this volume, and the anonymous reviewers who provided helpful comments. Dorothy Bishop should be singled out for her insightful comments on portions of an earlier draft. The writing of sections of the book in which I integrate some of my own recent research findings was supported in part by research grant number 5 R01 DC 00-458 from NIDCD.

PART I

Foundations

Chapter 1

Introduction

This book is about children with specific language impairment (SLI). These are children who show a significant limitation in language ability, yet the factors usually accompanying language learning problems—such as hearing impairment, low non-verbal intelligence test scores, and neurological damage—are not evident. This is a real curiosity, especially in light of the many language acquisition papers that begin with a statement to the effect that "all normal children" learn language rapidly and effortlessly. The only thing clearly abnormal about these children is that they don't learn language rapidly and effortlessly.

The prevalence of SLI is about 7%. SLI is more likely to be seen in males than in females, and children with SLI are more likely than other children to have parents and siblings with a history of language learning problems. There are areas of language that are especially difficult for most children with SLI, but the heterogeneity of language profiles in this population is nevertheless considerable. Treatment improves these children's language learning, but the deficits in language do not go away easily. Adolescents and adults who had been diagnosed with SLI as preschoolers often earn lower scores on tests of language ability than same-age peers with no such history. Children with SLI are also at risk for reading disorders when they reach school age.

The fact that SLI adversely affects the lives of children with SLI and their families constitutes the most pressing reason to uncover the mysteries surrounding this type of disorder. Through a better understanding of SLI, more effective methods of assessment, treatment, and prevention might be developed. However, there are also some good theoretical reasons to study SLI. On the face of it, SLI seems to constitute compelling support for certain theories of language development and structure; according to other theories, SLI shouldn't exist. In the pages of this book, we shall be considering all of these points in detail.

The study of children with SLI is not new. Unfortunately, terminology has changed greatly over the years, making it easy to miss some interesting details of the study of these children. For example, one of the early tests of Piaget's theories of the relationship between language and other symbolic abilities was conducted by Inhelder (1963), a frequent collaborator of Piaget's, who tested this relationship with a group of children with SLI. Menyuk's (1964) investigation of the grammatical deficits in children with SLI was one of the first to apply Chomsky's theory of transformational grammar to the study of language development. Bloom's (1967) first paper on the advantages of rich interpretation and the problems with pivot grammar was a critique of an investigation of the two-word utterances used by children with SLI. An abbreviated review of the history of the study of SLI will be provided in this chapter. However, we begin with a brief case study, for a more personal introduction to some of the characteristics of SLI.

One Child with Specific Language Impairment

Here are some brief excerpts from one English-speaking child with SLI, age four years, three months (4;3). The excerpts come from sessions that were part of a cross-linguistic investigation of SLI conducted by Leonard, Bortolini, Caselli, McGregor, and Sabbadini (1992). In (1), the child was being shown sets of sequence pictures and asked to make up a story for each.

(1) *Adult:*   Ok, ready?
     Child: Ready.
     Adult: This is Jim. Tell me a story about Jim.
     Child: Him going fishing. Jim hold...water. And go fish. And [unclear]
     Adult: I didn't hear this [last] one.
     Child: I don't know.
     Adult: Ok. How many more do you think we have?
     Child: I don't know.
     Adult: Ok, ready?
     Child: Ready.
     Adult: This is Kathy. Tell me a story.
     Child: Kathy brush teeth. Her eat. And her get clothes on.
     Adult: She must be getting ready to go to school, huh?

The child's productions in (2) and (3) were in response to pictures designed to create obligatory contexts for definite and indefinite articles, and regular and irregular past forms, respectively. Responses typical of normally developing children of the same age and obtained from the same study are provided in parentheses.

(2) a. *Adult:*   This is a woman and this is ....
         Child: Boy. (A man)
         Adult: The woman is washing dishes and....
         Child: Boy is painting. (The man is painting)
     b. *Adult:*   This is a baby and this is ....
         Child: A dog. (A dog)
         Adult: The baby is drinking milk and....
         Child: Dog chew bone. (The dog is chewing a bone)
     c. *Adult:*   This is a girl and this is ....
         Child: Boy. (A boy)
         Adult: The girl is throwing a ball and....
         Child: The boy hitting ball (The boy is hitting the ball)

(3) a. *Adult:*   He's catching the steer. What happened?
         Child: He caught him. (He caught him/it)
     b. *Adult:*   He's zipping his jacket. What did he do?
         Child: Zip jacket. (Zipped his jacket)
     c. *Adult:*   She's combing her hair. What did she do?
         Child: Comb hair. (Combed her hair)

d. *Adult:*   He's drawing a picture. What did he do?

    Child: Drawed picture. (Drew a picture; drawed a picture)

The child's utterances were quite short on average; in fact, in spontaneous speech they were characteristic of those produced by children more than a year younger. Omissions of grammatical suffixes and function words (grammatical morphemes) were rampant, even exceeding the degree of omission expected in short utterances. Yet, on occasion, errors of creativity were also seen, as in her production of *drawed* instead of *drew* in (3), an overregularization that we also saw in many of the normally developing children serving as age controls. As a conversationalist, this little girl was not assertive. Although she enjoyed interacting with others, she initiated verbal exchanges relatively infrequently. Communicative attempts were often abandoned if they were not understood on the first try.

Not obvious from the excerpts is the fact that her vocabulary was rather limited for her age, and she sometimes struggled to find words that she had occasionally produced in the past. In addition, her phonological abilities were below age-level expectations, though her use of final consonants was sufficient to rule out a purely phonological explanation for her omissions of grammatical inflections such as *-ed*. Formal test results obtained during the same period were consistent with the interpretation that this child's language abilities were quite limited for her age. Scores of language comprehension were higher than language production scores, but even some of these were below age level.

At the time of writing, this child was 11 years of age. She experienced academic difficulties in school, especially in reading. Her father had received professional attention for language problems as a youngster. He had some difficulties in school, though he earned his high school diploma. The child's mother had completed two years of college, and had never experienced language learning problems.

This child is not the most dramatic case that could be offered; some of these will be discussed later. However, she and her family illustrate some of the hallmarks of SLI that will be discussed more fully in this book. We shall return to this child later in this chapter.

A Brief Review of a Not-So-Brief History

On occasion, the study of SLI attracts a great deal of attention from the scientific community at large thanks to the appearance of an unusually important or interesting finding. Although the excitement surrounding such developments injects a freshness into the study of SLI, work in this area actually dates back to the first half of the nineteenth century. This section will provide a brief summary of the study of SLI. Additional details of work conducted in the twentieth century can be found in reviews by Myklebust (1971), Leonard (1979), Aram and Nation (1982), and Johnston (1988). P. Weiner (1986) provides an especially useful review of work done in the nineteenth century.

In 1822, Gall published a description of children who had clear problems in language but did not display the characteristics of other known disorders. One comment was the following (from the English translation of 1835):

> There are many children... who do not speak to the same degree as other children although they understand well or are far from being idiotic. In these

cases the trouble lies not in the vocal organs, as the ignorant sometimes insist, and still less in the apathetic state of the subject. Such children, on the contrary, show great physical vivacity. They not only skip about but pass from one idea to another with great rapidity. If one holds them and pronounces a word in their ear, they repeat it distinctly.   (p. 24)

By today's standards, Gall's description is a bit oversimplified (though not bad, considering that one of the treatments of the day was "curative tonic").

Gall's publication was followed by a smattering of case reports over the remaining years of the nineteenth century. These case studies, published in English, French, and, most often, German, were written by physicians. The authors emphasized the apparently normal nonverbal intelligence, seemingly good comprehension, and extremely limited speech output of these children. These studies included the reports of Wilde (1853), Benedikt (1865), Broadbent (1872), Waldenburg (1873), Clarus (1874), Bastian (1880), Uchermann (1891), Wyllie (1894), Lavrand (1897), and Moyer (1898). During this period, Väisse (1866) introduced the term "congenital aphasia," and applied it to these children. However, the German literature used the term (translated into English) "hearing mutism" (Coën, 1886). This term seemed apt in large part because the authors discussed children whose language output was severely restricted. Children producing utterances more than one word in length (even if the children were older) were considered to have production limitations attributable to severe phonological problems (Gutzmann, 1894; Treitel, 1893). The apparent reluctance to include grammatical difficulties in this clinical category continued until well into the twentieth century. For example, even after drawing parallels between agrammatism in adults with aphasia and the grammatical difficulties of an 11-year-old, Fröschels (1918) preferred the term "delayed speech development" as a descriptor of the child's deficit.

Liebmann (1898) may have been the first to discuss subtypes of children, but his descriptions covered only those children with severe output limitations. One subtype constituted a deficit that was motoric in nature. A second subtype concerned children who could succeed only in comprehending single words. The third subtype involved children who completely lacked the ability to comprehend language. This last subtype began to be called "congenital word deafness" (McCall, 1911); subsequently, the terms "congenital auditory imperception" (Worster-Drought & Allen, 1929) and "congenital verbal auditory agnosia" (Karlin, 1954) also were used.

Because neurological damage was not evident in these children, some authors proposed that the nature of the problem was functional (Coën, 1886). Limitations in attention and memory were hypothesized to play an important role (Treitel, 1893).

In the English and French literature of the early 1900s, the term "congenital aphasia" came into greater use. This term was applied more broadly than "hearing mutism," extending to children whose language output had progressed well beyond single-word utterances. Deficits in comprehension as well as production were included in this category. At the same time, it was recognized that earlier authors' efforts in ruling out deficits in hearing, oral motor ability, and nonverbal intelligence might have been too casual. Town (1911) offered the following definition:

Aphasia then is an inability, total or partial, to understand or to use language in any one or all of its forms, such inability being independent of any other mental capacity or of deformity or disease affecting the organs of articulation. (p. 167)

Additional advances were made concerning the bases of these difficulties. Ewing (1930) observed that the pattern of change in these children was consistent with a significant neurodevelopmental delay. During this same period, evidence for a hereditary basis for the expressive form of congenital aphasia emerged from a twin study by Ley (1929).

As the twentieth century progressed, the term "aphasia" was accompanied by new modifiers, though the basic meaning was unchanged. The highly respected child development team of Gesell and Amatruda (1947) used the term "infantile aphasia," as did several other authors (e.g., Van Gelder, Kennedy, & Lagauite, 1952). The term "developmental aphasia," first used in the second decade of the century (e.g., Kerr, 1917), became the preferred term by the 1950s. Influential studies by Morley, Court, Miller, and Garside (1955), T. T. S. Ingram and Reid (1956), Benton (1964), and Eisenson (1968) employed this term. At this time, authors began to use the terms "expressive developmental aphasia" and "receptive-expressive developmental aphasia" (sometimes "receptive developmental aphasia" was used instead of the latter) to distinguish between deficits centering on language production and those involving comprehension as well as production. An inspection of this literature shows clearly that children with grammatical problems were included in the category of developmental aphasia, often taking center stage.

Beginning in the 1960s, "dysphasia" began to appear along with "aphasia" (deAjuriaguerra, Jaeggi, Guignard, Kocher, Maquard, Roth, & Schmid, 1965; Inhelder, 1963; P. Weiner, 1969). By the 1980s, authors choosing one of these two terms were more likely to use "developmental dysphasia" (e.g., Chiat & Hirson, 1987; Clahsen, 1989; Wyke, 1978). Technical accuracy might have been one reason for this gradual change; the prefix a- implies the absence of language, whereas dys- implies only problems with language (e.g., Eisenson, 1972).

Although "developmental aphasia" gradually gave way to "developmental dysphasia," the most salient trend in the literature of the second half of the century has been a gradual shift away from both terms. There seem to be two reasons for this development. First, "aphasia" and "dysphasia" have a neurological connotation, due, of course, to their use as labels for language disruption caused by discrete brain damage such as that resulting from cerebral vascular accident (stroke). As late as the early 1960s, children with postnatal brain injuries were considered part of the category of developmental dysphasia (see Johnston, 1988). This is no longer the case. Children suffering such neurological insults are now described as "children with acquired aphasia" or "children with focal brain injury." The latter term is preferred when the damage occurs prior to the acquisition of language. When language disruption occurs with the onset of a convulsive disorder, the term "Landau-Kleffner syndrome" is used, after the researchers who first described the problem (Landau & Kleffner, 1957).

The second reason for change away from "aphasia" and "dysphasia" was emphasis. This era was one of intense description of the linguistic characteristics of these children's speech. Accordingly, most of the terms applied to these children contained the word "language," along with a descriptor that conveyed impairment. Unfortunately, for most of this period there was no consistency in the terms employed. To the uninitiated, the dizzying array of labels probably camouflaged the fact that the same general types of children participated in these studies. Since the 1960s, the following terms have appeared in the literature: "infantile speech" (Menyuk, 1964), "aphasoid" (Lowe & Campbell, 1965), "delayed speech" (Lovell, Hoyle, & Siddall, 1968), "deviant language" (Leonard, 1972), "language disorder" (Rees, 1973), "delayed language" (P. Weiner, 1974), "developmental language disorder" (Aram & Nation, 1975), "developmental language impairment" (Wolfus, Moscovitch, & Kinsbourne, 1980), "specific language deficit" (Stark & Tallal, 1981), and "language impairment" (Johnston & Ramstad, 1983).

Investigators studying the academic or preacademic skills of these children sometimes employed the term "language/learning-disabled" or "language/learning-impaired" (Tallal, Ross, & Curtiss, 1989b), presumably as a reminder that these children's spoken language deficits do not exclude them from being considered learning disabled (indeed, these deficits might be a central component of the learning disability). The term "specific language impairment" (e.g., Leonard, 1981), along with its abbreviation SLI (Fey & Leonard, 1983), is the most widely adopted term at present, especially in the research literature.

The clinical and educational world is also replete with alternative terms for SLI. The *Diagnostic and Statistical Manual of Mental Disorders, Fourth Edition* (DSM-IV) (American Psychiatric Association, 1994) uses the term "developmental language disorder," with the subtypes of "expressive" and "receptive and expressive." *The International Classification of Diseases, Ninth Revision, Clinical Modification* (ICD-9-CM) (United States Department of Health and Human Services, 1995) employs the terms "developmental aphasia" and "word deafness" as well as "developmental language disorder." Finally, some professionals in the clinical sphere have taken to heart guidelines from the World Health Organization (Wood, 1980) that make a distinction among "impairment," "disabilities," and "handicap." The first of these terms is used to refer to the abnormality itself; the second, to the functional consequences of the abnormality (e.g., the child can't communicate with peers); the third, to the social consequences of the abnormality (e.g., social isolation).

In summary, the study of children with SLI had its beginnings more than 150 years ago. The earliest emphasis was on children with severe output limitations. Gradually, the focus widened to include children who produced multi-word utterances; at this point, the significant grammatical deficits of children with SLI began to receive attention. Efforts were made to distinguish deficits of production from those involving both comprehension and production, though both types of problems were included in the category of SLI. Children with postnatal brain injury were eventually considered to fall outside of the category of SLI. The criteria for SLI were also tightened to ensure that children with demonstrable deficits in nonverbal intelligence, hearing, or oral motor skills were not included in this clinical category. The diagnostic boundaries resulting from these historical developments do not lead to a tightly homogeneous group of children. However, the characteristics they have in common are considerable, and often have been overlooked as a result of the excessive number of labels that have been employed.

Why Study SLI?

There are several important reasons to study SLI. The first, of course, pertains to clinical and educational concerns. With greater understanding of this type of disorder, more effective ways of assisting children with SLI and their families might be uncovered. If there were no other benefits to be gained, this possibility would be sufficient justification to pursue this work.

But there are other benefits. For example, children with SLI might contribute to a better understanding of language problems in other populations by providing a type of baseline of language impairment. Consider the case of deaf infants raised by deaf parents who communicate in the home through sign language. Presumably, these infants will develop a sign language in the customary fashion and at the customary rate. And if they don't? The existence of SLI, that is, a language impairment without any obvious causal factor, suggests that in principle some small percentage of native-signing deaf children, say 7%, will exhibit a significant language learning deficit. This deficit might have nothing to do with the fact of a signing environment.

The existence of SLI also has implications for children with both language deficits and other serious developmental problems. In the interest of parsimony, a single factor that can explain all of these deficits would be preferable. However, if language impairments can be found without these other developmental problems (as in SLI), they also, in principle, can co-occur with these problems yet have a separate source.

The study of SLI can also contribute to theories of language organization and development. If it is really the case that language is the sole problem area, or that some area within the domain of language such as morphosyntax is uniquely affected, then modular views of language would receive a significant boost. Data from children with SLI might help in another way. A theory might hold that a certain aspect of grammar is a prerequisite for other key attainments. Deficits in this area in children with SLI might be studied to determine if they are sufficient to steer the remainder of the children's grammatical development off course.

Theories that invoke maturational principles or critical periods are also obvious ones to be applied to children with SLI. One of the hallmarks of SLI is a slow rate of language development. It is natural to inquire, then, whether there are certain later-developing principles or parametric variations that remain inaccessible to these children.

It is not a stretch to argue that theories of language learning are obligated to consider SLI. Children with SLI pose a learnability problem, yet there doesn't seem to be a good basis for excluding them from consideration. What is most conspicuously atypical about these children is that their language development doesn't follow the usual blueprint. Thus, excluding them because they are not normal would be an exercise in circular reasoning, because it is their language development that makes them abnormal.

Let's look at the learnability problem by considering some of the empirical conditions that Pinker (1979) felt any satisfactory learnability theory should meet. Most of these conditions are now subsumed under Pinker's (1984) "learnability" condition, but the original distinctions serve as a useful framework for discussing this issue. The "time" condition requires a theory to contain mechanisms that account for the fact that children acquire language as rapidly as they do. However, children with SLI are slow in acquiring words and sentences. An issue raised by this fact is how the language learning mechanisms permit most children to learn language relatively quickly while they cause others to proceed very slowly.

The "developmental" condition requires a theory to predict the intermediate stages of development. If one confines analysis to individual features of language, the language of children with SLI resembles that of younger, normally developing children. However, across features of language, children with SLI do not have the same profiles as normally developing children at any point in time. Some features of language are weaker than others. The problem is how to account for this profile without proposing mechanisms incapable of yielding the usual relationships among the components of children's language.

The "learnability" condition—how the child actually succeeds in learning the language—is the core theoretical problem in language acquisition, according to Pinker (1984). Unfortunately, some children with SLI do not succeed. As adults, only a minority still exhibit obvious morphosyntactic as well as phonological errors. However, careful probing of adults who at one time committed such errors reveals that their linguistic knowledge continues to be quite restricted and certainly is not on a par with that of typical adults. These individuals, then, pose difficulties for the learnability condition, which hinges on the assumption that the child will reach the adult level.

We will revisit these ideas in later pages. To evaluate them properly, more information about these children is needed.

The Criteria for SLI

One of the banes of professionals who diagnose SLI is that it is a diagnosis based as much on exclusion as on inclusion. The inclusionary criterion of a significant deficit in language ability is the least problematic. A diagnosis of a language problem can usually be made with confidence. The trick is to distinguish SLI from other disabling conditions of which language problems are a part. Fortunately, the means available to rule out these conditions are more extensive and reliable in present-day clinical work than in the early days of SLI research.

Table 1.1 provides a summary of the areas that are usually considered before the term "specific language impairment" can apply.

The Language Deficit

Test Scores and Other Standardized Measures of Language   Most studies of children with SLI employ standardized tests of language ability. It is generally recognized that

Table 1.1
Criteria for SLI

Factor	Criterion
Language ability	Language test scores of −1.25 standard deviations or lower; at risk for social devalue
Nonverbal IQ	Performance IQ of 85 or higher
Hearing	Pass screening at conventional levels
Otitis media with effusion	No recent episodes
Neurological dysfunction	No evidence of seizure disorders, cerebral palsy, brain lesions; not under medication for control of seizures
Oral structure	No structural anomalies
Oral motor function	Pass screening using developmentally appropriate items
Physical and social interactions	No symptoms of impaired reciprocal social interaction or restriction of activities

even the best tests do not do justice to these children's language problems (e.g., Muma, 1986). Children with SLI show limitations in a wide range of language abilities, but areas of special weakness can be observed. The tests available usually do not cover the diversity of language details that can be problematic for these children, and where these details are covered, their representation may not adequately reflect their relative importance in the child's day-to-day functioning. As we shall see in chapter 3, standardized test scores serve only as the starting point. The work of actually describing and explaining these children's language functioning must then begin. In large part because this more detailed analysis is to be conducted, the standardized language tests used as inclusionary criteria can be quite broad in scope. Here we focus on two general standardized measures that are often used in this regard, the comprehensive language test score and the comprehension-production dichotomy.

Many studies of SLI require that the children included as participants earn low scores on a comprehensive language test. Such tests cover two or three areas of language (e.g., the lexicon, morphosyntax, phonology) in comprehension as well as in production. The Test of Language Development—Primary:2 (TOLD-P:2) (Newcomer & Hammill, 1991) is one such test. This test, standardized on children age 4;0 through 7;11, has been used as one of the selection criteria in several recent studies of SLI (e.g., Leonard et al., 1992; Tomblin, 1966b).

The TOLD-P:2 has three subtests that can be construed as grammatical in nature. Two of these involve production (sentence imitation, sentence completion), and one involves comprehension (pointing to the picture on a page that corresponds to a spoken sentence). Two other subtests tap lexical skills, one through production (in which the child says the meaning of each word provided by the examiner) and one through comprehension (pointing to the picture serving as the referent for the word produced by the examiner).

Finally, the TOLD-P:2 includes a phonological production subtest (in which the child's pronunciations during naming responses are scored) and a subtest assessing the child's ability to distinguish between phonologically similar words. Standard scores can be computed for each subtest, each area of language tested (vocabulary, grammar, phonology), and each modality (comprehension, production). And, of course, an overall or composite standard score can be calculated.

A composite score of 81 or lower (corresponding to at least 1.25 standard deviations [SDs] below the mean) has been employed as a criterion for SLI in some investigations. Speech-language pathologists have shown good agreement on the presence of a language disorder when composite scores are at this level or below (Records & Tomblin, 1994). The use of a composite score means that considerable variation is permitted in the subtest scores. This is illustrated in figures 1.1–1.3. Each of the composite scores shown in these figures is sufficiently low; however, the test profile in figure 1.3 reveals that standard scores for the comprehension modality and standard scores in the lexical area fall within 1 SD of the mean. Thus, children can meet the criterion if they are at age level on select areas or modalities of language, provided they are seriously deficient in others to bring the composite down to at least 1.25 SDs below the mean.

Some investigators have held out for a criterion that requires at least a mild deficit in comprehension as well as a deficit in production. The studies conducted by Stark and Tallal (see Stark & Tallal, 1988) are perhaps the best-known of the investigations employing such a criterion. The comprehension and production composites were

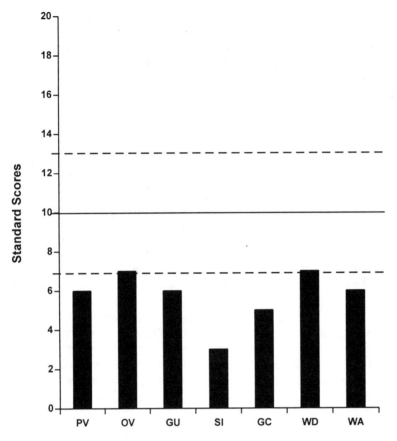

Figure 1.1
Profile of a five-year-old child with SLI with a composite score of 70 on the Test of Language Development—Primary:2. Five of the seven subtest scores were more than 1 standard deviation below the mean. PV, Picture Vocabulary; OV, Oral Vocabulary; GU, Grammatic Understanding; SI, Sentence Imitation; GC, Grammatic Completion; WD, Word Discrimination; WA, Word Articulation.

computed in terms of language ages instead of standard scores. (The latter are preferable—see Lahey, 1990; McCauley & Swisher, 1984.) A comprehension language age at least six months below expected levels was required along with both a production age and a composite (comprehension and production) age of at least one year below expectations. In later work by Tallal and her colleagues (see Tallal, Curtiss, & Kaplan, 1988), the criterion of separate deficits in comprehension was not employed, though this factor was monitored in case it had a bearing on the findings.

In the laboratories of Rice and her colleagues (e.g., Rice & Oetting, 1993), a comprehension as well as a production deficit is required, though the former is defined as a low standard score (at least 1 SD below the mean) on a separate test of vocabulary comprehension, the Peabody Picture Vocabulary Test—Revised (PPVT-R) (Dunn & Dunn, 1981).

Tests of language are not the only standardized measures of language ability that are employed in identifying children with SLI. Some measures are derived from samples of the child's spontaneous speech. The most widely adopted measure of this

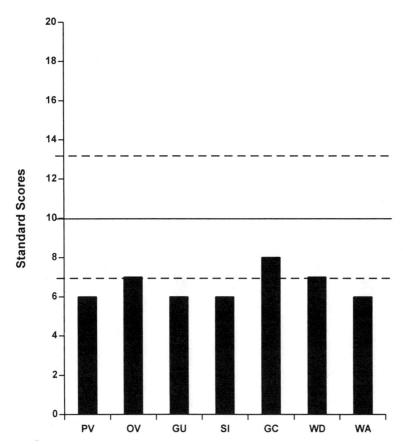

Figure 1.2
Profile of a five-year-old child with SLI with a composite score of 76 on the Test of Language Development—Primary:2. Four of the seven subtest scores were more than 1 standard deviation below the mean. PV, Picture Vocabulary; OV, Oral Vocabulary; GU, Grammatic Understanding; SI, Sentence Imitation; GC, Grammatic Completion; WD, Word Discrimination; WA, Word Articulation.

type is mean length of utterance (MLU). When MLUs are computed in terms of morphemes (e.g., Rice & Oetting, 1993), reference is usually made to the normative data of Miller and Chapman (1981). MLU computations using words rather than morphemes (e.g., Leonard et al., 1992) are compared with the normative data of Templin (1957). It would be rare to find MLU used as the sole language measure to identify children with SLI. Typically, it is used in conjunction with standardized tests. However, because spontaneous speech samples provide a different perspective from which to consider a child's language abilities, measures based on them can be of considerable service. In fact, Dunn, Flax, Sliwinski, and Aram (1996) found that measures based on children's spontaneous speech samples—including MLU—agreed more with clinical diagnoses of SLI than did results from formal tests.

An important point that must be raised about the criteria discussed thus far is that children with phonological disorders are included in the category of SLI only if they also perform poorly on other measures of language. By "phonological disorder," we refer to a sound system that is underdeveloped or otherwise deviates from that of the

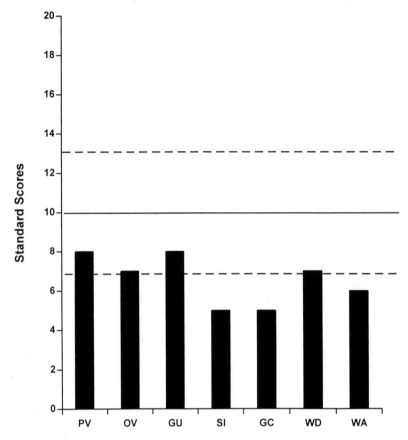

Figure 1.3
Profile of a five-year-old child with SLI with a composite score of 76 on the Test of Language Development—Primary:2. Although the composite score fell more than 1.25 standard deviations below the mean, the subtests dealing with comprehension (PV, GU, WD) and lexical abilities (PV, OV) were within 1 standard deviation of the mean. PV, Picture Vocabulary; OV, Oral Vocabulary; GU, Grammatic Understanding; SI, Sentence Imitation; GC, Grammatic Completion; WD, Word Discrimination; WA, Word Articulation.

ambient language. Children whose nonstandard pattern shows all of the contrasts of the target language but differs from it in the phonetic values for individual sounds (e.g., a lateral production for English /s/ and /z/) are not considered to exhibit a phonological disorder.

The exclusion of children with phonological problems only is peculiar in one sense. Phonology is a component of language, and therefore, if a child has a deficit in this area, the child, by definition, has a language disorder. This exclusion might be a holdover from the era when developmental dysphasia was sharply distinguished (without much empirical basis) from problems of which phonology was an obvious part. In any event, the exclusion of certain children with phonological problems does not mean that phonology is ignored. Because phonological abilities are so often limited in children otherwise meeting the criteria for SLI, this area of language receives a good deal of attention.

Social Devaluation and Other Considerations   We would expect children with SLI to exhibit low scores on standardized measures of language ability. However, such scores can be only part of the story. Imagine testing large groups of children on physical agility (or musical ability, or artistic ability; see Leonard, 1987, 1991). This is one of several abilities—like language—that can be elevated to the status of "intelligences," based on neurological and other evidence (see Gardner, 1983). That is, there is evidence that through brain damage, these areas can be selectively disrupted. And there are cases of persons with global deficits whose abilities in these particular areas are spared. So assume we test 1,000 first graders on their ability to perform tasks such as walking along a 15-foot-long board and running through a series of plastic hoops 12 inches in diameter. Performance on such tasks can be measured in both accuracy (making it across the board without stepping off, reaching the finish line without missing any of the hoops) and elapsed time to complete the tasks. For each measure, we can generate means and standard deviations, and, perhaps with some modifications of items, we will have a battery of tasks whose results from the first graders produce a bell-shaped curve, reflecting a normal distribution.

Now we administer our tasks to another group of first graders. How should we regard children who perform, say, 2 SDs below the mean? Should we treat them as motor-impaired? Only if we can find additional grounds for supporting such a view. Without such evidence, a more reasonable conclusion is that the children with low scores are simply not very coordinated.

Yet, the same type of conclusion seems odd when applied to the parallel case of low scores on a language test. An inspection of the manuals for most standardized tests of language ability reveals that the children making up the standardization sample were presumed to be developing normally; children with known handicapping conditions were not included. So shouldn't we expect a certain number of children to perform way below average, by definition? Certainly.

But this doesn't mean that we equate comparably poor scores in language and physical agility. Society places greater value on language ability, or, more accurately, we devalue poor language ability more than poor physical ability (Tomblin, 1983). Significant limitations in language ability do not augur well for a child's educational progress or future economic success. Although options remain, they will be limited to those not requiring academic accomplishment and verbal skill. Poor physical or musical ability, on the other hand, will limit the child's quality of life, but the restrictions on education and future professional endeavors will not be as great.

There is reason to believe that problems with language also limit a child's social well-being. Gertner, Rice, and Hadley (1994) studied peer preferences in a preschool for children with SLI and normally developing children. The children were asked to identify peers they enjoyed playing with. The normally developing children were selected more frequently than the children with SLI. In addition, the children with SLI entered into fewer mutual selections, in which they identified particular children as preferred playmates and the identified children in turn selected them. Fujiki, Brinton, and Todd (1996) found that a group of school-age children with SLI reported interacting with fewer peers in social activities than did same-age peers. Of course, factors other than language ability might have contributed to these findings. However, language is certainly the prime suspect, for limited language skills constituted the characteristic that the children with SLI shared. For this reason, the conclusion of Gertner et al. seems reasonable: children with SLI may be at risk for low social status among peers. In chapter 3, we will see a great deal of indirect evidence pointing to the same conclusion. It will be seen that children with SLI have a great deal of difficulty conversing with peers, and this fact is not lost on their interlocutors.

Peers do not seem to be alone in judging children with SLI as less adept in social skills. In the study of Fujiki, Brinton, and Todd (1996), teachers rated children with SLI as less socially skilled than normally developing children of the same age. Rice, Alexander, and Hadley (1993) asked adults to listen to audiotaped samples of the verbal interactions of children with SLI and children developing normally. On a variety of variables, including social maturity, leadership abilities, and popularity, the children with SLI were rated lower than the normally developing children. Because audiotapes were used, it might be argued that the children with SLI were put in the worst possible light; presumably, many of their stronger assets would be more apparent if visual information were available to the adult judges. However, at best this might have reduced the gap between the children with SLI and the normally developing children; the latter group, too, probably had strong attributes that could be detected by visual means.

We must be careful not to equate the judgments of others with the personal adjustment of the individuals in question. Records, Tomblin, and Freese (1992) administered a quality-of-life assessment instrument to young adults with a history of SLI and to age controls. Although the adults with a history of SLI continued to show low scores on measures of language ability, had completed fewer years of education, and received a lower rate of pay, their responses to the questions suggested the same generally positive attitude about their lives as was expressed by the controls. It is true that they conveyed feeling somewhat less in control of their lives, a finding that could have been related to their lower educational achievement, language ability, or type of employment obtained. The school-age children with SLI studied by Fujiki, Brinton, and Todd (1996) reported levels of satisfaction in their social relationships that were considerably lower than those reported by age controls.

It is rarely practical to obtain a direct measure of social devalue in the identification of children with SLI. More often, the type of evidence obtained is quite indirect. This typically takes the form of (1) parents and teachers referring the child to a speech-language pathologist or other professional with concerns about a child's functioning; and (2) the professional drawing the same conclusion, based on methods that go beyond standardized measures, such as observing and interacting with the child.

Nonverbal Intelligence

One of the most fundamental criteria in the diagnosis of SLI is a score on a nonverbal intelligence measure that is within age-appropriate levels. Usually this is defined as a nonverbal IQ of at least 85, or less than 1 SD below the mean. Nonverbal IQs can be more than 1 SD above the mean.

It is common to find children included in studies of SLI whose nonverbal IQs are around 90, and whose standard scores on language tests are in the high 70s. Some investigators have worried that such cases do not show a sufficient gap between nonverbal IQ and language score to place the child in the SLI category with confidence. They argue that a clearer discrepancy should exist between these two types of scores. In standard score terms, this discrepancy is usually set at greater than 15, the SD for each of these tests. However, in other projects, a discrepancy between mental age and language age of at least one year is used (Stark & Tallal, 1988; Tallal, Curtiss, & Kaplan, 1988).

There are some good reasons to be wary of such discrepancy criteria for defining SLI. For example, as pointed out by Lahey (1990), both nonverbal IQ tests and language tests have measurement error. Consequently, a definition of SLI that relies on differences between the scores on these tests runs the risk of compounding this error. We also have no clear evidence that children who do and do not show discrepancies are different in any meaningful way. For example, both types of children seem to show similar degrees of learning when presented with the same forms of treatment (Cole, Dale, & Mills, 1990; Fey, Long, & Cleave, 1994).

Use of a discrepancy criterion also introduces a social policy question that is best avoided. A significant percentage of the population with normal nonverbal IQs can be expected to show language scores at least 15 points lower than nonverbal IQ (Snyder, 1982). In fact, Aram, Hack, Hawkins, Weissman, and Borawski-Clark (1991) observed that over 18% of a group of typically developing children could be considered to have met a discrepancy criterion. There are children with above-average nonverbal IQs (e.g., 125) whose language scores fall squarely in the average range (e.g., 105). Instead of providing such children with a balanced educational program or offering them experiences designed to foster their strengths, should we instead concentrate on bolstering their lagging language skills?

Later in this book we review evidence suggesting that on certain nonlinguistic cognitive tasks, children with SLI do not perform as well as same-age peers. Thus, these children's age-appropriate scores on nonverbal tests of intelligence (a requirement for the label of SLI) should not be interpreted as meaning that all nonverbal cognitive operations in these children are above suspicion.

Hearing Sensitivity

SLI implies a language problem that cannot be attributed to impairments in hearing. Most investigators and clinicians feel comfortable assuming that a child's hearing is normal if the child passes a hearing screening. Typically, such a screening requires the child to detect pure tones presented at 20 dB in each ear at the frequencies 500, 1,000, 2,000, and 4,000 Hz. A full audiological assessment of a child's hearing is rarely conducted.

Otitis Media with Effusion

Otitis media with effusion (OME) is a disease in which fluid accumulates in the middle ear as the result of upper respiratory infection with poor functioning of the Eustachian tube. Often OME brings with it a mild and fluctuating conductive hearing loss. Although treatment is successful, OME often recurs. Because each episode of OME can lead to a period of impairment in hearing, and hearing ability can affect spoken language learning (Friel-Patti & Finitzo, 1990), there has been great interest in the role that OME plays in language disorders in children. It does not appear that children with SLI are more subject to OME than children without language problems (Bishop & Edmundson, 1986). Therefore, it seems unlikely that OME could be a major cause of these children's difficulties.

Nevertheless, it is important to document that the language problems observed in children with SLI cannot be attributed to recent bouts with OME. A hearing screening cannot ensure this, because the OME and associated hearing loss might have abated only days before the screening, and language might not have had sufficient time to resume its original course of development. For this reason, parents of children with SLI are asked about their children's history with OME. In some studies, if the children have had known episodes within the preceding 12-month period, they are not included in the research program (e.g., Loeb & Leonard, 1991).

Neurological Status

Several neurological conditions can lead to language disorders in children. These must be ruled out before a diagnosis of SLI is appropriate. There should be no evidence of focal brain lesions, traumatic brain injury, cerebral palsy, or seizure disorders. Some investigators allow for a brief period of febrile seizures during infancy, provided that the problem was resolved and the child was no longer on medication for the prevention of seizures (e.g., Stark & Tallal, 1988).

This neurological exclusionary criterion rules out certain groups of children who at one time were included in the category of developmental dysphasia. These are children with focal brain injury and children with Landau-Kleffner syndrome. Left in limbo are children fitting the description of congenital verbal auditory agnosia (congenital word deafness). The only hard neurological evidence of this condition comes from the postmortem examination of a single child. Results indicated bilateral damage to the auditory cortex and first temporal convolution (Landau, Goldstein, & Kleffner, 1960). Because such children are rare, it has not been determined whether all children showing the language symptoms of this disorder are in fact neurologically impaired. If this proves to be the case, these children might form a separate diagnostic category, distinct from SLI.

The neurological criterion for SLI does not rule out mild neuromaturational delays. As a group, children with SLI are more likely than age controls to exhibit clumsiness or slower motor responses typical of younger children (Bishop, 1990; Bishop & Edmundson, 1987b; Noterdaeme, Amorosa, Ploog, & Scheimann, 1988; Powell & Bishop, 1992; Stark & Tallal, 1988; Tallal, Dukette, & Curtiss, 1989). Other investigators have observed limitations in attention in children with SLI (Baker & Cantwell, 1982; Mackworth, Grandstaff, & Pribram, 1973; Tallal, Dukette, & Curtiss, 1989; Townsend, Wulfeck, Nichols, & Koch, 1995).

Oral Structure and Function

Children with abnormalities of oral structure that might impede normal language production are not included in the category of SLI. Equally important is the exclusion of children who exhibit problems in oral function. A screening of these abilities constitutes a regular part of any diagnostic battery. Volitional oral movements examined in the screening typically include rounding the lips, sealing the lips, biting down on the lower lip, protruding the tongue, and moving the tongue from one side of the mouth to the other. Although there is a developmental function for such movements, they are well controlled by age 3;6 (Robbins & Klee, 1987).

Interactions with People and Objects

Another important criterion for SLI is the absence of all of the symptoms of (1) impaired reciprocal social interaction and (2) restriction of activities listed in the DSM IV (American Psychiatric Association, 1994) criteria for autism and "pervasive developmental disorder not otherwise specified" (PDDNOS). Examples of symptoms of the first type are a marked impairment in eye contact and gestures to regulate interaction, and few or no attempts to share enjoyment, interests, or achievements with others. Examples of the second type are stereotyped and repetitive motor mannerisms, and inflexible adherence to specific nonfunctional routines.

It should be pointed out that the criterion of no symptoms of impaired social interaction and restriction of activities excludes children with "semantic-pragmatic" disorder from the category of SLI. These children show deficits in communicative ability marked by difficulties with comprehension in conversational contexts and especially severe problems in the semantic and pragmatic domains (Bishop & Rosenbloom, 1987; Rapin & Allen, 1987). Some of these children exhibit age-appropriate nonverbal IQ scores along with normal hearing and no evidence of frank neurological impairment. In this respect, they resemble children with SLI. However, their interactions with the social/material world are odd. Approximately 20% of such children do not show a sufficiently large number of social and behavioral problems to warrant the term "autism" (Rapin & Allen, 1987). However, as noted by Bishop (1989), a good case can be made for viewing these children as falling on the less severe end of an autism continuum. In terms of DSM-IV criteria, these remaining children would fall in the category of PDDNOS.

Some Limitations of the Criteria
The criteria in table 1.1 go a long way toward ruling out the inclusion of children whose language disorders are related to other disabling conditions. However, it must also be recognized that there are children with language problems who do not meet all of these criteria, yet do not easily fit into other diagnostic categories. Some of these children score in the 70–84 range on a nonverbal intelligence test; such scores are too low for SLI but too high to be considered a reflection of mental retardation. There are also children who have had a recent episode of OME. Often audiological testing rules out hearing impairment in these children. We must also add to this list those children who are regarded as having language learning difficulties but whose composite language test scores are not quite low enough for them to be included. These children also have needs that should be served; their exclusion from the category of SLI is primarily a product of research conservatism.

An investigation by Stark and Tallal (1981) offers a view into the difficulty of identifying children who fit each and every criterion for SLI. These investigators asked clinicians to refer language-impaired children in their caseloads whom they regarded as showing normal hearing and nonverbal intelligence, and adequate social and emotional adjustment. Of 132 children referred, only 39 met the criteria established for SLI. Fifty of the children exhibited nonverbal IQ scores lower than 85. Composite language scores that were too high constituted the basis for excluding 33 children. A few children did not qualify due to factors such as a recent episode of OME and evidence of frank neurological impairment.

An important point to make concerning the criteria for SLI is that the children who meet all of the criteria do not constitute a homogeneous group. There are common profiles, but some children do not show them. Later in this chapter, attempts to identify distinct subgroups of children with SLI will be discussed.

The Prevalence of SLI

It is not difficult to find estimates of the prevalence of SLI in the literature. For example, defining SLI (rather stringently) as a tested language age of no more than two-thirds of tested mental age, Tower (1979) estimated the prevalence of SLI to be about 1.5%. According to the American Psychiatric Association's DSM-IV (1994), the prevalence of children with SLI with production deficits only is approximately 5%; the prevalence drops to around 3% for combined comprehension and production deficits. But data supporting the prevalence estimates are much harder to come by.

Most of the prevalence studies that have been conducted were designed to answer a broader question, "What is the prevalence of language disorders in children?" (see reviews by Lahey, 1988; Silva, 1987). In most instances, only language ability was assessed. In a few studies (e.g., Stevenson & Richman, 1976), nonverbal intelligence as well as language tests were administered. However, because the full slate of tasks was not included, it cannot be determined from these studies how many of the children with documented language problems would have met the exclusionary criteria for SLI.

Tomblin (1996a, b) recently conducted a large-scale project on the prevalence of SLI in five-year-olds. More than 6,000 children participated. The children failing an initial screening were given a full battery of linguistic and nonlinguistic tests. The battery was also administered to 1,000 children who passed the screening, as a control. One of the advantages of Tomblin's method is that it provided a means of ensuring that the children exhibiting language problems but normal nonverbal IQs would meet the criteria for SLI in all other respects (e.g., display normal hearing). Validity studies were also conducted to ensure that the language test score criterion selected would closely agree with clinicians' judgments about the children's functioning (Tomblin, 1996a). Based on these steps, children were placed in the SLI category if they scored at least 1.25 SDs below the mean on two or more of five composite language measures: (1) vocabulary (comprehension plus production), (2) grammar (comprehension plus production), (3) narrative (comprehension plus production), (4) comprehension (vocabulary, grammar, plus narrative), and (5) production (vocabulary, grammar, plus narrative).

Using this method, the prevalence of SLI among five-year-olds was determined to be 7.4% (Tomblin, 1996b). This criterion had good sensitivity; 85% of children viewed as language-impaired by clinicians were identified by the criterion. And the specificity of the criterion was excellent: only 1% of children regarded as normal by clinicians showed scores that would erroneously identify them as children with SLI. Given that the language test criterion employed composite scores (each based on at least two individual test scores) and approximately 15% of children viewed by clinicians as language-impaired did not meet this criterion, there is no reason to believe that the prevalence of 7.4% is artificially high.

SLI is more prevalent in males than in females (e.g., Dalby, 1977; T. T. S. Ingram, 1959; Johnston, Stark, Mellits, & Tallal, 1981; Tallal, Ross, & Curtiss, 1989b). The ratio of males to females averages approximately 2.8:1 across studies (R. Robinson, 1987). However, ratios as high as 4.8:1 can be found in select settings, such as residential schools (Haynes, 1992).

The Long-Standing Nature of SLI

For many children with SLI, deficits of language persist. Gains in language ability are seen over time, but weaknesses in language often are still apparent in later childhood, adolescence, and, in some cases, adulthood.

Most studies documenting the persistence of language problems are retrospective. The individuals varied in age at follow-up from seven years to 25 years, and in the time from initial testing to follow-up from one year to around 20 years. Because these were retrospective studies and the investigators were dependent on the language testing instruments at hand when the children were first seen, some of the reports of the children's initial language status were not up to today's standards. Nevertheless, the results are quite convincing. Although most of the children showed improvement, and some appeared to be within normal limits at follow-up, the percentage of children who continued to have problems in language was very high.

For example, Aram and Nation (1980) found that 40% of preschoolers with SLI continued to have significant language problems four to five years later. Approximately the same percentage presented other learning problems; they were either held back at least one grade in school or were in classrooms for children with learning disabilities. In a study by Aram, Ekelman, and Nation (1984), 13 children who qualified as exhibiting SLI as preschoolers were tested again as teenagers. Eleven of these (85%) scored more than 2 SDs below the mean on a comprehensive test of language ability. Eight of the children (62%) repeated a grade, required tutoring, or were enrolled in special classes. R. Stark, Bernstein, Condino, Bender, Tallal, and Catts (1984) conducted a follow-up of 29 children with SLI originally seen when they averaged around six years of age. After four years, the children continued to perform well below age controls on measures of both language production and comprehension. Of the 29 children with SLI, 23 (79%) showed significant deficits in language ability four years later. The same number of children displayed significant problems in reading.

During later childhood and adolescence, the deficits seen in SLI are sometimes subtle. For example, Nippold and Fey (1983) examined the comprehension of metaphors in ten-year-old children with SLI whose language problems had first been diagnosed during the preschool years. These children performed significantly worse than a group of age controls whose scores on literal aspects of sentence comprehension seemed comparable.

Findings such as these are typical. Other retrospective studies that report high percentages of language problems at follow-up, and/or language performance at follow-up that is significantly below that of age controls, include deAjuriaguerra et al. (1965), Griffiths (1969), P. Weiner (1972), Garvey and Gordon (1973), Sheridan and Peckham (1975), Strominger and Bashir (1977), Kolvin, Fundudis, George, Wrate, and Scarth (1979), and R. King, Jones, and Lasky (1982). It appears that the degree of language deficit at six years of age and the degree to which language comprehension problems accompany production problems are important factors in language status at follow-up (Paul, Cohen, & Caparulo, 1983).

Case studies also have been informative. P. Weiner (1974) described a 16-year-old who had been first seen at four years of age. Although significant progress was seen in his language development, his skills were remarkably deficient for someone in his teens. Noun plural inflections were omitted in 77% of their obligatory contexts, copula *be* forms were omitted in 45% of their contexts, and the regular past *-ed* inflection never appeared. Other difficulties were also evident, as the examples in (4) illustrate.

(4)   a.   The grandmother look for son in room.
       b.   When the man plowing the field, her sister go to school that
             morningtime.

    c.   Now us have lot of snow at ... around this house.

    d.   That man in a dark room.

    e.   A little boy want to tell someone how he get hurt.

    f.   Those are businessmen talking for a building to build in that city.

    g.   Can I play with violin?

Perhaps the most telling retrospective studies are those documenting language problems in adults. For example, Hall and Tomblin (1978) found that nine of 18 individuals with SLI initially tested as youngsters continued to exhibit significant language difficulties in their early twenties. Only one individual out of a comparison group of 18 individuals with earlier phonological problems (only) showed similarly low scores on the language tests at follow-up.

Tomblin, Freese, and Records (1992) administered a series of tasks to a group of young adults with a history of SLI and to a group of age controls. Eleven of these were measures of spoken language ability. On all 11 measures, the adults with a history of SLI earned significantly lower scores. Several of the measures employed by Tomblin, Freese, and Records had been normed on adults; on these measures, the individuals with a history of SLI scored in a range that suggested their clinical status had not changed.

One of the case studies of an adult with persisting SLI is the report by Kerschensteiner and Huber (1975). At age 23 years, this young man continued to have great difficulty with grammatical constructions typically mastered by the end of preschool.

The most widely cited work on adults with language disorders in recent years is by Gopnik and her colleagues (e.g., Fee, 1995; Goad & Rebellati, 1994; Gopnik, 1990a; Gopnik & Crago, 1991; Piggott & Kessler Robb, 1994; Ullman & Gopnik, 1994). This group of investigators studied 30 members of a family, covering three generations. Sixteen of the family members had been diagnosed as language-disordered. Six of these showed nonverbal IQ scores below 85 (Vargha-Khadem, Watkins, Alcock, Fletcher, & Passingham, 1995) and thus would not fall in the category of SLI using conventional standards.

What is remarkable about this family is that such a large number of its members exhibited language problems, and to such a severe degree. Four of the five adults representing the second generation showed significant problems with language. Of the 11 members of the third generation with language problems, five were age ten years or older. Gopnik and her colleagues documented that the affected family members exhibited obvious problems in phonology and particular areas of grammar. Such problems are not seen in normally developing individuals at these ages. Indeed, some of the members of the same family showed significantly greater ability in these areas of language, suggesting that linguistic environment was not the primary factor involved. We shall have more to say about this interesting family in later chapters.

Prospective studies have also been carried out in recent years. Bishop and Edmundson (1987a) conducted a longitudinal investigation that included 68 children meeting the criteria for SLI at age 4 years. At age 5;6, a slight majority of children, 56%, continued to display poor scores in language. The same children were evaluated three years later, at age 8;6, by Bishop and Adams (1990). The children whose language difficulties appeared to resolve by age 5;6 continued to perform well on language tests; they differed from age controls on only two of several spoken language measures, and on none of the measures of reading employed. In contrast, the children continuing to show problems at age 5;6 lagged behind age controls at age 8;6 on all spoken and written measures except a single test of spelling. Similar findings were obtained in a prospective study reported by Beitchman, Wilson, Brownlie, Walters, and Lancee (1996). Children with SLI at age 5;0 continued to perform below the level of age controls on a wide range of measures when tested two years later.

Tallal, Curtiss, and Kaplan (1988) followed 101 children with SLI for five years, beginning at age 4;4. The children were administered standardized tests of spoken language during the first, third, and fifth years. Tests of reading were also employed in the final year. Although the children with SLI made gains across the five years, they earned significantly lower scores than age controls on all spoken language tests throughout the study. The spoken language test scores of the children with SLI were reminiscent of those seen in normally developing children one to two years younger. Their reading comprehension was also poorer than that of age controls when assessed in the final year.

It can be seen, then, that spoken language problems often persist, usually in subtle form though sometimes in dramatic fashion. In emphasizing problems with spoken language, we haven't done justice to the reading difficulties experienced by these individuals. This issue will be discussed more fully when we look at the relationship between SLI and specific reading disabilities in chapter 9.

The Search for Subgroups

In chapter 3, we shall see results from group studies that point to areas of special weakness in children with SLI. However, the common profiles observed are not found in all children. In many group studies there are children who do not fit the modal pattern. This fact has led some investigators to assume that the category of SLI is little more than a terminological way station for groups of children until such time as finer diagnostic categories can be identified. Subgroups must exist.

Or so logic tells us. But the task of identifying distinct and reliable subgroups has proven to be a formidable one. And we still don't have it right. Here we shall discuss the efforts that have been made toward this end.

As noted earlier, differences among children with SLI were being reported by the end of the nineteenth century, with special emphasis on the distinction between expressive problems and combined receptive and expressive difficulties. Although this practice has continued throughout the twentieth century, attempts to identify subgroups reached a more sophisticated level by the 1970s. Aram and Nation (1975) sought to identify subgroups of children with SLI on the basis of language test profiles. Forty-seven children from three through six years of age were administered comprehension, formulation, and repetition tasks dealing with semantics, syntax, and phonology. Six patterns were identified according to a factor analysis. One group performed relatively poorly on all tests. Another group performed as well on the production tasks (formulation, repetition) as on the comprehension tasks, though their levels of performance were in the moderate range relative to the other children participating in the study. A third group performed well on the comprehension tests but poorly on the production measures. The fourth group resembled the third group in that comprehension exceeded production, but the gap between the two was less dramatic. The fifth group identified by Aram and Nation showed significant problems in phonology, with a less marked problem in syntactic production. The final group exhibited moderate problems on most tests and language areas but were relatively strong in repetition ability.

Wolfus, Moscovitch, and Kinsbourne (1980) gave 19 children with SLI a battery of semantic, syntactic, and phonological tasks. Comprehension as well as production tests were included. A discriminant function analysis suggested two subgroups. The first could be characterized by deficits in the production of phonology and syntax. The second showed more global deficits in both comprehension and production.

The test profiles of 40 children were subjected to factor analysis in a study by Korkman and Häkinen-Rihu (1994). Three subgroups emerged, two of them applicable to children with SLI. Children in the first subgroup showed a deficit across a broad range of comprehension and production tasks. Children in the second subgroup seemed to experience difficulty in understanding sentences of greater length and complexity. Validation of these subgroups with a second group of 40 children revealed that 80% of the children could be classified according to the original subgrouping. Five of the remaining eight children seemed to fit a new category because they did especially poorly on naming tasks.

Other proposals of subtypes have been based on clinical judgments rather than statistical sorting procedures. The classification system of Rapin and Allen (1983, 1988) is the dominant system of this type. Three of the categories used by these authors are especially relevant, given the criteria for SLI discussed earlier. The subtype with the largest number of children according to Rapin and Allen is referred to as "phonologic-syntactic deficit syndrome." These children display mild comprehension problems and more severe deficits in the production of morphosyntax and phonology. Another subtype is "lexical-semantic deficit syndrome." The most salient aspect of this subtype is difficulties with word-finding. Sentence formulation is compromised by these children's difficulty in accessing the appropriate word as it is needed. In stretches of speech seemingly unaffected by such word-finding problems, morphosyntactic difficulties are only mild. Phonology is stronger in these children than in the first subtype. The third subtype is "verbal auditory agnosia" or "word deafness." These are children with extremely limited use of language, seemingly due to a severe comprehension deficit. Although such children have been described in the literature since the 1890s, they are rare.

Wilson and Risucci (1986) devised a scheme of subtypes of SLI that made use of both clinical judgment and cluster analysis. Ninety-three children were presented with a range of language, auditory processing, and visual processing tasks. Four subtypes of children with SLI emerged from the study. One group showed low scores on expressive language measures and relatively high scores on receptive language tasks. Another showed expressive language scores that resembled those of the first group but also displayed limitations in receptive language. The third group showed the greatest deficit on tasks requiring auditory memory and retrieval. The final group did very poorly on all measures.

At first blush, these studies do not seem to arrive at the same subgroups of children with SLI. However, closer inspection reveals particular subgroups that cut across investigations, terminological variation notwithstanding. All of the studies except that of Korkman and Häkkinen-Rihu (1994) appear to identify children whose limitations in the production of syntax and phonology are especially weak, and every study seems to identify another group who shows a flatter profile with comprehension difficulties approaching those seen in production. In some investigations, one or both of these two major groupings could be subdivided, according to the degree of comprehension-production gap (in the case of the first grouping) and the absolute severity of the global deficit (in the case of the second). These distinctions are generally accepted by researchers and clinicians as reflecting the most frequent ways in which children with SLI differ from one another. The lexical-syntactic deficit syndrome of Rapin and Allen (1983, 1988) seems to represent a profile that can be accommodated by one of the subcategories in each of the other studies reviewed, in that expressive skills are relatively depressed. Relative strength in repetition tasks and relative weakness in auditory memory tasks have been observed in individual children with SLI in studies in which such tasks constitute the dependent measure. However, the respective studies proposing them as distinct subtypes of SLI groups seem to be alone in doing so.

The contribution of these studies is that they formalize long-standing observations that children with SLI do not constitute a homogeneous group, and collectively, they confirm certain profiles as more common than others. Investigations are now needed to validate these as discrete subgroups (Aram, Morris, & Hall, 1993). Until this has been done, these subcategories should be useful in reminding us of the heterogeneity that exists among children with SLI, but they should not be taken as established diagnostic divisions. For this reason, we shall continue to use the term SLI in this book, recognizing it as the umbrella term that it is.

The Current Picture

Children with SLI experience significant limitations in language ability that cannot be attributed to problems of hearing, neurological status, nonverbal intelligence, or other known factors. Although such children have been described for over a century and a half, formal criteria for the identification of SLI have evolved only gradually. These criteria are rather stringent; there are children with language learning difficulties who do not meet all of the criteria, yet their symptoms do not fall into other established diagnostic categories. But those who meet the criteria are not few in number. Approximately 7% of five-year-olds appear to be children with SLI. Even though they meet rather strict criteria, children with SLI do not form a homogeneous group. They differ from one another in the areas of language posing the greatest difficulty, and in their comprehension abilities relative to their production abilities. Unfortunately, the problems of these children do not go away quickly. Spoken language problems can persist for years, and reading problems often emerge once academic instruction begins.

The child with SLI described at the beginning of the chapter illustrated some of the points raised. Her particular language profile was the most commonly reported one, characterized by greater weaknesses in syntax and phonology than in other areas, with comprehension (though not at age level) outpacing production. Word-finding problems were also noted. This child's difficulties persisted into the school years, when reading problems emerged. There are other details in this case study that warrant attention, such as the family history of language difficulties and her frequent omissions of grammatical forms that she is nevertheless capable of using in a creative manner. These will be taken up in later chapters.

Chapter 2

Characterizing the Language Deficit: Basic Concepts

Comparisons Between Children with SLI and Normally Developing Children: Choosing a Reference Point

With whom should children with SLI be compared? The answer to this question is not as simple as it seems. The choice of comparison group depends heavily on the type of information being sought.

Not surprisingly, most investigators who study children with SLI also recruit a group of normally developing children of the same chronological age. After all, it is not enough to document that children with SLI have difficulty with some aspect of language; unless typical children of the same age have fewer problems with the same aspect, the finding is largely irrelevant. Unfortunately, age matching has certain limitations.

It is not uncommon to find instances in which a group of children with SLI participating in a study show a nonverbal IQ range of, say, 90 to 105, whereas a group of age controls from the same community show a range of 100 to 115. To counter possible effects attributable to differences in nonverbal IQ, some researchers use a comparison group of normally developing children matched with the children with SLI on nonverbal mental age rather than chronological age (e.g., Kamhi, Ward, & Mills, 1995). This seems especially important if the dependent measures focus on nonlinguistic abilities with a clear conceptual basis. Studies of this type will be reviewed in chapter 5.

But mental age controls cannot solve another fundamental problem in the study of children with SLI. These children show weaknesses in many aspects of language. It is likely that some of these limitations strike closer to the heart of the basic problem than do others. However, because these children will differ from chronological age or mental age controls on almost any language measure selected, this fact might be obscured.

Assume that we wish to determine if one set of features of language is more problematic for children with SLI than another set. Any difference we observe in the children's comprehension or production of these two sets of features might be meaningful, or might instead reflect the fact that in normal development, one set is acquired before the other. If the chronological age or mental age controls are already approaching mastery on both sets, we won't know which of these two interpretations is correct. The solution to this problem that is most often adopted is the use of a group of younger normally developing children who are matched with the children with SLI on some measure of language ability.

The particular measure of language ability selected depends heavily on the nature of the investigation. In studies of the morphosyntax of children with SLI, MLU in morphemes is frequently selected as the means of matching the groups of children. The logic behind the use of this measure is straightforward. Consider the examples in (1) and (2).

(1)   Mommy putting juice
      You give to baby

(2)   Tammy putting food floor
      I give this to Mommy

The utterances in (1) and (2) suggest differences in the ability to express all of the arguments required in verbs such as *put* and *give*. But can we rule out the possibility that for the child in (1) the production of utterances five morphemes in length is a major chore, whereas for the child in (2) such utterances are effortless? If the use of the morphosyntactic feature of interest has an impact on utterance length, MLU matching serves to ensure that the failure to use the feature is not the result of length limitations.

However, there are instances in which a measure other than MLU in morphemes is preferable for matching groups of children. For example, if the focus of inquiry is on the use of grammatical morphology, MLU matching is somewhat risky. Because MLU is influenced by a child's use of grammatical morphemes, any failure to find a difference between groups of children might be due to the use of a matching technique that created a great deal of similarity between the groups on the variables of experimental interest. If the interest in grammatical morphology is restricted to grammatical inflections, computing MLU in terms of words will serve as a solution to this problem. However, if function words are among the grammatical morphemes of interest, matching according to MLU in words will still remove some of the differences the study was designed to examine. Johnston (1995) has noted that in this sense, the use of MLU matching is a highly conservative technique. If differences are observed (and they often are, as we will see in chapter 3), greater confidence can be placed in the findings.

Of course, there are instances in which a conservative approach is not ideal, as when an experimental question about certain grammatical morphemes is being posed for the first time. In these cases, measures such as the mean number of arguments expressed per utterance, the mean number of nonnuclear predicates produced per utterance (Johnston & Kamhi, 1984; Kamhi & Johnston, 1992), or the mean number of open-class words used per utterance (Rollins, Snow, & Willett, 1996) might be selected as the basis for matching the groups of children.

Measures of phonology, too, can serve as valuable tools for matching. For example, before it can be assumed that English-speaking children with SLI have special difficulties with free-standing forms (e.g., "*the* rat") and inflections (e.g., "go*es*"), it should be established that there is not a more general problem with weak syllables and word-final consonants. A comparison group might be employed that consists of younger normally developing children matched with the children with SLI on their ability to produce weak syllables and word-final consonants in monomorphemic contexts (e.g., "*giraffe*," "no*se*").

Expressive vocabulary can also serve as a matching criterion. In chapter 13, it will be seen that some theories of SLI require an assumption that children with this disorder remain at a point of using only single-word utterances for protracted periods.

To evaluate this assumption, children with SLI who are limited to one-word utterances can be compared with a group of younger normally developing children matched with the first group according to the number of different words in their expressive vocabularies. If the assumption is correct, multi-word utterances should be seen in the speech of the normally developing children.

Although the production abilities of children with SLI have captured the lion's share of investigators' attention, studies of these children's comprehension abilities are also important. For example, much is known about the production of morphosyntactic forms in children with SLI, but it is not yet clear whether some of the problems observed in production are attributable to a lack of knowledge of particular grammatical details. The likelihood of the latter would seem higher if problems in the comprehension of these details were found.

Studies of the comprehension abilities in children with SLI usually require the use of comparison groups matched on some comprehension measure. Matching on the basis of production measures can be problematic because children with SLI are often poorer in production than in comprehension. The use of a production measure, then, might result in a significant mismatch in general language comprehension that favors the children with SLI. Of course, the comprehension measure used as the basis for matching should provide some useful control for the comparison of interest. For example, if grammatical constructions such as relative clauses (e.g., "the rabbit is kissing the duck that is sleeping") or full passives (e.g., "the rabbit is kissed by the duck") are under investigation, children with SLI and normally developing children might be matched on their ability to comprehend structurally simple sentences whose lengths approximate those of the constructions of interest (e.g., "touch the green car and the blue boat").

Some scholars have questioned the wisdom of using language measures as a basis for matching. Several different types of criticism have been leveled against the use of such measures. Some have to do with the way the measure has been characterized; in other cases the criticism is more fundamental. One of the problems with MLU in morphemes as a matching technique stems from the fact that many investigators employing it have treated it as a rough measure of grammatical development (after R. Brown, 1973). Although MLU in morphemes serves as a good predictor of morphosyntactic skills when MLU is below 3.0, the correlation between the two becomes weaker thereafter (Klee & Fitzgerald, 1985; Scarborough, Rescorla, Tager-Flusberg, Fowler, & Sudhalter, 1991).

Johnston and Kamhi (1984) observed differences between children with SLI and MLU controls on two quite different measures of morphosyntax. Such a finding seems difficult to reconcile with the view that MLU in morphemes represents a particular stage of grammatical development. Even at low MLU levels, it is easy to identify instances in which two utterances of the same length reflect different morphosyntactic contributions. Rollins and her colleagues (Rollins, 1995; Rollins, Snow, & Willett, 1966) provide a number of examples in which two children seemed to achieve similar MLUs through different means. In some of the pairs of children studied, for example, one child showed more advanced propositional complexity, whereas the other exhibited greater morphological complexity within noun phrases.

If grammatical development is the intended basis for matching, a more direct measure of grammatical complexity should be used. Matching for grammatical complexity makes a great deal of sense for many research questions. For example, if the hypothesis is advanced that children with SLI are extraordinarily limited in their use of conversational speech acts (e.g., promising, warning) and such acts require the children's control of particular grammatical constructions, it would be important to ensure that the normally developing children with whom they are compared are not more sophisticated in their grammatical abilities.

However, it is one thing to have a measure of grammatical complexity that can serve as a control for the experimental question of interest, and quite another to have a number that can be taken to represent a stage or level of grammatical development. At present, there is no single measure that can be interpreted in this way.

Another criticism directed toward MLU is that its psychometric properties are less than ideal. In fact, there are some psychometric limitations, though few if any developmental measures have fared as well. At lower MLUs in morphemes, there is a significant correlation between MLU and age of at least .75 in both normally developing children and children with SLI (Klee, Schaffer, May, Membrino, & Mougey, 1989; Miller & Chapman, 1981; Scarborough, Rescorla, Tager-Flusberg, Fowler, & Sudhalter, 1991). Beyond age 3;0, the relationship becomes weaker (Scarborough, Wyckoff, & Davidson, 1986). Interexaminer reliability is high (Klee & Fitzgerald, 1985); however, because most MLU computations are now performed by computer software programs, the more relevant interexaminer reliability concerns the transcription that is entered into the program. Transcription reliability, too, appears good (Klee, Schaffer, May, Membrino, & Mougey, 1989). However, a major weakness of MLU in morphemes seems to be its high variability when MLU rises above 2.50; two blocks of 100 utterances obtained during the same sample can vary considerably in MLU (Klee & Fitzgerald, 1985).

There are also data on the psychometric properties of MLU computed in terms of words. This measure, called "mean length of response" in the earlier literature, was largely replaced by MLU computed in morphemes beginning in the 1970s, except in crosslinguistic studies, where it continues to prove useful. This measure is significantly correlated with age (Templin, 1957) and adults' judgments of children's language proficiency (Shriner, 1967), and shows good intra- and interexaminer reliability (G. Siegel, 1962). For samples of at least 100 utterances, MLU in words appears to be quite stable (Darley & Moll, 1960). Minifie, Darley, and Sherman (1963) observed that samples obtained from the same children on two consecutive days produced similar values. Unlike MLU computed in morphemes, MLU in words appears to retain its positive psychometric properties until approximately age 5;0 (Shriner, 1967). However, this might be due to the fact that samples analyzed in terms of MLU in words were usually obtained through a more structured sampling procedure of asking children to describe pictures.

Another criticism is a more sweeping one, applying to language measures of any type. Normally developing children matched with children with SLI on almost any language measure will be significantly younger than the children with SLI. Thus, the normally developing children might differ in many ways from the children with SLI, some of which may have more to do with the general developmental differences between the groups than with differences pertaining to the details of language under investigation (Plante, Swisher, Kiernan, & Restrepo, 1993).

The solution to this problem is not clear. For example, it is far from obvious how correlational techniques using only children with SLI and chronological age controls can correct this problem, because differences in the relationships among variables that will arise from such comparisons are also seen when comparing two groups of normally developing children who differ in chronological age. As a case in point, suppose we compare four-year-old children with SLI and chronological age controls on their use of the regular past inflection -ed in obligatory contexts. We will find that the children with SLI show much lower percentages (e.g., Leonard, Bortolini, Caselli, McGregor, & Sabbadini, 1992; Rice, Wexler, & Cleave, 1995). We will probably find another difference as well. The age controls' errors on irregular past forms will usually be overregularizations (e.g., *throwed*), whereas the errors of the children with SLI will usually be uninflected forms (e.g., *throw*). Employment of percentage of correct use of regular past -ed as a covariate to remove the effects of skill level will not remove the difference in types of error produced on verbs requiring irregular past. This might tempt us to conclude that the children with SLI could not be described as falling on a continuum with the age controls, and thus were atypical in their developmental pattern. But such an interpretation would be erroneous; younger normally developing children also produce uninflected forms in place of irregular past forms when their percentages of use of -ed in obligatory contexts are low. Overregularizations are not customarily seen until percentages of regular past -ed approach 50 (Marcus, Pinker, Ullman, Hollander, Rosen, & Xu, 1992).

Because relationships between different measures often change as a function of age, it does not seem appropriate to dispense with younger normally developing children as additional controls. Though age controls are a crucial part of any investigation, for some experimental questions they are not sufficient.

Five Ways to Differ from Normally Developing Peers: Improving on the Delay-Deviance Dichotomy

For approximately 30 years, investigators have attempted to determine whether the linguistic characteristics of children with SLI are like those seen in younger normally developing children, or are altogether different from anything seen in normal development. Studies by Menyuk (1964), Lee (1966), Leonard (1972), and Morehead and Ingram (1970, 1973) are among the earliest of this type. Eventually, the cover terms "delay" and "deviant" were adopted to represent these dichotomous possibilities. Despite questions raised about the value of continuing the delay versus deviant debate (e.g., D. Ingram, 1987b), discussion of this issue can be found even in the current literature.

Certainly, attempts to characterize the linguistic abilities of children with SLI are not misguided; the problem is that the delay-deviant dichotomy does not adequately capture the various ways in which children with SLI can differ from normally developing children. We shall review each of these types of differences here.

Delay

The common use of the term "delay" suggests a late start and nothing more. If one leaves home 15 minutes late, the destination is reached approximately 15 minutes later than the original expected arrival. Indeed, in this sense of the term "delay," it might even be possible to make up for lost time, say, by running to the destination instead of walking. Either of these possibilities might hold for children who begin

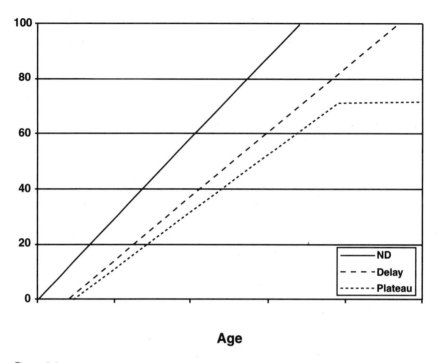

Figure 2.1
Course of language development of children exhibiting a language delay and a delay with a plateau relative to the course of development seen in normally developing (ND) children.

speaking late but are clearly at age level by around three years of age. As we will see in chapter 9, such children are not likely to be diagnosed as exhibiting SLI.

For many children with SLI, a delay involves not only the late emergence of language but also slower than average development of language from the point of emergence to the point of mastery. This pattern is illustrated in figure 2.1. Here it can be seen that the children exhibiting a delay show both a later emergence than the normally developing children and protracted development from emergence to mastery. This protraction can be seen in the shallower slope of development for these children than for the normally developing children. The gap between the two groups of children widens across time.

Plateau

In chapter 1 it was noted that SLI can be a long-standing problem, continuing into adulthood in some individuals. The language development of these individuals cannot be considered to be delayed only, for mastery levels are never reached. This pattern of development is also illustrated in figure 2.1. Like the pattern of delay, there is late emergence and protracted development, but in this case, a plateau is reached at some point before certain aspects of language are mastered. The basis for such arrested development is presumed to be biological age. We will have more to say about this issue in chapter 9.

Normal Development

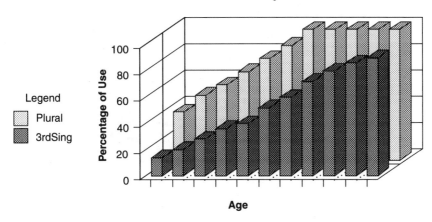

Figure 2.2
Acquisition pattern seen in normally developing children, in which progression toward mastery of the noun plural -s inflection precedes that of the third-person singular -s verb inflection.

Profile Difference
The picture of delay that has been painted thus far makes it seem that a child with SLI at age four years might look just like two-year-old normally developing children; the same child at ten years of age might be identical to six-year-old normally developing children, and so on. However, this picture sidesteps the fact that at any given age, children show greater ability with some features of language than others. For example, two-year-old normally developing children acquiring English will show greater ability with the noun plural inflection -s than with the present third-person singular verb inflection -s. In fact, plural -s seems to emerge earlier and reach mastery levels sooner than third-singular -s. This pattern of development is illustrated in figure 2.2.

Accordingly, if children with SLI were really delayed in the sense described above, not only should there be later emergence, slower development, and later mastery of each feature of language, but the relationship between features should match that seen for normally developing children. Figure 2.3 provides a view of this delayed pattern.

And if the relationship between features of language does not match that seen in normally developing children? Figure 2.4 provides an example. Such a pattern would be seen, for instance, if children with SLI were functioning like normally developing children one year younger in their use of plural -s but like normally developing children two or three years younger in their use of the third-singular -s verb inflection. Technically, such a pattern might constitute different degrees of delay across features.

However, patterns of this type should not be lumped together with the uniform delay illustrated in figure 2.3 because in the present case, the profile simply does not match that of younger normally developing children at any age. For any given level of plural -s use, say 80%, the gap between the use of plural -s and the use of third-person singular -s is wider for the children with SLI than for the normally developing children.

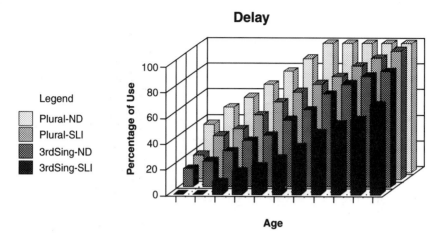

Figure 2.3
Acquisition pattern of children with SLI reflecting a language delay, in which the relationship between two grammatical features is the same as seen for normally developing (ND) children.

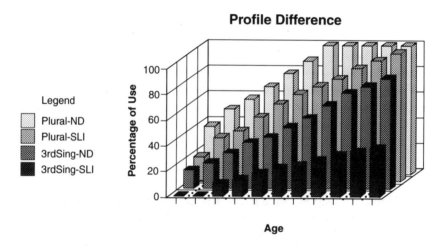

Figure 2.4
Acquisition pattern of children with SLI reflecting a profile that differs from the profile of normally developing (ND) children.

The type of profile difference between children with SLI and normally developing children that is illustrated in figure 2.4 is quite common in the literature, as we shall see. However, it shows up most often when the two groups are matched according to a language measure such as MLU. (As noted earlier, in such a comparison, the children with SLI are older than the normally developing comparison group.) Here, a pattern of delay would be a case where the degree of use of a given form, such as third-person singular -s, is the same for the two groups at the same level of MLU. In other words, the relationship between MLU and third-singular -s is the same in the two groups. An uneven profile, in contrast, would look like figure 2.5. The children with SLI show lower percentages of third-singular -s than would be predicted by their MLU, based on what we observe in normally developing children.

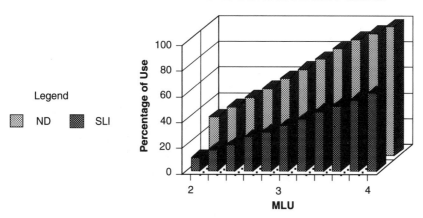

Profile Difference: MLU Match

Figure 2.5
Acquisition pattern of children with SLI reflecting a profile that differs from the profile of normally developing (ND) children as defined by the relationship between the acquisition of the third-person singular verb inflection and mean length of utterance (MLU).

Abnormal Frequency of Error
Another pattern is seen when children with SLI exhibit a particular type of error that can also be seen in the speech of younger normally developing children, but never with such high frequency. One error that might fit this example is the use of an accusative case pronoun in contexts calling for a pronoun of nominative case. Examples include *Me want the dolly* and *Him pushing car*. Younger normally developing children produce errors of this type. However, some (though not all) investigators have found that this error occurs more frequently in the speech of children with SLI. This type of difference is illustrated in figure 2.6. It can be noted in this figure that errors persist for a longer period in the speech of children with SLI, though this fact might be attributable to the slower rate of development seen in these children. More noteworthy is the observation that at no point in development is the percentage of errors for the normally developing children as high as it is for the children with SLI.

Qualitative Difference
The term "deviant" implies something out of the ordinary, something not seen in normally developing children; yet there are very few examples in the literature on SLI that warrant such a description. Occasionally one sees commission errors of the type *They likes ice cream*, but it turns out that young normally developing children commit such errors as well. There are examples of *brooming* and *barefeeting* used as verbs, but these might easily be based on lexical items such as *brushing* and *elbowing*, respectively; similar if not identical creations are found in the speech of young normally developing children.

Unusual phonological patterns are sometimes seen in children with SLI. For example, one child reported in the literature produced [s] in the final position of all words that did not require a labial in this position. Thus, *home* was produced appropriately but *cook* was produced as [kus] and *blue* was produced as [bus]. Of course, patterns of this type are not typical even of children with SLI. One could not hope to conduct a

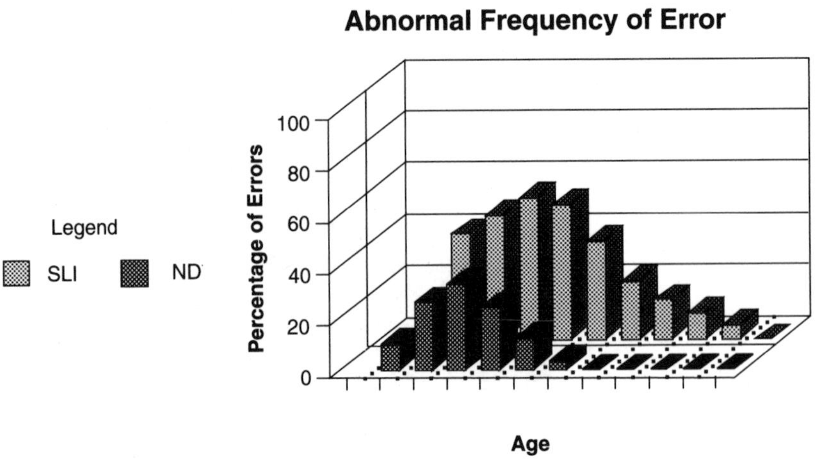

Abnormal Frequency of Error

Figure 2.6
Acquisition pattern reflecting a frequency of error in children with SLI that is higher than the error frequency seen at any age in normally developing (ND) children.

comparative study of, say, 30 children with SLI and 30 normally developing children and expect to find that 12 children showed this pattern, all of them children with SLI. It is more likely that no child in either group would show this pattern.

Even the finding that individual children with SLI can exhibit a unique phonological pattern must be interpreted with caution. Reports of highly unusual patterns can be seen in the literature on "normal" phonological development as well. For example, one otherwise normally functioning child described in the literature placed all word-initial strident continuants in final, rather than initial, position. Thus, *soap* was produced as [ops], *zoo* was produced as [uz], and *phone* was rendered as [onf]. As we shall see in chapter 3, unusual phonological patterns seem to be more frequent in the speech of children with SLI, but it is not yet clear that such patterns are unique to these children.

The possibility of five different patterns—delay, plateau, uneven profiles, abnormal frequency of error, and qualitative difference—makes it apparent that the delay-deviance dichotomy is an oversimplification, and can even be misleading. After all, which of the patterns described above constitutes a deviance? If deviance is defined as a linguistic characteristic never seen in normal development, only the last pattern qualifies. If it is defined as holding anytime the linguistic details seen in children with SLI fail to match an earlier stage of normal development, all patterns but the first represent a deviance.

Moving Beyond Description

The proper description of how children with SLI differ from normally developing children is an important start. However, a true understanding of these children's problems will not be achieved until we can explain how the different strengths and weaknesses of these children fit together. If a child with SLI has problems with regular past -ed and third-person singular -s, are these separate problems, or are they meaningfully connected in some way? Correlational techniques can give us some leads, especially when they are part of a longitudinal design. However, investigators in the area of SLI have begun to adopt an alternative and rather powerful tool for investigating these relationships—a treatment design. The basic idea is this: If we believe that problem A and problem B are related because they both require a certain type of linguistic knowledge (or require a certain type of comprehension strategy, or a certain articulatory maneuver), and that problem C is an independent problem, we can provide treatment focusing on problem A and should see improvement as well in problem B but not in problem C. This design permits an exploration of possible causal relationships.

In the case of children with SLI, of course, it has an added benefit. These children are usually enrolled in treatment programs, and if the features of language under study are key areas of difficulty for the children, they would probably be the focus of intervention efforts anyway. However, in order to use treatment as a means of testing hypotheses about how several linguistic difficulties fit together, some additional provisions are necessary.

We can look at some of these provisions by considering an example. Assume we wish to compare two different types of explanations for the serious difficulties with verb morphology that are seen in English-speaking children with SLI. Both of these accounts will be presented in greater detail in later chapters, but they can be introduced briefly here for illustrative purposes. One account, which we'll call the "surface" account, attributes the problem to a difficulty with the phonetic properties of English inflections (e.g., -ed, -s) and auxiliaries (e.g., is, are). According to this account, the underlying grammars of children with SLI are not defective; rather, the problem rests with detecting the morphemes of relatively short duration (which represent the great majority of the grammatical morphemes pertaining to verbs in English) and hypothesizing their grammatical function before they are erased from memory.

A competing account, which is called the "extended optional infinitive" account, places the problem in the underlying grammar. In this case, it is assumed that children with SLI go through an extended period of time before they learn that tense marking is obligatory in main clauses. What links verb inflections and auxiliaries in this type of account is the fact that they require tense. Thus, when these children produce utterances such as *Michelle like coffee* and *Mommy drinking tea*, they are treating the main verb as a tenseless form—in essence, as an infinitive. The children are capable of expressing tense, and on occasion they will use verb inflections and auxiliaries correctly. Their problem is that they don't know that tense must always be expressed in these contexts.

To set up the treatment design, we can rely on some of the assumptions of each account. According to the surface account, children with SLI can incorporate grammatical morphemes into their grammars provided these are salient. Fortunately, some of the morphemes whose durations are usually relatively short can be placed in sentence contexts in which they are significantly lengthened. For example, Swanson and Leonard (1994) noted that in mothers' speech to young children, uncontracted copula *is* averaged approximately 60 ms in sentence-initial (e.g., *Is the milk good?*) and sentence- (and clause-) medial position (e.g., *His fish is big*), whereas the duration of this form in final position (e.g., *Who's making that sound? The mouse is*) averaged around 250 ms.

One might create a treatment regimen in which the first phase consisted of presentations of a story with multiple instances of copula forms in utterance-final position. To ensure that these forms are grammatically instructive as well as salient, copula *is*, *am*, *are*, *was*, and *be* could be used at the outset, for these would help the child learn that the choice of forms varies with person, number, tense, and finiteness. A second phase would involve placing these forms in utterance-medial position but deliberately producing them with longer durations. In a third phase, the forms could be retained in medial position but their durations could be shortened to approximate those seen in typical adult-to-child speech.

According to the surface account, treatment of this type should assist the child in the acquisition of copula forms, but should have no appreciable effect on the use of verb inflections. (To acquire inflections, the child would need to have multiple occasions to hear inflections in salient contexts, as in *There she goes* and *I know what she sees*.) Thus, one test of this account would be to determine whether treatment on copula forms resulted in significant gains only on these and phonetically and positionally identical forms (e.g., auxiliary *is*, which, like copula *is*, immediately follows the preverbal noun phrase).

A second test of the surface account might be conducted by adding a treatment condition in which a different group of children with SLI heard the story presentations with copula forms only in medial position with typical duration values. According to the surface account, the first group of children should show greater gains than the second.

To evaluate the extended optional infinitive account, another component should be added to the procedure. The principal assumption of this account is that children with SLI understand tense but do not know that it is obligatory in main clauses. Thus, the procedure as described above might simply reinforce what the children already know—that tense can be expressed in main clauses—but might fail to teach them that tense is required in these contexts. To accomplish the latter, we might embellish the activity by incorporating an additional character, say, a puppet, who requests clarification whenever the storyteller uses a nonfinite form where a finite form is required. For example, the storyteller could proceed with sentences as above, but on occasion deliberately produce utterances such as *Is the girl happy now? I think she be.* These would be corrected by the puppet (e.g., *You meant to say, "Is the girl happy now? I think she is!"*).

The predictions of the extended optional infinitive account differ in two ways from those of the surface account. First, the children should do equally well whether copula forms first appear as utterance-final, lengthened syllables or in utterance-medial position throughout the treatment program. Second, if gains are seen in the use of copula forms as a result of treatment, gains should also be seen in the use of other morphemes requiring tense, such as third-singular *-s* and past *-ed*. That is, if gains are seen in the children's use of copula forms, it should be because the children recognized that tense is obligatory in main clauses. This insight should carry over to finite verb inflections.

The example is not yet complete. If the children showed the broad gains consistent with the predictions of the extended optional infinitive account, skeptics might claim that they were the result of a maturational spurt rather than of newfound knowledge that tense is obligatory. In fact, if there were a slight lag in gains made on the verb inflections following treatment on copula forms, this argument might seem even more plausible. Yet, such a pattern could still be consistent with the extended optional infinitive account. Even if the children learned from treatment on copula forms that tense is obligatory, there might not be an instantaneous application of this knowledge across the grammar. To guard against the maturation interpretation, measures should also be obtained of the children's pre- and posttreatment use of grammatical morphemes theoretically unrelated to tense. For example, the children's use of the genitive 's inflection might also be monitored. If this inflection shows smaller gains than inflections involving tense, it would be difficult to attribute the findings to general maturation.

The Evolution of Experimental Design in the Study of SLI

The language abilities of children with SLI lag behind those of same-age peers in many ways. For purposes of careful description, comparison groups in addition to age controls might be necessary. The choice of comparison groups should be tailored to the details of language of primary interest. By using more than one comparison group and measuring more than one feature of language, it is usually possible to determine more precisely how children with SLI differ from normally developing children. However, to understand the relationships that hold among these features of language—and hence understand the nature of the impairment—methods of observing these relationships in action must be adopted. Treatment designs represent one such method.

It will be clear from the chapters to follow that research in the area of SLI has been an evolutionary process. Across time, comparisons have become more sophisticated, and the hypotheses that have been advanced have been put to a stronger test. The methodological considerations outlined in this chapter will be apparent. We turn now to a detailed look at the descriptive data.

PART II

Describing the Data: Linguistic and Nonlinguistic Findings

Chapter 3

The Language Characteristics of SLI: A Detailed Look at English

In this chapter, we review data on the production and comprehension of English by children with SLI. The amount of evidence on this subject is considerable, and much of it will be covered here. The evidence comes mainly from standard dialects of British and American English. However, in all cases, the reference point is the child's own dialect group. From the data presented, a picture of the linguistic strengths and weaknesses of English-speaking children with SLI should emerge. Just as important, a significant portion of this evidence will be put to use in evaluating the competing accounts of SLI reviewed in later chapters.

The data are organized in terms of the dimension of language studied (e.g., the lexicon, morphosyntax). However, the problems of children with SLI do not always fall neatly into one of these categories. For example, in a study of 256 children with SLI, Miller (1996) found that 107 fell more than 1 SD below the norm in their (frequent) use of pauses, and 98 of the children were more than 1 SD below the norm in the number of words produced per minute. Such fluency and rate problems are probably tied to difficulties in sentence formulation, but their precise source is unclear.

Lexical Abilities

The Early Lexicon

Children with SLI appear to be late in acquiring their first words. This fact, long taken for granted, has an unexpectedly small database. Much of the available evidence comes from early case studies. For example, Bender (1940) observed a child who failed to produce words until after age 4;0. A child studied by Werner (1945) did not begin using words until after age 5;0. In one of the few investigations dealing with more than a single child, Morley, Court, Miller, and Garside (1955) noted ages of first-word acquisition ranging from 1;6 to 5;0 among 15 children with SLI. Nice (1925) reported one child's lexical development beyond the first word. Five words were acquired by age 2;0; by age 3;0, the child had not yet acquired 50 words. Weeks (1974) reported a child whose production lexicon did not reach 50 words until age 2;4.

More recent work with larger numbers of children confirms the impression given by earlier case studies. A retrospective study of 71 children with SLI based on parental report by Trauner, Wulfeck, Tallal, and Hesselink (1995) indicated an average age of first words of almost 23 months, compared with an age of almost 11 months reported by parents of normally developing children.

The finding that children with SLI acquire their first words at a later-than-expected age is not surprising, but it is not logically necessary. It could have been possible, for example, for first words to appear on schedule but for lexical development thereafter to flounder, or for lexical development to proceed normally up to the point of word combinations. That this is not the case suggests that signs of language problems are evident at least from the point when communication is expected to take a verbal form.

The types of words used by children with SLI during the early period of language development seem to match the types observed in the speech of young normal children. General nominals (e.g., names of objects, substances, animals) constitute approximately 55% of lexical types, whereas words referring to actions and to properties each represent about 12% (Leonard, Camarata, Rowan, & Chapman, 1982). As has been found in normally developing children, children with SLI seem to vary in whether they use many general nominals and few personal-social words (e.g., *please, thank-you, peek-a-boo*), or the converse (Weiss, Leonard, Rowan, & Chapman, 1983). Form class errors are not at all common. Early case studies described a few such errors, as in *brooming* and *barefeeting* (Weeks, 1975). However, these are sensible overgeneralizations of rules for deriving verbs from other parts of speech. Furthermore, Rice and Bode (1993) found that form class errors occurred only seven times out of 2,998 verb tokens in the speech of three preschoolers with SLI.

By the time children with SLI begin to produce multi-word utterances, their lexical abilities are not so easily characterized as matching those of younger normally developing children. Verbs, in particular, begin to show deficiencies that seem to go beyond the general lag in these children's lexical abilities. Fletcher and Peters (1984) found that one of the variables that served to differentiate preschool-age children with SLI from age controls in a discriminant function analysis was verb type frequency; higher frequencies were seen for the age controls.

Watkins, Rice, and Moltz (1993) found that preschool-age children with SLI used a more limited variety of verbs than MLU controls as well as age controls, though the verbs that they used with high frequency (e.g., *do* as a main verb, *go, get, put, want*) were also among the most frequently used by both control groups. In a study of three preschoolers with SLI, Rice and Bode (1993) reported a similar list of frequently used verbs. These kinds of discrepancies between children with SLI and MLU controls have not been reported for other types of words; indeed, it appears that certain word types, such as locative terms, can be more advanced than expected on the basis of MLU (Stockman, 1992).

Lexical Learning in the Preschool Years
In a series of studies, Leonard, Schwartz, and their colleagues used an experimental task to examine lexical learning by young children with SLI (Leonard, Schwartz, Allen, Swanson, & Loeb, 1989; Leonard, Schwartz, Chapman, Rowan, Prelock, Terrell, Weiss, & Messick, 1982; Leonard, Schwartz, Swanson, & Loeb, 1987; R. Schwartz, 1988; R. Schwartz, Leonard, Messick, & Chapman, 1987). Children participated in play sessions during which nonsense names of novel objects and actions were provided by the experimenter. These names were presented several times per session for up to ten sessions, depending upon the study. The children's comprehension and production of the words were then tested.

Comparisons were drawn between three- to four-year-old children with SLI at the single-word production level and younger normally developing children with comparable lexical sizes and utterance lengths. Across studies, the two groups performed in a similar manner. Both groups comprehended and produced more object words than action words, and comprehension was superior to production. Action words referring to intransitive actions were, if anything, learned better by children with SLI than by the younger controls. This advantage was not apparent for words referring to transitive actions. The children with SLI were less able to extend newly learned object names to unnamed but appropriate exemplars in a comprehension task.

Chapman, Leonard, Rowan, and Weiss (1983) employed a similar design but expanded the referents used during testing to those for which the newly introduced words would be inappropriate from the adult standpoint. The use of inappropriate extensions by the children with SLI resembled that seen for younger normally developing children. In the majority of cases, the referent receiving the inappropriate label was both perceptually and functionally similar to the true referent of the word.

The findings of these studies were surprising. Earlier studies had suggested that during the period of development in question, lexical acquisition in children with SLI was slow. Yet the children with SLI in these studies acquired as many words as the control children. In one study, they did not extend these words to new referents to the degree seen in the control children, but some appropriate extensions were observed; and, in another study, so were overextensions. Measures obtained from these children one year later indicated that they still exhibited significant language deficits. At this point, however, the acquisition of morphosyntax appeared to be their biggest obstacle.

In the studies conducted by Leonard, Schwartz, and their colleagues, each new word was presented numerous times across sessions, in the same few play contexts and in the same few sentence frames. It is not known if only a few presentations would have produced the same similarities between the two groups of children. This question has been pursued in studies examining "fast-mapping" ability in children with SLI, that is, the ability to form an initial association between a word and its referent after only one or two exposures of the word.

Dollaghan (1987) seems to have been the first to apply a fast-mapping paradigm to the study of children with SLI. She studied a group of four- to five-year-old children with prominent deficits in the production of morphosyntax. The children were found to be comparable with a group of age controls in correctly associating the nonsense name *koob* with an unfamiliar object on a comprehension task. However, the children with SLI performed below the level of the control children in their production of this word.

Rice, Buhr, and Nemeth (1990) employed a task in which five unfamiliar names from each of the categories of objects, actions, attributes, and affective states were presented to five-year-old children with SLI in a television story format. The words were incorporated into stories and appeared ten times each. The children with SLI showed poorer overall mapping ability on a comprehension task than did both age controls and MLU controls. The names of actions were especially difficult for each group of children.

In a study of object and attribute names only, Rice, Buhr, and Oetting (1992) found that children with SLI were more limited than age controls in associating these words with their referents on a comprehension measure. Rice, Oetting, Marquis, Bode, and Pae (1994) observed that children with SLI showed no evidence of learning a set of new object and action names after only three exposures of each word. Learning was apparent for words presented ten times each, but the gains seen for action names were not retained. The action names showing a drop in retention were presented in contexts requiring the past tense inflection *-ed*. Rice, Oetting, Marquis, Bode, and Pae speculated that the children might have had difficulties recalling the action terms because features such as tense were not adequately stored, and thus the representation of these words in memory was insufficiently elaborate.

Verbs differ widely in the types of meanings they convey. Thus, it is possible that differences between children with SLI and controls might have as much to do with the distribution of the types of verb meanings employed in the study as with the fact that they were verbs. Kelly and Rice (1994) obtained preliminary evidence of this type by examining children's preferences for interpretation of novel verbs enacted in videotaped scenes. The distinction between change-of-state verbs (e.g., *break*) and motion verbs (e.g., *jump*) was of primary interest. Novel actions corresponding to these meaning types were presented simultaneously on a split screen, and the children heard nonsense words presented in a short sentence frame that was not biased toward either type of meaning. The children selected the scene that they felt corresponded to the sentence. In another condition, the children labeled the scenes. A group of five-year-olds with SLI and a group of MLU controls showed no preference according to meaning type. In contrast, a group of age controls showed a clear preference for a change-of-state interpretation.

Investigators have also examined whether the manner in which the novel words are presented plays an important role in lexical learning. Words appearing as bare stems in sentence-final position appear to be learned most easily by children with SLI (e.g., Leonard, Schwartz, Chapman, Rowan, Prelock, Terrell, Weiss, & Messick, 1982). Adding a pause before the novel word at the end of the sentence does not provide any additional benefit (Rice, Buhr, & Oetting, 1992). Words presented in inflected form in varying sentence contexts (Haynes, 1982) and words appearing in sentences presented at a rapid rate (Ellis Weismer & Hesketh, 1993) have a more detrimental effect on children with SLI than on control children.

It can be seen, then, that the lexical abilities of preschoolers with SLI generally parallel those of MLU controls, with verbs in even greater peril. We will return to verb acquisition in the sections on argument structure and morphosyntax.

Lexical Abilities During the School Years

Experimental studies of word learning have been extended to the school years. For example, Oetting, Rice, and Swank (1995) reported findings for learning as a function of word type that were very similar to those seen for preschool children with SLI. Of particular importance was the finding that the school-age children with SLI learned object names almost as well as did age controls, but their learning of action names fell well below that of their same-age peers.

The lexical limitation most frequently identified in the literature on school-age children with SLI is a "word-finding" problem, that is, a problem in generating the particular word called for in the situation (e.g., German, 1987; McGregor & Leonard, 1995; Rapin & Wilson, 1978; P. Weiner, 1974). These difficulties have been variously described as "lexical look-up" problems (Menyuk, 1975, 1978) and problems involving "delayed speed of word retrieval" (E. Schwartz & Solot, 1980). According to the clinical literature, the chief symptoms of word-finding problems are unusually long pauses in speech, frequent circumlocution, and/or frequent use of nonspecific words such as *it* or *stuff*. Naming errors also occur rather frequently. Usually substitutions are semantic in nature, such as *shoes* for "pants," but phonological substitutions such as *wrangler* for "ankle" are not infrequent (Casby, 1992; Fried-Oken, 1984; McGregor, 1994; Rubin & Liberman, 1983).

Earlier descriptions of word-finding difficulties—as the very term suggests—assigned the blame to problems of retrieval. Children with SLI showing a greater number of naming errors than did age controls gave evidence of comprehending the target word's meaning when tested in a picture-pointing task (Rubin & Liberman, 1983; Wiig, Semel, & Nystrom, 1982). Thus, it seemed possible that the words were present in memory but the children used an inefficient or inappropriate means of accessing them.

However, another possibility exists. Words are not represented in memory in an all-or-none fashion. Some words have a richer network of associations and/or a stronger set of associations in memory than other words. Many of these connections are semantic in nature, but others are grammatical and still others are phonological. In effect, two words can be known but one can be known better than the other. The richer and stronger the network of associations, the more readily the words can be retrieved, much like the claw that is guided to pick out a toy in an amusement park game. The claw will have a better chance of latching onto a toy that has several firm appendages than an otherwise similar toy with no appendages. In the former case, retrieval can tolerate a margin of error; chances are good that what is grasped will lead to the target. Search in the latter case is more likely to be unsuccessful.

This is readily demonstrated with normally functioning adults. By including names with low frequencies of occurrence in a picture-naming task, adults will occasionally make errors on items they can identify in a picture-pointing task. Presumably, less frequent words have been encountered less often and have a correspondingly weaker network of associations. But a more sensitive measure is response time. Even when correct responses only are considered, adults show slower response times in naming pictures with low-frequency names than those of higher frequency. To continue the analogy with the amusement park claw, successful grasping of the object without appendages requires more precision, and hence more time.

If one imagines SLI as a type of filter such that some but not all experiences with a word are registered in semantic memory, then it seems reasonable to suspect that the strength and number of associations in the semantic memories of children with SLI are weaker and fewer than is the case for age-mates. The resulting network of associations would be akin to that seen in younger normally developing children and comparable with fewer toys with appendages and fewer appendages on the toys that have appendages. Although this would have a detrimental effect on retrieval, the problem is not one of retrieval.

Anderson (1965) may have been the first investigator to examine naming response times in children with SLI. A group of school-age children with SLI named pictures of common objects with slower response times than a group of age controls. Similar findings were observed by Wiig, Semel, and Nystrom (1982) with stimuli that included color and shape names. Katz, Curtiss, and Tallal (1992) found that children with SLI were slower than age controls in the number of pictures they could name in a 60-second period. Leonard, Nippold, Kail, and Hale (1983) employed pictures whose names varied in frequency of occurrence. Children with SLI named pictures less rapidly than age controls but more rapidly than a group of younger controls matched according to composite language test score. For all groups, naming times were significantly related to frequency of occurrence.

Kail and Leonard (1986) studied naming times in three conditions. In one condition, the presentation of the picture was preceded by a sentence that the name of the picture logically and grammatically completed. In a second condition, the picture again completed a sentence, but the sentence appeared in a larger text. The third condition was the standard condition of picture naming with no prior linguistic context. Response times for children with SLI were faster with increasing amounts of supportive linguistic context. However, in all conditions, the children with SLI were slower than age controls. These findings indicated that the children with SLI used linguistic information to guide retrieval, as did the control children. The fact that they were generally slower didn't seem to be tied to a specific deficit in retrieval.

Ceci (1983) employed a different type of paradigm to examine picture naming in children with SLI. Pictures were preceded by three types of prompts: one that should have facilitated retrieval of the correct name (e.g., "Here's an animal" prior to presentation of a picture of a horse), one that was neutral ("Here's something you know"), and one that should have started retrieval down the wrong path (e.g., "Here's a fruit" prior to presentation of the horse picture). Ceci varied the ratio of helpful to harmful prompts at different points in the study. The response times of the children with SLI varied with the type of prompt, but not to the same extent as for age controls.

Other types of lexical abilities have been examined by using response time. For example, Kail and Leonard (1986) found that children with SLI were slower than controls in making judgments about whether a presented picture had been included in a series of pictures shown a moment before. Sininger, Klatsky, and Kirchner (1989) reported slower response times for children with SLI compared with age controls in a similar task employing digits. In another experiment by Kail and Leonard (1986), children with SLI were slower than controls in responding to picture pairs according to whether the two pictures were physically identical or shared the same name.

Montgomery, Scudder, and Moore (1990) used a word monitoring task in which children pressed a panel as soon as they heard a particular target word. The words appeared in grammatical and normally produced sentences, grammatical sentences produced in a monotone, and monotone strings in which words were in scrambled order. The response times of the children with SLI were similar to those of younger children matched according to language test score, though once response times were corrected for simple motor-response time differences between the groups, the children with SLI were slower. However, the two groups showed the same systematic increase in response time as a function of the degree to which the stimuli departed from normal sentences. R. Stark and Montgomery (1995) found that the response times of children with SLI became faster when target words appeared in sentences as opposed to word lists. As with age controls, the response times of the children with SLI were faster when the target word appeared later rather than earlier in the sentence. Both groups' response times became faster when the sentences were time compressed, and slower when the sentences were low-pass filtered, allowing only information below 2 kHz to appear. Despite the similar pattern of performance, the response times of the children with SLI were slower than those of the age controls.

In all of these studies, the children with SLI showed slow response times, but their responses varied with the presentation condition (e.g., number of items in a comparison set, amount of linguistic information) in the same way as the responses of the control children. Nothing implicated retrieval as a special source of difficulty.

Recent evidence suggests that the response time data obtained in these studies might not be representative of all children with SLI. Lahey and Edwards (1996) and J. Edwards and Lahey (1996) have shown that on picture-naming tasks and tasks requiring judgments of whether a phonetic string is a real word, only children with SLI who have some degree of comprehension problem in addition to a production deficit seem to respond with longer latencies. Because most of the children with SLI participating in earlier studies had both comprehension and production problems, this point might have been missed.

Word recall tasks also have provided data consistent with the idea that known words are insufficiently elaborate in the memories of children with SLI rather than retrieved improperly. Lists of familiar words are recalled less well by children with SLI than by age controls (Kirchner & Klatsky, 1985; Sommers, Kozarevich, & Michaels, 1994); when words can be grouped into categories (e.g., clothing, animals), performance improves, though it falls below levels seen for control children in the same condition (Kail, Hale, Leonard, & Nippold, 1984). In a task in which children listen to stories and attempt to recall a word in the story that immediately followed a probe word, recall is poorer by children with SLI than by age controls, but the two groups' responses show similar effects attributable to the number of words intervening between the target word and the recall prompt, and the type of clause (main or subordinate) in which the target word appeared (Kail & Leonard, 1986).

In free recall tasks requiring children to generate as many items from a category (e.g., animals, items of furniture) as they can think of, children with SLI generate fewer items than age controls and the items generated reflect fewer subcategories (e.g., farm animals, jungle animals) than the responses of their age-mates. However, the organization of their responses, as defined by the order in which items are generated and the pauses between items from different subcategories, gives every indication that the retrieval process is the same as that of normally developing children (Kail & Leonard, 1986). McGregor and Waxman (1995) reported similar semantic organization reflected in the responses of children with SLI and control children. The groups differed primarily in the tendency of the children with SLI to rely on morphological cues to generate subordinate terms. For example, the control children could provide an appropriate subordinate term such as *sunflower* whether the cue was *rose* or *cornflower*. The children with SLI, in contrast, were significantly aided by the latter, presumably because of the commonalities in form.

Early Word Combinations

In the 1960s and 1970s, much work in normal language development was devoted to describing the nature of the early word combinations used by children. (Reference lists at the time were heavily weighted in the letter "B"—Bloom, Braine, Brown, and Bowerman.) The kinds and breadth of relational meanings reflected in these utterances were a source of considerable debate. For certain kinds of notions, such as recurrence and rejection, a single word (*more, no*) often combined with another word (e.g., *more juice, no syrup*). For other notions, pertaining to people performing actions on objects, or the attributes or locations of objects, the means used for expression varied. In many instances, cover terms such as agent + action, attribute + object, and locative state seemed appropriate. In some instances, the words used in combination were more restricted; agents were limited to the words *mommy* and *daddy*, or attributes were limited to *big* and *little*, for example. Still other cases suggested that broader terms might be appropriate, representing full-fledged grammatical categories such as noun and verb.

The work on the semantic notions reflected in normally developing children's early word combinations sparked interest in the word combinations of children with SLI. Most of this work was performed from the mid-1970s to the mid-1980s, though important information has appeared since that time.

Not surprisingly, age of first word combinations appears to be later in children with SLI than in normally developing children. Trauner, Wulfeck, Tallal, and Hesselink (1995) found average ages of almost 37 months and 17 months for the two types of children, respectively.

Case studies have reported word combinations that reflect rather narrow meanings in children with SLI (Leonard, 1984; Leonard, Steckol, & Panther, 1983). For example, Leonard, Steckol, and Panther observed a child with SLI age 3;11 whose word combinations could be divided into a small number of combinatorial rules narrower in scope than notions such as agent + action, attribute + object, and the like. Utterances of the first type were limited to *me* as the agent (e.g., *me do, me make, me reach*); utterances of the second type were best described as reflecting size + object, for all of the combinations involved *big, little,* or *baby* (e.g., *baby rabbit*). Another child, age 3;2, used locative utterances (e.g., *cup box, ring block, dog paper*) only when one object was physically on top of another object. Narrow meanings such as this are probably not a diagnostic marker of SLI. As noted earlier, many normally developing children show word combinations of this type (e.g., Braine, 1976; D. Ingram, 1979).

Leonard, Steckol, and Schwartz (1978) found that preschoolers with SLI were generally similar to MLU controls except that agent + action and action + object (two early notions) were used more frequently by the children with SLI, whereas experience + experiencer (e.g., *scare Heather*), a later notion, was used more frequently by MLU controls. Leonard, Bolders, and Miller (1976) and Freedman and Carpenter (1976) found no differences in the semantic relations reflected in the utterances of children with SLI and MLU controls.

The evidence on early word combinations seems to indicate considerable similarity between children with SLI and MLU controls. However, the findings of Leonard, Steckol, and Schwartz (1978) raise the possibility that there is a slight lag in the development of relational meanings relative to MLU in children with SLI.

The Lexicon Meets Syntax: Argument Structure

In the section on the lexical abilities of children with SLI, we saw that verb learning seemed most precarious. One possible reason for this is that the learning of verbs involves extra obstacles, which might be especially difficult for children with SLI to overcome. Gleitman and her colleagues (e.g., Gleitman, 1990; Gleitman & Gleitman, 1992) have demonstrated that the meaning of many verbs cannot be learned on the basis of simple exposure to events and the verbs that describe them. Rather, it appears that the learner must also be provided with the sentence frame in which the verbs appear. Through such "syntactic bootstrapping," the child can refine the meaning of the verb, eliminating other plausible interpretations.

Unfortunately, the process of interpreting meanings based on sentence frames might play into one of the major weaknesses of children with SLI. Grammatical limitations in these children are significant. As Rice (1991) has put it: "The end result would be the opposite of bootstrapping. Instead of using one area of language to build another, SLI children would be left without a solid strap to hang onto" (p. 455).

However, grammatical information comes in different forms. One aspect of grammar that might play a large role in contributing to children's discovery of verb meanings is argument structure. For example, if children can identify the noun phrases in a sentence such as *Carol zimmed the dog in the river*, they could guess that *zimmed* refers to an action involving transfer of an object to a particular location. (Additional sentence frames would help, of course. It might be that *zimmed* refers to an action such as washing. A sentence containing the same verb without the location could clear this up.) We begin with a review of studies designed to examine how well children with SLI produce obligatory arguments in their sentences. Then we will consider studies that looked at these children's ability to deduce argument structure on the basis of syntactic information.

The Use of Arguments in Spontaneous and Elicited Utterances

One of the earliest studies involving argument structure was conducted by Lee (1976). She compared preschoolers with SLI and younger normally developing children matched according to a measure of syntactic development. The children with SLI and the control children were similar in their use of argument structures. The less developed children in each group were more likely to omit obligatory arguments (e.g., *Doggie get*; *He put his finger*), but the two groups of children did not differ in this respect.

Rice and Bode (1993) found very few instances of omitted arguments in the spontaneous speech of three preschoolers with SLI. Subject omissions were even less frequent than object omissions. These investigators noted that the speaking contexts allowed for considerable subject elision, and elided subjects were not treated as cases of subject omission.

J. Roberts, Rescorla, and Borneman (1994) compared three-year-old children with SLI and age controls on the use of arguments. The vast majority of utterances (approximately 85%) produced by the children with SLI lacked an obligatory argument or the main verb itself. This number exceeded the number seen for the age controls (approximately 60%). This finding is not especially surprising, given the MLU differences between the two groups. The MLUs of the age controls averaged 4.1 morphemes, whereas the mean for the children with SLI was 2.6.

Grela and Leonard (1997) compared preschoolers with SLI and MLU controls on the arguments reflected in their spontaneous speech. They focused primarily on the presence of subject arguments as a function of the kind of verb used. Two factors seemed relevant. One was the number of arguments. It seemed possible that verbs requiring a greater number of arguments (e.g., transitives versus intransitives) might be more likely to be used with the subject missing, due to constraints on length or complexity. In addition, within the class of intransitives, verbs can be unergative or unaccusative. The latter kind of intransitive (seen in verbs such as *fall*) employs subjects that are themes rather than agents. In most constructions, themes are in postverbal rather than preverbal position. In contrast, the subjects of unergative verbs (e.g., *sing*) are agents, which are usually in preverbal position. Grela and Leonard (1997) asked whether themes located in subject position would cause problems for children with SLI. If so, subject omissions would be more frequent for unaccusative verbs than for unergative verbs. For both the children with SLI and the MLU controls, the number of arguments of the verb had no effect on the children's likelihood of producing subject arguments. However, the children with SLI produced fewer subject arguments with unaccusative verbs than did the MLU controls. This difference was not seen for the unergative verbs.

Determining the argument structure of certain verbs is complicated by the fact that some morphemes, such as *in, on,* and *over,* can be either prepositions (e.g., *Jump over the table*) or particles (e.g., *Push over the table*). Watkins and Rice (1991) found that children with SLI used these forms as prepositions to the same degree as did MLU controls (though less than age controls); their use of the same forms as particles was more limited than that of the MLU controls. Though ungrammatical, omission of particles does not alter argument structure, whereas omission of prepositions can change goals to themes or patients. Object noun phrases and verbs were also omitted more frequently by the children with SLI than by the children in the other two groups.

By school age, argument errors are infrequent in the spontaneous speech of children with SLI, at least in absolute number. G. King and Fletcher (1993) observed that such errors occurred in fewer than 3% of the verb tokens used. This percentage was not significantly higher than the percentage seen for a group of MLU controls. The control children's argument errors were usually limited to omissions of the object or location argument for the verb *put.* The children with SLI also had some difficulty with this verb, but those making argument errors attempted fewer instances of *put* than was true for the MLU controls. The children with SLI showed a greater diversity of verbs with argument errors. The examples in (1) are representative of those noted in the speech of the children with SLI.

(1) a.   you can take over there
     b.   don't let out
     c.   there it's something white but I can't find

In a comparison of children with SLI age six to nine years and normally developing five-year-olds, Fletcher (1991) reported higher frequencies of errors described as formulation errors on the part of the group with SLI. These included argument errors.

Some investigators have studied children's control of verb alternation to examine the children's ability to use the same verb with different argument structures, or the same argument structure in different syntactic arrangements. In a study by G. King, Schelletter, Sinka, Fletcher, and Ingham (1995), school-age children with SLI and younger (vocabulary-matched) controls watched a video designed to elicit descriptions that tapped two different types of verb alternations. One of these was the causative/inchoative alternation in which the subject is either the agent (e.g., *The boy is bouncing the ball*) or the theme (e.g., *The ball is bouncing*). The other alternation applies to locative and contact verbs where the direct object carries the role of the theme and the prepositional phrase expresses the goal (e.g., *She's loading the bricks onto the toy truck; He's scraping the shovel along the wall*) or the goal serves as direct object and the theme appears as a prepositional phrase (e.g., *She's loading the truck with bricks; He's scraping the wall with a shovel*).

The children with SLI produced relatively few alternations; they were more likely to use one argument order even when the events of the video promoted a change in order. However, the numerical difference between these children and the younger controls was not statistically significant. When G. King, Schelletter, Sinka, Fletcher, and Ingham (1995) compared the data from the two groups of children with the data from adults, the adults produced significantly more alternations than the children with SLI but not more than the control children.

Loeb, Pye, Redmond, and Richardson (1994) also examined causative alternations. In addition to verbs permitting the causative/inchoative distinction (e.g., *We broke the window; The window broke*), these investigators examined intransitive verbs that require a periphrastic construction to appear in transitive contexts (e.g., *The baby cried; We made the baby cry*) and transitive verbs that had to be passivized to appear in an intransitive context (e.g., *We put the frog in the bathtub; The frog was put in the bathtub*). Children with SLI age five to seven years were comparable with age controls on the causative/inchoative distinction but had greater difficulty than the control children on the periphrastic and passive responses required.

It appears that the causative/inchoative alternation is a productive rule, such that after hearing alternations such as *The window broke* and *We broke the window*, children sometimes generate utterances such as *We cried the baby* along with *The baby cried*. Loeb, Pye, Redmond, and Richardson (1994) found evidence for such overgeneralizations in the responses of several children with SLI and several age control children. However, this similarity between the two groups was deceptive because the age controls were more facile with the correct (periphrastic) constructions for the verbs in question and as a result had fewer opportunities to overgeneralize. When Loeb, Pye, Redmond, and Richardson examined the data from a younger control group, they found that overgeneralizations were more abundant than for the other two groups. This finding suggested that the rules of the children with SLI were not as productive as their correct use of the causative/inchoative alternation might suggest. Much of their correct use might have been built up through exposure to the specific verbs in each construction.

Unlike arguments, adjuncts are substantive elements of a sentence that are not grammatically obligated. Nevertheless, they play an important role in specifying meaning. Fletcher and Garman (1988) examined adjuncts that express information about time, location, or manner of action, as in "We washed the car *yesterday/in the driveway/in haste.*" These forms, often called adverbials, were examined in the spontaneous speech of seven- to nine-year-old children with SLI and normally developing three-, five-, and seven-year-olds. The children with SLI most resembled the five-year-old controls in the structure of their adverbials (most were single words such as *yesterday*) and the sentence locations of these adverbials (usually they followed the verb).

However, although the children with SLI resembled the five-year-olds in the percentage of adverbials that expressed time, the two groups differed in an important respect. When the context did not provide cues that specified reference time, the younger control children used a temporal adverbial nearly three times out of four. The children with SLI, in contrast, provided a temporal adverbial only about one-quarter of the time. The children with SLI did not seem to appreciate that an utterance such as *We left* was not helpful without context; a specification such as *yesterday, at three o'clock,* or *last Saturday* was needed. Wren (1980) found adverbials (in addition to inflections) to be one of the major elements that distinguished a group of children with SLI from age controls. Johnston and Kamhi (1984) found adverbials to be used less frequently by a group of children with SLI than by a group of MLU controls.

The findings on argument structure and related phenomena point to subtle but identifiable problems in this area for children with SLI. Most argument structures seem to be reflected in these children's speech, though consistency is lacking. Elements that add substantively to the message without being obligated by the verb (viz, adjuncts) seem to constitute a real stretch for these children, and are often left out in places where normally developing children would include them.

Interpreting Argument Structure from Syntactic Information

Once the process of grammatical development is well under way, children with SLI seem capable of using a new verb with the appropriate arguments if they have had the benefit of observing events for which the verb is appropriate (van der Lely, 1994). That is, if the children hear a new word produced in isolation (e.g., *sloodge*) in the context of a new action being performed with familiar characters and objects (e.g., the mother drop-kicking a can into a recycling bin), they can construct an appropriate sentence for the occasion (e.g., *Mommy sloodging can in recycling*). This ability is helped considerably, according to van der Lely (1994), because once the arguments are known (through observing the event), the possibilities for syntactic organization are fairly limited. Agents are usually subjects, themes are usually objects (especially if there is an agent), and goals are usually prepositional phrases. In fact, this characterization of the event probably does not do justice to the information actually available to the child. An agent is causing the object to move through the air (in end-over-end fashion if the drop-kick is well executed) toward and into a container as the result of movement of the leg and contact between foot and object.

The narrowness of this characterization is only slightly exaggerated; Pinker (1989) has proposed that correct (and only correct) application of the dative alternation rule in (2) requires knowledge of narrow-range lexical rules of the type shown in (3).

(2)   to cause X to go to Y (e.g., *Jill throws the ball to Danny*) → to cause Y to have X (e.g., *Jill throws Danny the ball*)

(3)   instantaneous imparting of force to an object, causing ballistic physical motion → causing someone to possess an object by means of instantaneously imparting force to it

Narrow-range lexical meanings that do not share many of the semantic details with the verbs that participate in the rules are prevented from being included in these rules. If this is the case, the possible sentence frames for events observed by the child (assuming the events themselves are not ambiguous) must be rather limited.

In an experimental task that examined this kind of ability, van der Lely (1994) found that school-age children with SLI performed as well as younger control children matched according to language test scores. However, when the task was changed so that visual information was eliminated, the results were quite different. In this task, the children only heard a new word in a sentence (e.g., *The car fims the train to the lorry*) and had to act out relationships among a set of objects in a way that seemed sensible. The children with SLI performed significantly more poorly than the control children. This deficit was not seen if the children had already seen the verb acted out with another set of props.

The reason for the poor performance when the children were asked to respond to a sentence containing a verb never previously associated with a specific action, according to van der Lely (1994), is that it is much more difficult to select an argument structure from a syntactic frame than vice versa. The same syntactic frame can accommodate dozens of relationships of the sort described by Pinker (1989). Therefore, the child must have a more detailed representation of the syntactic frame, one that specifies all the arguments within the frame. According to van der Lely (1994), children with SLI lack this detailed knowledge of structural relationships.

This does not mean that children with SLI ignore syntactic frames when responding to verbs. For example, Hoff-Ginsberg, Kelly, and Buhr (1995) found that five-year-old children with SLI would modify their interpretation of known verbs upon hearing them in novel frames (e.g., *The giraffe falls the camel*). Kouri, Lewis, and Schlosser (1992) observed that five-year-old children with SLI could use syntactic information to decide whether a novel verb was causal. According to O'Hara and Johnston (1997), a critical factor might be limitations in processing capacity. They obtained findings essentially identical to those of van der Lely (1994), using a similar procedure.

However, inspection of the errors committed by the children led them to believe that the children's knowledge was underestimated by the scoring procedure. The children with SLI included critical components such as causation, contact, and object movement according to the sentence frame requirements. In contrast, details that were more vulnerable to working memory limitations were more likely to be lost. For example, errors on the noun phrase showed recency effects (e.g., greater accuracy on *lion* than on *monkey* in *The doctor fets the monkey to the lion*) and the sentences were sometimes acted out with use of an incorrect object that shared most of its semantic features with the appropriate choice (e.g., choosing a toy bear instead of a toy cow in *The farmer voofs the cow and the boy*). The error types of the children with SLI were the same as those of the control children; they were simply more numerous.

Morphosyntax

Many of the points made in the discussion of argument structure were made more easily because we took morphosyntax for granted. For example, differences between *She's loading the bricks onto the truck* and *She's loading the truck with bricks* are difficult to fathom if one has no knowledge that word order matters or that prepositional phrases differ from noun phrases. In this section, morphosyntax is the focus of attention. For convenience, we divide the discussion of the data on morphosyntax into syntactic structure and grammatical morphology. The former is concerned with structural relationships between constituents. These are the facts of language that dictate word order, permit proper movement of constituents (e.g., *Is the boy who is on the beach your brother?* and not *Is the boy who on the beach is your brother?*) and substitution through pronominalization (e.g., *I need my shoes, and I need them now!* and not *I need my shoes and I need my them now!*), among other operations.

Grammatical morphology pertains to the closed-class morphemes of the language, both the morphemes seen in inflectional morphology (e.g., "play*s*," "play*ed*") and derivational morphology (e.g., "fool," "fool*ish*"), and function words such as articles and auxiliary verbs. As put so well by R. Brown (1973, p. 249), grammatical morphemes share the characteristic of representing a sort of ivy growing up between and upon nouns and verbs.

However, the division between syntactic structure and grammatical morphology is somewhat artificial. For example, auxiliary verbs are needed for framing *wh-* questions and using the passive voice. In turn, the form that pronouns take is dictated by their structural position. Pronouns serving as the object of the verb assume the accusative case, for example, (*Mary saw him* and not *Mary saw he*). These inter-relationships make it clear that problems with grammatical morphology will have ramifications for syntactic structure, and vice versa. Weaknesses in one of these areas might be more serious than weaknesses in the other, but it seems implausible that one area could be deficient and the other problem-free. As we will see, difficulties in both of these interrelated areas are evident in children with SLI.

Syntactic Structure

Probably the first systematic study of the morphosyntax of children with SLI was made by Menyuk (1964). In what proved to be the prototype for future studies, Menyuk collected spontaneous speech samples from children with SLI age three to five years and normally developing children matched according to age. The samples were then analyzed for evidence of the transformations assumed in then-current theories of grammar. These included the use of auxiliary inversion in questions, formation of the passive voice, insertion of negative particles between the auxiliary and main verb, and use of an infinitival complement.

Deviations from the adult grammar in terms of substitution, redundancy, or omission also were examined. The results indicated that a greater number of age controls showed evidence of transformations, whereas a greater number of children with SLI deviated from the adult grammar. Omissions were the most common type of deviation. Lee (1966) compared the spontaneous speech of a normally developing child age 3;1 and a somewhat older (age 4;7) child with SLI in terms of evidence of different types of sentence constructions. The child with SLI did not use some of the constructions seen in the speech of the normally developing child, suggesting that his grammar possessed a more restricted set of syntactic rules.

Leonard (1972) also employed an age control design but examined the frequency with which children with SLI and control children used particular types of utterances as well as the number of children using them. Although the number of children showing evidence of transformations was larger among the age controls, the most striking difference between the two groups (in the same direction) was in frequency. Errors of omission, on the other hand, were much more frequent on the part of the children with SLI. Leonard also analyzed the children's samples in terms of a system developed by Lee and Canter (1971; see also Lee, 1974) that assigned developmental weight to particular morphemes such as conjunctions, indefinite pronouns, and *wh-* words. Again frequency differences favoring the age controls were seen. However, for several of the morpheme categories, the mean developmental level of the morphemes used by the children with SLI was as high as the level for the control children. Degree of use appeared to be the most dependable basis for distinguishing the groups of children.

Morehead and Ingram (1970, 1973) seem to be the first investigators to have employed a comparison group matched according to MLU. The children with SLI ranged from 5 to 8 years of age; the MLU controls were, by comparison, quite young, ranging in age from 20 months to just under 3 years. The two groups were quite similar in the syntactic rules reflected in their speech, with a few important exceptions. The children with SLI did not use major syntactic categories (e.g., noun, verb, embedded sentence) in as many different sentence contexts, on average, as the MLU controls.

Leonard, Sabbadini, Volterra, and Leonard (1988) reported evidence consistent with the Morehead and Ingram (1970, 1973) finding that major syntactic categories were present in the spontaneous speech of children with SLI. They applied criteria for categories such as noun phrase and prepositional phrase based on Valian (1986) and found that preschoolers with SLI met the distributional criteria for all categories examined. For example, nouns appeared with determiners in preverbal (e.g., *The man going home*) and postverbal positions (e.g., *I get the toys*), and following prepositions (e.g., *Put this in the box*); they were also replaced as a unit by pronouns in appropriate contexts (e.g., *I want the ball, I want it*). However, the criteria applied by Leonard, Sabbadini, Volterra, and Leonard (1988) were rather broad; had they employed the detailed system of defining sentence contexts used by Morehead and Ingram (1970, 1973), limitations in the range of application of these categories might have been detected.

Johnston and Kamhi (1984) measured grammatical structure in terms of a propositional complexity index. Structures that influenced this count included embedded propositions (e.g., "We knew *he was in trouble*") and propositions that served as modifiers of elements in the main proposition (e.g., "We looked at the *crying* baby"). Children with SLI averaging five years of age produced utterances with a lower number of propositions than a group of MLU controls.

D. Ingram (1972a) observed that when questions were categorized in terms of the type of *wh-* word used, children with SLI used the same types of questions as younger normally developing children with similar MLUs. Questions with *what* and *where* were more likely than *why* questions, with *who* and *when* questions the least frequent. Leonard (1995) analyzed the use of auxiliary inversion in the *wh-* questions used by children with SLI and MLU controls (e.g., *What can we make?*). He found that the children with SLI produced a higher percentage of questions with the auxiliary in declarative sentence position (e.g., *What we can make?*). Smith (1992) found that when using *wh-* questions containing embedded clauses (e.g., *What do you think Evelyn broke?*), children with SLI often produced a second *wh-* word in the embedded clause (e.g., *What do you think what Evelyn broke?*) or filled the slot serving as the presumed origin of the *wh-* word with a lexical item (e.g., *What do you think Evelyn broke something?*).

Given the test profiles of children with SLI, it would be reasonable to expect that these children's comprehension of syntactic structure would exceed their production abilities. As noted in earlier chapters, one common subdivision of SLI is between production problems and production plus comprehension problems. And even among children with limitations in both areas, it is likely that production problems will be more severe than comprehension problems. Nevertheless, it is not safe to conclude that children with SLI who are considered to have production deficits only are actually free of comprehension limitations. These subdivisions are based in part on standardized test scores; consequently, the content of the test and the degree to which the children under study match the demographic characteristics of the standardization sample are critical issues. Bishop (1979) has provided some valuable information in this regard. She administered vocabulary and grammatical comprehension tests to a group of school-age children with SLI classified as production-impaired (only) and a group of controls matched on a subsection of a nonverbal intelligence test. The controls scored higher on each type of comprehension test.

Bishop (1979) also compared the grammatical comprehension of a larger group of children with SLI (including the production-impaired subgroup) and that of a group of age controls and a group of younger controls matched according to vocabulary comprehension score. The grammatical comprehension items permitted an inspection of the children's understanding of many different aspects of grammar. The children with SLI performed more poorly than the age controls on all aspects of grammar, but lower than the vocabulary comprehension controls only on items assessing reversible passives such as *The cow is pushed by the boy*. The children with SLI performed like their younger controls on other items requiring an understanding of word order, such as reversible active sentences (e.g., *The boy chases the dog*) and prepositional phrases in which each noun possessed the attributes to play either role (e.g., *The cup in the box*).

In a subsequent study, Bishop (1982) examined the grammatical comprehension performance of a group of production-impaired children with SLI age 8 to 14 years. These children performed in a manner that closely resembled the performance of a group of eight-year-old normally developing children, the oldest control group employed. There was no evidence that the children with SLI consistently misinterpreted reversible passives.

Paul and Fisher (1985) described the comprehension of word order in a younger group of children with SLI, averaging 46 months of age. The children's scores on a general language comprehension test revealed mild deficits. The children were presented reversible active and passive sentences, some reflecting probable events (e.g., a girl carrying a baby), and some improbable events (a baby carrying a girl). The children performed better on active than on passive sentences, and better on sentences reflecting probable events than on those reflecting improbable events. However, the responses on the latter indicated that they had reached the point at which a word order strategy (Bever, 1970) was replacing a probable event strategy (R. Chapman, 1978).

A study by van der Lely and Harris (1990) revealed comprehension difficulties in which a word order strategy did not seem to apply. The children with SLI in their study ranged in age from four to seven years. Age controls and younger controls matched according to language production and comprehension test scores served as subjects. The sentences of interest included reversible active and passive sentences—referred to as canonical and noncanonical, respectively—as well as canonical and noncanonical locative sentences (e.g., *The cup is in the box; In the box is the cup*) and dative sentences (*Give the boy to the girl; Give the girl the boy*). Although the canonical sentences were generally easier for all of the children than were the noncanonical equivalents, the children with SLI performed more poorly than the language test score controls as well as the age controls even on canonical active sentences and dative sentences.

One possible explanation of these findings, according to van der Lely and Harris, is that the children with SLI had great difficulty assigning roles such as agent and theme or theme and goal on the basis of syntactic structure alone. Support for this interpretation came from a later study by van der Lely (1996). Children with SLI not only had greater difficulty with full passives (e.g., *The teddy is mended by the girl*) than did language test score controls, but also were more likely than controls to interpret similar sentences without the prepositional phrase (*The teddy is mended*) as an adjectival construction (choosing a picture of an already mended teddy seated in a chair with no agent in view). The latter interpretation is not incorrect, of course; however, the difference between children suggests a possible avoidance of a passive interpretation on the part of the children with SLI.

Finally, Gopnik and Crago (1991) administered comprehension tasks to members of the three-generation family discussed in chapter 1. On tasks that tapped knowledge of syntactic structure—such as the distinction between active and passive constructions—the individuals with SLI showed appropriate comprehension.

Grammatical Morphology

Many of the studies of grammatical morphology in children with SLI were shaped significantly by the work of R. Brown and his colleagues (e.g., R. Brown, 1973; de Villiers & de Villiers, 1973), who studied a set of grammatical morphemes that seemed to make their way into normally developing children's speech at a relatively young age. Over time, the scope of the research on grammatical morphology expanded, and included pronouns, modal auxiliaries, and complementizers, among others. In recent years, the presumed interrelationships between particular grammatical morphemes and their connections to broader linguistic achievements have been the basis of a great deal of research.

Extraordinary Limitations in Grammatical Morphology   In this section, we review evidence suggesting that grammatical morphology—or at least a significant portion of grammatical morphology—constitutes a relative weakness in children with SLI. However, we begin with an early study whose findings suggest just the opposite. In the ground-breaking investigations of Morehead and Ingram (1970, 1973) discussed earlier in this chapter, the use of grammatical inflections by the children with SLI and their MLU controls was also examined, as determined by word-morpheme ratios. No differences were seen between the two groups overall. However, at lower MLU levels (up to 3.50 morphemes), it was the children with SLI who showed higher ratios. From the analysis of syntactic structure used by Morehead and Ingram, it was also possible to identify the degree of use of select function words. One of these, auxiliary *do*, was used less frequently by the children with SLI than by the MLU controls.

Examining the same database, D. Ingram (1972b) found that the children with SLI used both auxiliary and copula *be* forms with lower percentages in obligatory contexts than did the MLU controls. This was true for both contractible and uncontractible contexts. However, Ingram noted that these differences were due principally to the nine children in each group with MLUs above 3.00 morphemes. For the remaining children, percentages were actually somewhat higher for the children with SLI, though values at these lower MLU levels were only around 10% to 30%.

Subsequent studies of grammatical morphology in the speech of children with SLI appeared in abundance. Kessler (1975) examined these children's use of many of the grammatical morphemes (noun plural *-s*, progressive *-ing*, past tense *-ed*, copula *be* forms) studied in young normally developing children by R. Brown (1973) and de Villiers and de Villiers (1973). A ranking of these morphemes according to percentage of correct use in obligatory contexts yielded a sequence that fell well within the range of sequences reported for typical children.

In a study of 287 children with SLI, Johnston and Schery (1976) reported much the same finding but also examined the MLU levels at which mastery (90% correct in obligatory contexts) occurred. According to their data, children with SLI required higher MLU levels before mastery than had been reported for normally developing children in previous investigations. Khan and James (1983) also found higher-than-expected MLU levels before adequate use of grammatical morphemes was seen in the speech of children with SLI. Longitudinal case studies by Trantham and Pedersen (1976), Cousins (1979), and Eyer and Leonard (1995) yielded similar findings.

Steckol and Leonard (1979) compared children with SLI and younger MLU controls on their use of several grammatical morphemes and found higher percentages of use by the MLU controls. Albertini (1980) employed a similar design, but followed the preschool-age participants for six months, beginning when their MLUs ranged from 1.5 to 2.1 morphemes. The morphemes tracked for this period were the progressive -ing, plural -s, in, on, and genitive 's. At the end of six months, the children with SLI had mastered only in and on, whereas some of the control children had mastered plural -s and -ing as well as in and on. Even when mastery was not attained, percentage differences were quite large between the children in the two groups. Mean percentages for the normally developing children and the children with SLI were, respectively, 87 and 10 for plural -s, and 63 and 3 for possessive 's. Of course, this difference is probably enhanced by the fact that after six months, the normally developing children's MLUs were probably higher than those of the children with SLI. As in previous studies, the apparent sequence of development of the morphemes was similar in the two groups of children.

Other studies that have reported differences favoring MLU controls have ranged from focusing on particular grammatical morphemes such as articles (Beastrom & Rice, 1986) and past tense (Oetting & Horohov, 1997) to composite scores reflecting a collection of grammatical morphemes. Johnston and Kamhi (1984) combined data for noun plural -s, past -ed, auxiliary and copula be forms, infinitival to, and the complementizer that. Bliss (1989) used two composites (differences favoring the MLU controls were seen for each). The first was a composite for noun morphology, involving articles, noun plural -s, and pronouns; the second was a verb morphology composite, consisting of past -ed, third-person singular -s, irregular past, infinitival to, auxiliary and copula be forms, and auxiliary do forms. Leonard (1995) employed a composite consisting of genitive 's, nonthematic of (e.g., glass of milk; piece of candy), auxiliary be forms, auxiliary do forms, and infinitival to.

Leonard, Eyer, Bedore, and Grela (1997) compared preschoolers with SLI, age controls, and MLU controls on their use of two sets of morphemes. The first set consisted of finite verb inflections and copula be forms; the second consisted of a collection of inflections and function words that were unrelated to finiteness: noun plural -s, genitive 's, infinitival to, and articles. The children with SLI showed lower percentages of use of both sets than did the MLU controls and the age controls.

In other recent studies, separate analyses have been performed for each morpheme examined. Leonard, Bortolini, Caselli, McGregor, and Sabbadini (1992) found higher percentages of use in obligatory contexts by MLU controls for the noun plural inflection, regular past inflection, third-person singular inflection, and copula be forms. Differences in the same direction for articles and irregular past forms failed to reach statistical significance. Rice, Wexler, and Cleave (1995) found significant differences favoring MLU controls for regular past, third-person singular, copula and auxiliary be forms, and auxiliary *do* forms. Hadley and Rice (1996) observed more limited use by children with SLI of copula and auxiliary *be* and auxiliary *do*. Cleave and Rice (1995) also noted differences favoring MLU controls on copula and auxiliary *be* forms. According to these studies, development of copula *be* precedes that of auxiliary *be*, which in turn outpaces auxiliary *do*, for both groups of children. Cleave and Rice also found greater use of *be* in contractible (e.g., *Ken's funny*) than uncontractible contexts (e.g., *Chris is funny*), a pattern that is not evident in D. Ingram's (1972b) earlier data on these forms.

Additional findings of lower use of grammatical morphology by children with SLI than by control children will be included below, in discussion of related issues. These findings come from about a dozen additional samples of children (and, in one case, adults as well) in English-speaking countries. Not all grammatical morphemes have yielded differences, and a few morphemes have produced differences in some studies (always in the same direction) but not in others. The earlier findings of Morehead and Ingram (1970, 1973) must still be taken into account (and will be discussed in chapter 11), but the dominant finding is clearly one of a relative weakness in grammatical morphology in children with SLI.

This finding of extraordinary problems in grammatical morphology is itself extraordinary, for two reasons. First, Lahey, Liebergott, Chesnick, Menyuk, and Adams (1992) found a great deal of variability in the use of grammatical morphemes by young normally developing children. It appears that children with SLI who used certain grammatical morphemes with lower percentages than did MLU controls in one study would not have been lower had they been compared with normally developing children with similar MLUs participating in another study. In a sense, younger normally developing children who might serve as controls represent a moving target. However, if the findings of group differences had been simply a function of chance matchups, one must wonder how so many differences were seen and, crucially, why these differences were not approximately evenly divided in terms of which group had the higher percentages. As many differences favoring the children with SLI as favoring the MLU controls should have been witnessed.

The second remarkable aspect of the findings for grammatical morphology is that the measure most often used for subject matching—MLU—is influenced by grammatical morpheme use (as noted in chapter 2). This means that the adoption of MLU for matching constituted a highly conservative research strategy, for the likelihood of observing differences was reduced. The fact that so many differences were seen renders the data from this area quite convincing.

Grammatical Morpheme Productions Are Appropriate   In the early literature on SLI, specific reference to children's use of grammatical morphemes in inappropriate contexts (e.g., *They likes milk*) was not frequent. What little mention was made of this type of error suggested that children with SLI do not usually insert grammatical morphemes in the wrong places. Menyuk (1964) had included categories that represented commission errors of this sort and found that children with SLI made the same number of errors (and for some categories fewer errors) than age controls. Leonard (1972) also employed Menyuk's categories and reported the same finding. D. Ingram (1972b) reported that most of the errors on copula and auxiliary *be* by the children with SLI in his study were omissions. However, occasionally *is* replaced *are* and *was* replaced *were*.

Table 3.1
Percentages of Inappropriate Use of Grammatical Morphemes

	Leonard et al. (1992)		Rice et al. (1995)	
	SLI	MLU	SLI	MLU
Articles	<1	1		
Plural	2	2		
Third singular	8	8	0	0
Regular past	0	0	0	0
Copula *be*	4	5	7	9
Auxiliary *be*			4	5

SLI, children with specific language impairment; MLU, normally developing children matched according to MLU.

A different impression emerged from a case study by Gopnik (1990b). The eight-year-old with SLI participating in this study exhibited errors on many inflections and function words involving tense, definiteness, person, number, and gender. Errors were not limited to omissions of forms from obligatory contexts; inappropriate productions of these morphemes were also seen, as in *You got a tape recorders, The Marie-Louise look at the bird*. These findings led Gopnik to suspect that the child had no knowledge of the grammatical role of these morphemes.

More recently, investigators have reported, for each grammatical morpheme studied, the number of productions of the morpheme in inappropriate contexts as well as the number of productions in obligatory contexts. The findings from two of these studies are summarized in table 3.1.

Clearly, when these grammatical morphemes are produced, they usually appear in appropriate contexts. Some commission errors occur, but, as can be seen in table 3.1, they occur just as frequently in the speech of MLU controls. And what is the nature of the inappropriate productions that do occur? Leonard, Bortolini, Caselli, McGregor, and Sabbadini (1992) found that the third-person singular inflection was occasionally used with third-person plural subjects. However, most of these errors occurred on sentence completion items in which the subjects were invariant in number, such as *sheep, deer*, and *fish*. The children (in both groups) may have treated the experimenter's use of these words (e.g., "But here the deer …") as instances of the singular. The absence of such production errors in the data of Rice, Wexler, and Cleave (1995) is consistent with this possibility. Commission errors on copula and auxiliary *be* were usually cases of the third-person singular *is* replacing the third-person plural *are*. Some, though not all, of these instances in the Leonard, Bortolini, Caselli, McGregor, and Sabbadini study seemed attributable to the same invariant noun problem just described. The nonfinite form *be* almost never appeared when a finite form was required, a finding also reported in the study of Hadley and Rice (1996).

Productivity   Two major types of evidence are used to argue that the use of a grammatical morpheme reflects the child's grammatical knowledge rather than the result of rote learning. One type of evidence is overregularization. If a child produces utterances such as *runned* and *sheeps*, for example, it is likely that past *-ed* and plural *-s* are operative in the child's grammar. The other type of evidence is the use of the grammatical morpheme with lexical items that are new to the child. This is especially clear in the case of the use of a grammatical morpheme with a nonsense word provided by the investigator in a game fashioned after Berko's (1958) classic task (e.g., *plicked, the spax, to lootch*).

Gopnik and her colleagues (Crago & Gopnik, 1994; Gopnik & Crago, 1991; Ullman & Gopnik, 1994) tested the use of grammatical morphology by 20 members of the three-generation family described earlier. Seven of the individuals seemed to be acquiring language normally, whereas the remaining individuals exhibited deficits in language ability. The individuals with language impairment were significantly poorer in grammatical morphology than the controls from the same family. Deficits were seen in the use of noun plural -*s*, regular past -*ed*, and third-singular -*s*, among other morphemes.

A task modeled after that of Berko (1958) was employed with mixed results. Instances of affixation with nonsense words were seen in the responses of the individuals with language impairment. However, fewer responses of this type were provided relative to the number produced by the family members serving as controls. Furthermore, some of the stem-plus-affix responses that were provided by these individuals seemed very deliberate and forced. According to Gopnik and her colleagues, the low frequency of grammatical morphemes suggests that these individuals had learned inflected words as separate lexical items, much as irregular forms such as *men* and *sang* are learned. The automatic application of a productive rule was unavailable to these individuals.

The sporadic success these individuals had on the nonsense word task was taken as the occasional application of an affixation rule that had been learned in school as a conscious, metalinguistic rule. Gopnik and her colleagues interpreted this evidence as consistent with the dual mechanism model of Pinker (1991), wherein irregular forms are built up through an associative learning process and regular rules are automatically applied as a default in cases where no irregular forms have been stored. The regular rules apply without regard to frequency of exposure.

Data reported by van der Lely and Ullman (1995) might be interpreted as consistent with the data of Gopnik and her colleagues. These investigators compared the past-tense productions of school-age children with SLI and younger controls matched according to language test score. The children with SLI showed lower degrees of use of regular past -*ed* with both real verbs and nonsense verbs. Only minimal use of affixation with nonsense words was observed, approximately 5%. For the children with SLI, correct use of the past inflection with real words was related to the frequency of occurrence of the word; this effect was not seen in the control children, and should be reserved for irregular verbs, according to the model adopted by Gopnik and her colleagues. The children with SLI were also more likely than the controls to form the past of a nonsense word by using a pattern seen among irregular verbs, such as *strank* as the past of *strink*.

In other studies, evidence of productivity is not uncommon. Overregularization of plurals has been reported by Albertini (1980) and Rice and Oetting (1993). Overregularization of past has been documented by Eyer and Leonard (1994, 1995), G. King, Schelletter, Sinka, Fletcher, and Ingham (1995), Leonard, Bortolini, Caselli, McGregor, and Sabbadini (1992), Leonard, Eyer, Bedore, and Grela (1997), Oetting and Horohov (1997), and Smith-Lock (1995), among others. The degree to which overregularizations are seen in the speech of children with SLI varies. Rice, Wexler, and Cleave (1995) found that they occurred only rarely in their data. On the other hand, Marchman, Wulfeck, and Ellis Weismer (1995) found that overregularizations constituted 35% of the noncorrect productions of the children with SLI in their investigation, in contrast to 50% for a group of age controls. Of the 27 irregular verbs studied by Marchman, Wulfeck, and Ellis Weismer, all were overregularized by at least one of the 41 children with SLI in their study. Interestingly, overapplication of an irregularization pattern was also observed (e.g., *brang*) in the responses of both groups of children.

Marchman, Wulfeck, and Ellis Weismer (1995) found that the tendency toward overregularization was related to factors such as frequency of occurrence of the past forms, whether the stem ends in /t/ or /d/, and the number and frequency of occurrence of other phonologically similar stems that undergo the same types of changes from present to past tense. Bortolini, Leonard, and McGregor (1992) found that the overregularizations of a group of children with SLI was related to whether the vowel contained in the present or past form was more dominant, according to the proposals of Stemberger (1993) for young normally developing children.

The findings of Marchman, Wulfeck, and Ellis Weismer (1995) suggest another interpretation of some of the findings of van der Lely and Ullman (1995) and of Gopnik and her colleagues. Instead of assuming dual mechanisms with one mechanism unavailable to individuals with SLI, Marchman, Wulfeck, and Ellis Weismer assumed that there is a single mechanism that simply operates less efficiently in individuals with SLI.

Evidence of productivity from nonsense word tasks is also easy to locate in the literature. For most of the grammatical morphemes of interest in the Leonard, Eyer, Bedore, and Grela (1997) investigation, items were included in which the children had to produce the morpheme with a nonsense word (e.g., *flaxed*, *to reeb*). The degree of application of the morphemes to the nonsense words appeared somewhat more limited in quantity by the children with SLI, though the small number of items employed did not permit statistical comparison. However, each of the children with SLI produced one or more grammatical morphemes with a nonsense word. Importantly, analysis of listeners' judgments of these productions suggested they were no different from the productions of MLU controls.

Other studies that found evidence of grammatical morphemes used with nonsense words were those of Oetting and Horohov (1997) for regular past, and Bellaire, Plante, and Swisher (1994) for several different grammatical inflections, including the noun plural inflection, possessive '*s*, third-person singular -*s*, and regular past -*ed*. In both of these investigations, the children with SLI did not apply the grammatical morphemes to nonsense words to the same degree as the control children.

The lower frequency with which children with SLI showed overregularization or affixation to nonsense words in these studies is understandable, considering the fact that in most cases these children used the morphemes in obligatory contexts with lower percentages than did control children. Indeed, some of the children showing this type of creative application of the morphemes seem to have had no business doing so. Overregularization was observed in children with SLI producing regular past inflections in as few as 10% of obligatory contexts (Leonard, 1994). In the study by Albertini (1980), overregularization of the plural was observed in the speech of a child with SLI who used plurals in only 7% of their obligatory contexts. When children with SLI and normally developing children are matched according to their use of grammatical inflections with real words, children with SLI perform as well as the control children in adding inflections to nonsense words (Smith-Lock, 1995).

More on Past Tense   Thus far, two observations have been made about past tense. First, the regular past inflection is used with lower percentages in obligatory contexts by children with SLI than by MLU controls. Second, in spite of this reduced level of use, many of these children show evidence of productivity. Here, we discuss additional details concerning these children's command of past tense.

In the investigations of Leonard, Bortolini, Caselli, McGregor, and Sabbadini (1992), Leonard, Eyer, Bedore, and Grela (1997), and Oetting and Horohov (1997), children with SLI did not differ from MLU controls in the use of irregular past forms even though they showed lower percentages of use for the regular past *-ed*. Johnston, Miller, Tallal, and Curtiss (1994) also found that children with SLI were similar to younger controls in their use of irregular past; however, they also found no differences for regular past. These investigators attributed the latter to the basis for matching the two groups of children. The groups were matched on the basis of scores on an expressive language test that focused on grammatical morphology as well as syntactic structure. Therefore, potential differences between groups were reduced by matching them on a measure that is affected by the same ability. (As discussed earlier, the same problem exists for MLU but perhaps to a lesser extent.)

Oetting and Horohov (1997) compared the regular and irregular past use of six-year-old children with SLI, age controls, and MLU controls. As usual, the age controls' use was superior to that of the other two groups. The children with SLI were similar to the MLU controls in their use of irregular past, but used regular past forms with lower percentages. The children with SLI were most limited in their use of regular past with words of low frequency of occurrence. The regular past use of both the children with SLI and the MLU controls was influenced by the phonological characteristics of the verb. Oetting and Horohov also introduced the children to denominal verbs—verbs that are derived from nouns, as in *ring (the city)* and *fly (out to center field)*. These verbs take the regular past. The particular denominal forms chosen have verb homophones that require the irregular past (e.g., *ringed the city; rang the bell*). The children were introduced to one or the other type of verb and their use of the forms in the past was evaluated. The children with SLI, like the control children, were more likely to apply the regular past to the denominal verbs and the irregular past to their homophones. Thus, although the children with SLI showed more limited use of regular past, they gave considerable evidence of knowing how and when it is used.

Moore and Johnston (1993) asked whether the difficulties with past forms in five-year-old children with SLI derived exclusively from the fact that past time must be marked on the verb. They devised tasks in which children were obligated to complete sentences with past verb morphology (regular or irregular) or with a temporal adverb such as *yesterday* or *last night*. The children with SLI resembled a group of three-year-old controls in their use of past forms and a group of four-year-olds in the use of temporal adverbs. According to Moore and Johnston, this difference between verb morphology use and temporal adverb use is even more noteworthy considering the possibility that adverbs are more complex from a semantic standpoint. Whereas inflections encode the fact that the event occurred prior to the moment of speech, temporal adverbs provide, in addition, a specification of the time prior to the moment of speech when the event occurred. The greater difficulty with past verb forms was probably due to special difficulties with verb morphology rather than notions of time. It should be noted that because irregular past forms were included among the verb morphology items, the Moore and Johnston findings might constitute an exception to earlier findings that irregular past forms are used by children with SLI and younger controls to a comparable degree.

Some Grammatical Morphemes Are Disproportionately Difficult   Although group studies (e.g., Leonard, Bortolini, Caselli, McGregor, & Sabbadini, 1992; J. Roberts, Rescorla, & Borneman, 1994) and case studies (Crystal, Fletcher, & Garman, 1976; Eyer & Leonard, 1995) have reported problems with noun phrase morphology (e.g., noun plural inflections) as well as verb-related morphology, there is reason to believe the latter is more problematic. However, this is not an easy fact to establish. Earlier it was seen that the sequence of grammatical morpheme development of children with SLI approximates that of their normally functioning peers. Because some of the earliest appearing grammatical morphemes are affiliated with the noun phrase, it follows that children with SLI, too, will look more advanced in their use of these morphemes relative to grammatical morphemes pertaining to the verb phrase. For the same reason, ceiling effects could mask real differences between children with SLI and MLU controls in the use of noun phrase-related morphemes. Clearly, the age and MLU level of the children must be considered carefully in drawing conclusions from the data.

The most consistently observed differences between children with SLI and control children have been for finite verb inflections and copula and auxiliary forms requiring agreement. Many of these studies have already been noted; a few additional studies will be discussed here because they suggest that problems with verb-related morphology are diagnostic of SLI.

Two types of findings are illuminating. First, the use of verb-related morphology seems to emerge from discriminant function analyses designed to determine which factors are successful in distinguishing normally developing children from children with SLI. Fletcher and Peters (1984) found that the use of auxiliaries and verb inflections contributed significantly to distinguishing preschool-age children with SLI from age controls in a discriminant function analysis. The children with SLI used fewer instances of these forms. Gavin, Klee, and Membrino (1993) found that the variable that contributed most to the distinction between preschoolers with SLI and age controls was the high frequency of verb phrase errors by the children with SLI, such as omission of copula and auxiliary *be* forms.

Another important type of finding comes from studies reporting deficits in verb-related morphology that continue in the school years and beyond. Examples from Gopnik and her colleagues (e.g., Gopnik & Crago, 1991) and van der Lely and Ullman (1995) have already been noted. In addition, Haber (1982) observed that a group of school-age children had greater difficulty with auxiliaries than a group of normally developing four-year-olds. G. King and Fletcher (1993) found that school-age children with SLI showed lower percentages of use of the third-person singular inflection in obligatory contexts than did MLU controls. G. King, Schelletter, Sinka, Fletcher, and Ingham (1995) observed that school-age children with SLI used regular past -*ed* and third-singular -*s* less frequently than younger controls. Rice and Wexler (1995b) followed a group of children with SLI from age five to six years, and a group of MLU controls from age three to four years. At each observation point, the younger controls showed greater use of the regular past and third-singular inflections, copula and auxiliary *be* forms, and auxiliary *do* forms. By age four years, the control children were approaching mastery levels. In contrast, approximately half of the children with SLI showed no change from five to six years of age.

Thus far, the term "verb-related morphology" has served as a useful cover term for the areas of particularly serious difficulty in children with SLI. However, this term is too imprecise. The progressive inflection *-ing* is not only one of the first grammatical morphemes to be acquired by children with SLI, it also seems to be acquired at about the same MLU level as for normally developing children. This morpheme is nonfinite in nature, but another nonfinite form, infinitival *to*, does appear to give children with SLI more problems than MLU controls. Modal auxiliaries have not received much attention from investigators, owing in part to the difficulty in identifying obligatory contexts for these forms. Recourse to frequency counts or to the number of children in a group using the form is usually required.

Menyuk (1969) observed that modals are not used as extensively by children with SLI as by age controls. However, differences between children with SLI and MLU controls are more subtle. Past tense modals such as *could* and *should* appear to be used by fewer children with SLI, but the present form *can* is just as likely to be used by these children as by MLU controls (Eyer & Leonard, 1995; Leonard, 1995). Sturn and Johnston (1993) found evidence that children with SLI use modal expressions with the same distribution of meanings (e.g., possibility, necessity) as do age controls, but use them only as frequently as younger controls matched according to language measures.

Although identification of the boundaries of the extraordinary difficulty with grammatical morphology is helped by excluding forms such as *-ing* and *can*, other pockets of special difficulty seem to fall outside the domain of verbs. Longitudinal case studies and composite measures used in group studies indicate problems with genitive *'s* and nonthematic *of* (Eyer & Leonard, 1995; Leonard, 1995; Leonard, Eyer, Bedore, & Grela, 1997). Furthermore, several studies have identified differences between children with SLI and MLU controls for articles and/or noun plural *-s* (Albertini, 1980; Beastrom & Rice, 1986; Eyer & Leonard, 1995; Leonard, Bortolini, Caselli, McGregor, & Sabbadini, 1992; Leonard, Eyer, Bedore, & Grela, 1997). The noun plural inflection is among the earliest grammatical morphemes to be acquired by normally developing children, and though it is acquired earlier than most verb inflections by children with SLI as well, its development is often slow. For example, J. Roberts, Rescorla, and Borneman (1994) found that three-year-old children with SLI resembled age controls in their degree of use of grammatical morphemes that had been mastered by the control children (e.g., progressive *-ing*), with the exception of the noun plural inflection. On this morpheme, the children with SLI lagged behind their same-age peers (68% versus 93%).

However, the status of noun plural *-s* is a matter of some dispute. Rice and Oetting (1993) compared children with SLI and MLU controls on the use of noun plural *-s* and third-person singular *-s*. The control children showed significantly higher percentages for the third-person singular inflection. Significant differences in the same direction were also seen for the noun plural inflection. However, percentages were quite high even for the children with SLI (83%, versus 93% for the MLU controls), and clear evidence of productivity, in terms of both the number of different nouns that were used with the inflection and the appearance of overregularizations, such as *foots*, was seen on the part of the children with SLI. Further inspection of the data indicated that the children with SLI were more likely to omit the plural inflection if a quantifier preceded the noun (e.g., *two cat*).

In an especially thorough test of plural use, Oetting and Rice (1993) found no differences between children with SLI and MLU controls. The two groups of children were similar in their use of the plural inflection with both real and nonsense words. Oetting and Rice also explored the children's use of plural forms during compounding. In English, nouns with irregular plural forms can participate in compounds in either singular or plural forms, as in *mouse-eater* and *mice-eater*. However, nouns taking *-s* in the plural must remain in singular form; hence, *rat-eater* is grammatical, but *rats-eater* is not. Oetting and Rice observed that children with SLI, like normally developing children, obeyed this distinction.

Pronouns   Children with SLI are slow to develop certain pronominal forms. Reports of difficulties with accusative pronouns (e.g., *Don't push me; Mommy kissed him*) are scarce. However, studies revealing weaknesses in the use of other types of pronouns are quite common in the literature. For example, Schelletter (1990) used a measure that assigned developmental scores to different indefinite and personal pronouns. The pronouns used by seven- to nine-year-old children with SLI were dominated by earlier developing forms, in contrast to those used by seven-year-old controls. The pronouns used by the children with SLI more closely resembled the pronouns used by five-year-olds. Later developing forms such as *anything*, *everybody*, and *herself* were used relatively infrequently if at all.

Nominative case pronouns (e.g., *I, he, she, they*) have received the greatest attention from investigators. Loeb and Leonard (1988), Leonard (1982a), Lee (1966), and Menyuk (1964) all reported instances of accusative for nominative case pronouns (e.g., *Him eating popcorn*) that seem higher in frequency than is reported for younger normally developing children. Loeb and Leonard (1991) compared the nominative case pronoun use of preschoolers with SLI and MLU controls, and found greater use of accusative for nominative forms by the children with SLI. These differences were correlated with problems the same children had with verb agreement morphology, such as the third-person singular inflection and auxiliary *be* forms. Subsequent studies by Loeb and her colleagues also reported higher percentages of case errors in children with SLI than MLU controls, both when these errors are studied longitudinally (Loeb & Mikesic, 1992) and through sentence repetition tasks (Loeb, 1994). In a case study by Ramos and Roeper (1995), a close relationship was observed between case errors and limited verb morphology.

More recent work has cast some doubt on the reliability of these findings. Most notably, Moore (1995) did not obtain the same results. The five-year-olds with SLI in her study made a greater number of pronominal case errors than did age controls, but not more than a group of three-year-olds matched according to a measure of syntactic development. In the speech of the child with SLI studied longitudinally by Eyer and Leonard (1995), many instances of accusative pronouns in place of nominative pronouns were noted. Early in the study, this pronoun use occurred in sentences with no verb morphology (e.g., *Me put that up; Me like doughnut place*). By the end of the study, this pronoun pattern was still evident even though verb morphology was more extensive. Unexpected co-occurrences such as *Me drinked it all* were also seen.

Moore (1995) reported that case errors involving the third-person feminine pronoun (e.g., *her sleeping*) were more common than those involving the masculine (e.g., *him going*). This last finding was explored in greater depth by Ogiela (1995). Of particular interest was the pattern of substitution error as a function of the degree to which the nominative, accusative, and genitive forms of the same person and gender shared phonetic material. Rispoli (1994) had proposed that, in normal language development, when a pronoun shared no phonetic material with the corresponding pronouns of different cases, it was unlikely to serve as a substitute in replacing these other pronouns during instances of error. Conversely, the same pronoun was more vulnerable to replacement by a pronoun that shared phonetic properties with the remaining pronouns because phonological material cannot be recruited to assist in retrieval. Ogiela (1995) found evidence consistent with this view in a group of children with SLI. These children were more likely to produce *her* in contexts requiring *she* than to produce *him* for *he*. Ogiela also observed occasional instances in which a nominative pronoun replaced an accusative form. These substitutions were also in keeping with Rispoli's proposals. The nominative form *he* sometimes replaced *him*, but *she* never replaced *her*.

Data collected by Webster and Ingram (1972) are also consistent with the notion of shared phonetic material, but seem to extend to gender errors as well as case errors. In the speech of the children with SLI studied by Webster and Ingram, *he* and even *him* replaced *she*, but the converse was never observed.

Variability in a Grammatical Morpheme's Appearance   Some investigators have attempted to gain a better understanding of the inconsistent use of grammatical morphemes by children with SLI. Bishop (1994) found evidence that inflections and omissions sometimes occurred with the same stem; it was not the case that the inflections were consistently produced with certain words and consistently absent from others. Leonard, Eyer, Bedore, and Grela (1997) found that all nine children with SLI who participated in their investigation showed this pattern for at least one grammatical inflection; for most children, the majority of inflections were distributed in this way.

Even though inflectional variability often occurs with the same lexical material, there might be characteristics of lexical items that make them more or less likely to be inflected. Some of these are semantic and apply during the early stages of inflectional use. For example, Johnson and Sutter (1984) found that children with SLI, like normally developing children, were more likely to use the regular past *-ed* inflection with verbs of momentary duration (e.g., *jumped*) and only later used it with verbs of continuative duration (e.g., *played*). Marchman, Wulfeck, and Ellis Weismer (1995) observed that the same frequency and phonological factors that related to overregularizations of past by children with SLI were also important predictors of whether the past inflection would appear or would be omitted.

Learning New Grammatical Morphemes   Several studies have examined the learning of novel suffixes by children with SLI. In an investigation by Swisher and her colleagues (Swisher, Restrepo, Plante, & Lowell, 1995; Swisher & Snow, 1994), English-speaking children with SLI and age controls learned a nonsense affix *-u*, corresponding to a derivational morpheme with the meaning "large size," to be used with nonsense nouns that also were taught. The children with SLI had greater difficulty applying the morpheme, but this was significantly related to how many of the nonsense nouns had been learned. It was not related to performance on nonverbal tasks of hypothesis testing, even though the children with SLI also had more difficulty with the latter than did age controls (Restrepo, Swisher, Plante, & Vance, 1992). The children with SLI were not assisted by a method in which the rule was verbalized explicitly by the experimenter. In fact, they performed better with a method that relied on implicit discovery of the rule.

Bellaire, Plante, and Swisher (1994) found that school-age children with SLI were less able than age controls in their ability to learn and apply nonsense affixes to novel material, though the degree of application seen was sufficient to conclude that a productive rule had been acquired. The same children also could use English inflections with nonsense nouns and verbs.

There is evidence that the learning required for children with SLI to grasp the meanings of new grammatical morphemes is not always sufficient to ensure the production of these forms. In studies by Connell (1987) and Connell and Stone (1992), children with SLI showed evidence of comprehending novel suffixes representing derivational morphemes (e.g., -um with the meaning "broken") through a procedure in which the rule was illustrated visually and described verbally. However, unlike age and language comprehension test score controls, these children failed to produce the suffixed forms unless they also had practice in producing the forms through imitation.

Johnston, Blatchley, and Olness (1990) examined grammatical morpheme learning in a larger context. These investigators taught two miniature languages to a group of children with SLI. The novel nouns of the language referred to fantasy animals that played either an agent or a patient role. Novel verbs were used to refer to the actions that the agent performed on the patient. A novel suffix was attached to the noun serving as the patient. One of the languages possessed the word order of agent + patient + suffix + action; the other language had the order action + agent + patient + suffix. Johnston, Blatchley, and Olness found that the children with SLI had a tendency to direct most of their limited resources toward one characteristic of the language. For the second of these languages, it was the perceptually salient suffix that appeared at the end of the sentence. This characteristic seemed to be learned at the expense of word order. For the first language, the reverse was true.

Comprehension and Grammaticality Judgments   Although most investigations have dealt with the production of grammatical morphology, there have been several efforts to determine how well children with SLI understand and process this aspect of grammar. Fellbaum, Miller, Curtiss, and Tallal (1995) used a picture identification task in which the selection of the appropriate picture depended on the understanding of a single grammatical morpheme. The morphemes were divided into sets according to their acoustic duration. Consonantal inflections such as the noun plural and the regular past were among the brief-duration set, for example, whereas object pronouns were among the longer-duration set. Six-year-old children with SLI and age controls participated. The performance of the children with SLI was consistently lower than that of the age controls on the morphemes of brief duration. Most of the longer-duration morphemes were understood equally well by the two groups.

Grammaticality judgments also have been studied in children with SLI, though these judgment tasks have been reserved for children of school age. Although the focus of these studies was on grammar in general, inspection of the items used reveals that the manipulations concerned grammatical morphology. Relative to chronological age controls and mental age controls, children with SLI had greater difficulty identifying errors involving grammatical morphemes (e.g., *John and Jim is a brother; They throwing a stick*) than violations of lexical restrictions (e.g., *The dog writes the food; Jill eats cards*) (Kamhi & Koenig, 1985; Liles, Schulman, & Bartlett, 1977).

Certain types of grammatical morpheme judgments might be especially difficult for children with SLI. In a study by van der Lely and Ullman (1995), children with SLI were more likely than language test score controls to accept sentences with missing past tense inflections in clearly past tense contexts. However, when children with SLI are compared with control children matched according to their ability to use past tense inflections, no difference in judgment ability is seen (Smith-Lock, 1995).

Wulfeck and Bates (1995) used response time as well as accuracy measures to examine grammaticality judgments by children with SLI age 7 to 14 years, and a group of age controls. The children with SLI accepted a greater number of ungrammatical sentences than did the control children, and were slower in their response time in making grammaticality judgments. However, the pattern of their responses was generally similar to that of the controls. Errors appearing later in the sentence were more readily detected than earlier errors, and errors involving determiners (e.g., omission, agreement error, or transposition error) were detected more quickly than errors involving auxiliary *be* forms.

Phonology

Problems with the sound system of the language are, by definition, language problems. In this chapter, we emphasize studies of children whose phonological limitations are accompanied by deficits in other areas of language. However, no sharp distinction can be drawn among groups of children with limitations in phonology. During the preschool years, if children exhibit deficits in morphosyntax and lexical skills, they almost invariably show weaknesses in phonology as well. If children are identified first on the basis of phonological problems, a majority will also show problems in other areas of language (e.g., Paul & Shriberg, 1982; Ruscello, St. Louis, & Mason, 1991; Shriberg, Kwiatkowski, Best, Hengst, & Terselic-Weber, 1986). Approximately one-third will score below age level on measures of language comprehension, but approximately 80% will display deficits in language production (Shriberg & Kwiatkowski, 1994). For children with moderate to severe phonological problems during preschool, improvement continues into the school years, but the rate of phonological gain shows a deceleration between eight and nine years of age (Shriberg, Gruber, & Kwiatkowski, 1994).

Relatively few studies in the area of phonology have employed comparison groups of younger normally developing children matched on some language measure. Most have employed either age controls or no control group. Usually in the latter case, investigators have compared their findings with those reported in the literature for younger normally developing children. A great deal of insightful work is apparent in this literature, but these design features make conclusions about the relative status of phonology in children with SLI quite tentative. The general impression is that these children show many of the phonological characteristics seen in younger normally developing children, with only a few areas that pose unusual difficulty.

The models of normal phonological development that have been applied to children with SLI run the gamut. Some investigators have employed a framework in which it is assumed that the adult form is represented in the children's underlying phonological system but that rules are applied that alter the output (Compton, 1970,

1976; Lorentz, 1976; Oller, 1973). (In some of these studies, it was concluded that the assumption of an adultlike underlying system is not warranted. See Dinnsen, 1984; Maxwell, 1979.) Other investigations have used a model in which the underlying form is assumed to be the result of perceptual-encoding rules; a separate set of production rules is then applied to yield the output (e.g., Chiat, 1983; McGregor & Schwartz, 1992).

Nonlinear approaches also have been applied. The details of the nonlinear frameworks employed have varied from study to study, but they share the property of representing prosodic as well as segmental information, and doing so in a manner that permits study of both their independent functions and their interaction. Application of this general approach is seen in the work of Bernhardt and her colleagues (Bernhardt, 1992; Bernhardt & Gilbert, 1992; Bernhardt & Stoel-Gammon, 1994), as well the work of Chin and Dinnsen (1991), Gandour (1981), Leonard (1992a), and Spencer (1984).

Segments

The measures used to examine phonology have varied considerably. The earliest work looked at segment accuracy, that is, the accuracy of each consonant and vowel. Not surprisingly, children with SLI are late in acquiring the segments of the language. Segments that are acquired early by normally developing children (e.g., /n/, /m/, /b/, /w/) are likewise the first to be acquired by children with SLI, albeit at a later age. Segments that are acquired later by normally developing children (e.g., /s/, /v/) can continue to be difficult for children with SLI well into their school years (e.g., Farwell, 1972). In contrast to the work on consonants, few investigations have dealt with the use of vowels. The studies that have been conducted have not yielded identical results (e.g., Hargrove, 1982; Pollock & Keiser, 1990; Stoel-Gammon & Herrington, 1990). However, there appears to be agreement that the vowels that are used with only limited accuracy by children with SLI are typically those that are difficult as well for young normally developing children.

Distinctive Features

An obvious shortcoming of segment analyses is that each consonant and vowel is treated as if it were independent of all others. Yet, clinicians and researchers alike recognize that some segments are more similar than others, and hence problems with more than one segment might easily be due to a problem with the shared characteristic. To capture this information, investigators began to apply distinctive feature analyses by the early 1970s. The most frequently adopted distinctive feature system was that of Chomsky and Halle (1968). In this system, the similarities and differences between segments are expressed by the use of features with binary (+/−) values. Thus, the similarity between /t/ and /s/ is represented through the shared values for a number of features: [− vocalic], [+ consonantal], [+ anterior], [+ coronal], [− voice], and [− nasal]. Differences are limited to the fact that [t] is [− continuant] and [− strident], whereas [s] is [+ continuant] and [+ strident]. Importantly, error patterns can be described by means of distinctive features. For example, errors such as [bi] for *pea*, [di] for *tea*, and [gi] for *key* can be characterized as difficulty with [− voice].

For the most part, the distinctive feature patterns reflected in the speech of children with SLI resemble those seen in the speech of younger normally developing children. However, there are two possible differences. Menyuk (1968) observed that once segments carrying [+ strident] appear in the speech of normally developing children, this feature is relatively resilient, even if other features in the target sound are in error. However, data from Leonard (1973) and McReynolds and Huston (1971) suggest that children with SLI retain [+ strident] least often. Thus, whereas errors such as [su], [tsu] or [tʃu] for *shoe* might be most likely for typical children, errors such as [tu] for *shoe* might be seen more often among children with SLI.

The best-documented difference between children with SLI and younger normally developing children concerns [+ voice]. However, the direction of this difference is not the usual one. Although children with SLI are not as proficient as their same-age peers in their ability to produce voicing contrasts (e.g., *coal-goal*) (Catts & Jensen, 1983), this aspect of phonology seems to be a relative strength in these children, at least in word-initial or prevocalic position. In a comparison of children with SLI and younger normally developing children having similar consonant inventory sizes, D. Ingram (1981) observed that prevocalic substitution of [+ voice] for [− voice] (e.g., [gol] for *coal*) was a frequent error only in the normally developing group. Schwartz, Leonard, Folger, and Wilcox (1980) found evidence for such errors in the speech of both children with SLI and younger normally developing children; however, unlike the case for the latter group, substitution of [+ voice] for [− voice] was never the most frequent error for the children with SLI.

Similar findings have been reported when the voicing contrast was measured in terms of voice onset time (VOT), the time between the release of energy in the vocal tract and the vibration of the vocal folds. In English, VOT is longer for sounds carrying the [− voice] value. Farmer and Florance (1977) found that a group of children with SLI showed VOT values for word-initial stops that approximated those seen for normally developing children of the same age. In a study of children with SLI and younger normally developing children matched for both vocabulary size and consonant inventory size, Leonard, Camarata, Schwartz, Chapman, and Messick (1985) found that VOT differences between word-initial voiced and voiceless stop consonant targets were more frequent in the productions of the children with SLI. A study conducted by Bond and Wilson (1980) had a somewhat different outcome. Children with SLI showed a pattern of VOT development resembling that of a group of MLU controls; however, the children with SLI showed a greater tendency toward prevoicing, that is, initiating vocal fold vibration prior to the release of air from the vocal tract.

Finally, D. Ingram (1990) studied the voicing contrast question through an analysis of the types of distinctions that were reflected in the consonant inventories of a large group of children with SLI and a comparable group of younger normally developing children. Ingram observed that whereas place of articulation distinctions (e.g., *tea-key*) appeared before voice distinctions in the speech of the younger comparison group, the reverse was often true for the children with SLI.

This unexpected advantage shown by children with SLI in contrasting [+ voice] and [− voice] consonants might have its basis in neuromotor development. It has been proposed that voiced English consonants (with smaller VOT values) permit considerable variability in the coordination of oral and laryngeal movements, whereas voiceless consonants (with larger VOT values) require a higher degree of neuromuscular coordination (Gilbert, 1977; Kewley-Port & Preston, 1974). Even though children with SLI might show slower maturation than their same-age peers, as we saw in chapter 1, in these studies of voicing, the children with SLI were older than the

Table 3.2
Some Phonological Processes

Phonological Process	Examples
Consonant cluster reduction	snow [no], blue [bu]
Final consonant deletion	boat [bo], soup [su]
Liquid gliding	read [wid], lake [wek]
Prevocalic voicing	two [du], key [gi]
Stopping	food [pud], soap [top]
Velar assimilation	dog [gag], take [kek]
Word-initial weak syllable deletion	away [we], banana [nana]

normally developing children. It seems likely that they had attained a higher level of neuromotor development as well.

Phonological Processes
Methods of analyzing children's phonology were significantly changed with the appearance of Stampe's work (1969, 1973). The most noteworthy change was the use of phonological process analysis to capture generalizations about a child's speech. Phonological processes are systematic sound changes that affect classes of sounds or sound sequences. Some examples are provided in table 3.2. It can be seen that some processes are literally recastings of distinctive feature patterns. Prevocalic voicing is the substitution of [+ voice] for [− voice] sounds in prevocalic position; stopping is the substitution of [− continuant] sounds for [+ continuant] sounds; and so on. Other processes can be described in distinctive feature terms, but only awkwardly, and still others seem to evade a distinctive feature characterization. Weak syllable deletion, for example, involves the loss of a weak syllable, but the feature composition of the syllable is irrelevant.

Comparisons between children with SLI and younger normally developing children show a great deal of similarity in the processes reflected in the children's speech. Word-initial weak syllable deletion might be more prevalent in the speech of children with SLI (D. Ingram, 1981), and we have already discussed prevocalic voicing differences in distinctive feature terms. Otherwise, the similarities are striking. For example, consonant cluster reduction, liquid gliding, final consonant deletion, and word-initial weak syllable deletion still occur with relatively high frequency in two-year-old children, with the first two of these persisting for some time. These processes are also the most prevalent in the speech of children with SLI (M. L. Edwards & Bernhardt, 1973; Hodson & Paden, 1981; D. Ingram, 1976, 1981; Leonard, 1982b; Schwartz, Leonard, Folger, & Wilcox, 1980).

A minority of individuals with SLI continue to exhibit certain phonological processes even as adults. Fee (1995) examined the phonological characteristics of 12 of the members of the family studied by Gopnik and others. Eight of these individuals (ranging from 7 to 46 years of age) exhibited language impairments. Data were collected at two points in time, 17 months apart. Fee found that the phonological abilities of the older individuals in the group were better than those of the younger, and the later testing sessions showed better performance than the testing 17 months earlier. Nevertheless, even some of the older individuals continued to show evidence of final consonant deletion. The most frequent process was consonant cluster reduction in word-final position.

Implicational Laws
Dinnsen, Chin, Elbert, and Powell (1990) asked whether the presence of particular phonetic distinctions in a child's inventory implied the presence of another type of distinction. These investigators reported evidence suggesting that the implicational laws of a group of children with SLI matched those reflected in the speech of younger normally developing children. For example, whenever there were distinctions between the lateral sonorants [r] and [l], there were already distinctions between the nasal and oral sonorants [n] and [l]; before the latter distinction was seen, distinctions were observed between continuant and stop consonants such as [f] and [p].

The Dinnsen, Chin, Elbert, and Powell (1990) study examined phonetic distinctions. However, children can have a wide variety of sounds at their disposal yet still have serious phonological difficulties if they don't use these sounds in a contrastive manner. For example, it doesn't matter a great deal if the child produces both [s] and [t] if these two sounds are equally likely to be used in words such as *sea, tea, sew,* and *toe*. Gierut, Simmerman, and Neumann (1994) adopted a procedure that employed phonemic contrasts and obtained findings somewhat different from those of Dinnsen et al. For example, a voicing distinction among fricatives implied a voicing distinction among stops. When the latter condition held, children with SLI also had nasals as well as stops in their phonemic inventories.

Subphonemic Distinctions
Some of the changes taking place in young children's pronunciations over time are not apparent from perceptual analysis. Sometimes, changes that more closely approximate the adult system are noted only at a subphonemic level through acoustic analysis. This seems to be true for children with SLI as well. For example, as with younger normally developing children, the productions of word-initial voiced and voiceless stop consonant targets by children with SLI sometimes differ in VOT even when judges transcribe them all as having been produced with a voiced stop (Forrest & Rockman, 1988; Gierut & Dinnsen, 1986; Maxwell & Weismer, 1982; Tyler, Edwards, & Saxman, 1990). In this case, the voiceless targets are produced with longer VOTs than the voiced targets. Subtle but reliable VOT differences can also be found between stop (e.g., /t/) and fricative (e.g., /s/) targets that appear to be produced as stops (Tyler, 1995).

Vowels that precede final voiced stop consonants have longer durations than vowels that precede final voiceless stops. For some young normally developing children and (somewhat older) children with SLI, these same differences can be detected even when the final consonants themselves seem to be omitted (Weismer, Dinnsen, & Elbert, 1981) or produced as a glottal stop (Smit & Bernthal, 1983). Other reported differences that are consistent with developmental changes include differences in frequency and transition rate of the second formant of /r/ targets versus /w/ targets (Huer, 1989), differences in rise of the second formant in vowels preceding targets with a final velar stop versus no final consonant (Weismer, 1984), and spectral differences between the first 40 ms of word-initial /t/ versus word-initial /k/ targets (Forrest, Weismer, Hodge, Dinnsen, & Elbert, 1990).

Avoidance

During the early period of phonological and lexical development, some young normally developing children avoid the use of words whose adult forms contain certain sounds or syllable shapes. In some of these studies, young children were presented with novel words whose phonological characteristics either resembled those of words already used by the child or showed little similarity to the characteristics of the words that the child had attempted in the past. The findings indicated that the children produced a greater number of novel words that conformed to their existing phonologies than novel words that were inconsistent with their phonologies. Leonard et al. (1982) found precisely the same thing for a group of children with SLI in the single-word-utterance period of development.

Unusual Errors

It appears that children with SLI are more likely than younger normally developing children to produce errors of an unusual nature (Leonard, 1985). Some of these are cases of presumably later-developing sounds replacing presumably earlier-developing sounds (e.g., [v] for /d/) (M. L. Edwards & Bernhardt, 1973; Grunwell, 1981; Leonard & Brown, 1984; Lorentz, 1974; F. Weiner, 1981). Another form of unusual error is seen when the child's production constitutes an addition to the adult form. For instance, D. Ingram's (1976) analysis of data first reported by Hinckley (1915) revealed instances of nasals added to the initial position. Pollock (1983), too, has observed similar use. Both M. L. Edwards and Bernhardt (1973) and Grunwell (1981) observed children who added nasals to final position or inserted nasals before alveolar stops (e.g., [tond] for *toad*). In some cases, unusual productions are seemingly promoted by the appearance of some other type of sound in the word (Grunwell, 1981; Willbrand & Kleinschmidt, 1978). For example, Leonard and Leonard (1985) described one child who produced [k] or [g] in word-initial position whenever the adult form contained a nonlabial consonant in initial and final position. Productions such as [go] for *door*, [ka] for *sock*, and [gad] for *tired* were seen. In contexts free of this constraint, as when nonlabial consonants appeared in initial position only (e.g., *do*, *see*, and *tie*), productions were usually accurate.

Some unusual errors involve the use of consonants not found in the ambient language. Examples from English-speaking children with SLI include the use of nasal friction and/or snorts (Beebe, 1946; M. L. Edwards & Bernhardt, 1973; Grunwell, 1992; Hall & Tomblin, 1975), alveolar affricates (Fey, 1985), lateral fricatives (M. L. Edwards, 1980; Grunwell, 1992), ingressive lateral fricatives (Grunwell, 1981), and ingressive alveolar fricatives (D. Ingram & Terselic, 1983).

It would be easy to draw the conclusion from these reports of unusual errors that children with SLI possess bizarre phonological systems. However, two important facts indicate that the findings are not as dramatic as they might appear. First, these have been case studies or individual cases identified from larger groups of children with SLI. If a prospective study were carried out with the idea of measuring the frequency of, say, nasal snorts or [v] for /d/, it is doubtful that a single data point would be registered. Second, unusual errors can also be found in the literature of normal language development. Individual children who reached every major language milestone on schedule—including most phonological milestones—have been observed doing things like moving all sibilants to word-final position or adding nasals to the ends of word-final voiced consonants. The safest conclusion seems to be that unusual errors are documented more frequently in the literature on SLI than in the child language literature more generally.

Unsystematic Application

Variability is one of the most conspicuous characteristics of young children's phonologies, and the phonologies of children with SLI are no exception. It is often possible to find some reasons for the variability, suggesting that the children's use is, in spite of appearances, systematic. For example, some studies of children with SLI have revealed that the adjacent vowel (Camarata & Gandour, 1984; Grunwell, 1981; Leonard, Devescovi, & Ossella, 1987) or the prosodic structure and syllabification of the word (Chiat, 1983, 1989) are determining factors.

However, children with SLI seem to be more likely than younger normally developing children to show variability even when the phonetic contexts of the words and the children's own consonant inventories provide no rationale for it (Grunwell, 1981, 1992; Grunwell & Russell, 1990; Martin, 1991). This is also seen in controlled studies in which new words (that do or do not conform to the children's existing phonological patterns) are introduced to the children. In such studies, children with SLI produce a larger percentage of the new words in an exceptional manner than do younger control children matched according to consonant inventory size (Leonard, Schwartz, Swanson, & Loeb, 1987).

Phonology and the Lexicon

D. Ingram (1987a) has advanced the notion that the degree of a child's phonological deficit is a consequence of the inverse relation between the child's phonological abilities and the number of words in the child's expressive vocabulary. This idea has not been put to an adequate test, but it does appear that many children with SLI have phonological skills that are even further behind age level than their lexical abilities. This type of gap constitutes another way in which these children differ from normally developing children.

Pragmatics

In chapter 1, it was noted that children with SLI incur a significant social penalty for their difficulties. They are accorded lower social status by peers, and are viewed by adults as less mature and less likely to assume a leadership role. To some degree, commission errors that are morphosyntactic, phonological, or lexical in nature might contribute to these reactions from others. There are instances when confusions of word meanings, for example, can have a significant negative impact. Brinton and Fujiki (1995) described one child whose meaning of the word *liar* extended to persons who unintentionally provided incorrect information. Calling someone a liar in such instances is not exactly a way to win friends. However, these children's social standing is probably more often influenced negatively by what they are unable to communicate with language.

The communicative uses of language fall within the domain of pragmatics. Although clinical texts have long emphasized the importance of the communication skills of children with SLI, systematic study of the pragmatic abilities of these children is only about 20 years old. The first work of this type appeared shortly after the early studies of pragmatics in normal child language by Bates and her colleagues (Bates, 1976; Bates, Camaioni, & Volterra, 1975).

Given the criteria for SLI, it would be natural to assume that any pragmatic difficulties observed in these children were secondary to problems of linguistic form or content. These children score at age-appropriate levels on nonverbal intelligence tests and show social and emotional behaviors that are safely outside the boundaries of disorders such as autism. Furthermore, many children with SLI exhibit language comprehension skills that exceed their production abilities. These facts might suggest that when children with SLI fail during communicative exchanges, it is because they don't have the requisite morphosyntactic or lexical form. Indeed, some of the evidence of pragmatic difficulties in children with SLI is of this type. In other instances, however, the basis of the problem is not so clear.

Pragmatic abilities can be divided in different ways, depending on the framework adopted. In an earlier review, Fey and Leonard (1983) adopted the scheme introduced by Bates (1976), expanding it somewhat to accommodate new findings. This expanded scheme will be used here to review the findings for children with SLI.

Speech Acts

Speech acts are the functions that utterances are designed to serve, such as requesting, naming, thanking, warning, and congratulating. Some of the most basic functions, such as requesting and naming, are not strictly dependent on grammar, in the sense that they can be reliably identified in the productions of young children with speech limited to single-word utterances. In fact, precursors to verbal requesting and naming can be observed in the form of gestures. In what was probably the first modern-day pragmatic study of children with SLI, Snyder (1975, 1978) reported that children with SLI at the single-word utterance level could express both requestive and declarative functions but usually did so through gestural means. In contrast, a group of younger normally developing children similarly restricted to one-word utterances were more likely to convey these functions through word use. Rowan, Leonard, Chapman, and Weiss (1983) conducted a similar study, controlling for the vocabulary required for the requestive and declarative tasks administered, and found that the two groups were comparable in their ability to use words to express these functions.

Leonard, Camarata, Rowan, and Chapman (1982) employed an expanded set of 16 communicative functions to examine speech acts by children with SLI and younger controls at the single-word utterance level. These speech acts included describing, refusing, and answering as well as naming and requesting. The children with SLI produced a greater number of answers and fewer spontaneous instances of naming than did the control children.

The speech acts of children with SLI who produced multi-word utterances have been examined by several research teams. The most common design involved children with SLI, age controls, and younger normal controls matched according to MLU or comprehension test score (Ball & Cross, 1981; Fey, Leonard, Fey, & O'Connor, 1978; Rom & Bliss, 1981). In other studies, younger normally developing children served as a comparison group but were not strictly matched on any measure with the children with SLI (Prinz, 1982). Deficits relative to age controls were seen in these studies, though the speech acts used by the children with SLI resembled those of the younger control children. Prinz (1982) found that the requests of children with SLI were similar in type to younger normally developing children, though the latter were more likely to produce grammatically complete renditions of these requests (see also Prinz & Ferrier, 1983). Fey, Leonard, Fey, and O'Connor (1978) found that the children with SLI would have more closely resembled age controls if it were not for the fact that certain speech acts required grammatical constructions that were seemingly out of reach for these children. For example, one speech act, termed "rules," required modal or semiauxiliaries, as in *You hafta stay over there.*

These group studies suggest that these children's speech act abilities fall behind peers no more than might be expected given their morphosyntactic abilities. However, there are case studies of children with SLI who apply seemingly routinized forms to perform speech acts for which the forms are not suited (e.g., Blank, Gessner, & Esposito, 1979; Gallagher & Craig, 1984). The child studied by Gallagher and Craig, for example, used *it's gone* as a means of initiating an interchange with other children. His goal was to recruit the children's participation in a disappearance game, but this means of conveying his intent was, not surprisingly, unsuccessful.

Sensitivity to the speech acts of others is sometimes lacking in children with SLI. Shatz, Bernstein, and Shulman (1980) found that a group of children with SLI responded appropriately to indirect requests (e.g., "Can you open the door?") but had great difficulty responding to similar utterances with an informing ("yes" or "no") response when the situation was altered to promote such responses. Younger normally developing children with comparable MLUs could make this change.

The comprehension of figurative language such as metaphors requires a type of speech act ability in that the listener must recognize that an interpretation beyond the literal meaning of the utterance must be sought. Nippold and Fey (1983) observed that school-age children with a history of SLI did not understand metaphors (e.g., *My head is an apple without any core*) as well as peers did, even though the two were matched on their performance on literal aspects of language.

Conversational Participation and Discourse Regulation
Conversational participation and discourse regulation refer to a broad range of abilities including conversational initiations and replies, topic maintenance, turn taking, and repairing utterances based on listener feedback or interruption. The degree to which children with SLI initiate an interaction with a coconversationalist is often taken to be an indication of how conversationally assertive they are. When speaking to adults, children with SLI are less likely to initiate conversations than are same-age peers, and more likely to restrict their participation to utterances that acknowledge the prior message or indicate that the message was understood (Sheppard, 1980; Siegel, Cunningham, & van der Spuy, 1979; Stein, 1976; Watson, 1977). Similar differences have been reported for comparisons between children with SLI and younger control children, though exceptions are evident. Johnston, Miller, Curtiss, and Tallal (1993) found that children with SLI around six years of age were more likely to answer adults' questions with elliptical responses than were younger control children matched according to language production test score. On the other hand, Prelock, Messick, Schwartz, and Terrell (1981) observed that children with SLI were as likely as younger MLU controls to initiate conversational turns.

Leonard (1986) compared the conversational replies of children with SLI and younger normally developing children. The two groups showed similar lexicons and speech limited to single-word utterances. Although replies can be viewed as relatively unassertive conversational behaviors, they vary in terms of whether or not they are obligated by the preceding speaker turn, and whether or not they add information to the previous turn. For example, the replies in (4) and (5) (termed "affirmative answer" and "alternative," respectively, by Leonard) are both obligated by the adult's question, yet (5) adds information.

(4)   *Adult*:   Was that a wreck?
       Child: Yeah.

(5)   *Adult*:   Is she dirty?
       Child: Clean.

The replies in (6) and (7) (called "affirmation" and "expansion," respectively) are not obligated. However, (7) adds new information.

(6)   *Adult*:   He's thirsty.
       Child: Yeah.

(7)   *Adult*:   You have a cat.
       Child: Home.

Leonard found that the children with SLI produced a greater number and variety of replies than the younger controls. These differences included the nonobligated, information-bearing replies. In a study of children with SLI who produced word combinations, Van Kleeck and Frankel (1981) also observed the use of replies that added information. Together, these investigations suggest that the characterization of children with SLI as passive conversationalists has clear limits.

In fact, individual cases have been described that show rather assertive conversational activity on the part of children with SLI. Fujiki and Brinton (1991) described one child who produced over 60% of the utterances that occurred in a dialogue with an adult. This child's outgoing conversational style often led to overestimations of his linguistic ability by others.

Most studies have examined children with SLI interacting with adults. Fey and Leonard (1984) studied children with SLI interacting in dyads with peers and toddlers in addition to adults. Age and MLU controls also interacted with these three types of coconversationalists. The results showed that when speaking to younger listeners, the children with SLI were as conversationally assertive as age controls were. The language skills of the children with SLI relative to the conversational partner seem to play a major role in findings such as these. Fey, Leonard, and Wilcox (1981) found that children with SLI took greater conversational initiative when interacting with MLU-matched listeners than when speaking with age-mates. Jacobs (1981) reported that children with SLI were more conversationally assertive when speaking to other children with SLI than when talking to normally developing children of the same age. Craig and Gallagher (1986) examined the nonobligated replies of a child with SLI in dyadic interaction with same-age and younger control children. This child was clearly responsive, though the degree to which his utterances contributed substantively to the conversation was influenced greatly by factors such as whether there was shared reference in the discourse.

Although children with SLI have been shown to be active conversationalists when speaking with other children, their abilities seem to drop off sharply when they must participate in conversations with more than one child. Craig and Washington (1993) and Craig (1993) found that children with SLI had great difficulty entering established conversations between normally developing children. Some children made attempts, though often by nonverbal means, whereas others chose to wait in the wings, either because of previous failures or because they didn't have a clear sense of how to enter in. Rice, Sell, and Hadley (1991) found that in classroom settings, children with SLI were more likely to initiate interaction with adults than with peers, whereas normally developing age-mates were more likely to initiate interactions with peers than with adults (see also Schuele, Rice, & Wilcox, 1995). What is especially interesting about this finding is the fact that in dyadic interaction, children with SLI are less conversationally assertive with adults than with other children (Fey & Leonard, 1984). Clearly, settings with multiple participants constitute a conversational hazard for children with SLI.

Children with SLI also have difficulties resolving conflicts in a verbal manner. When involved in conflicts, these children seem more likely than normally developing children to become physically aggressive or to withdraw (Baker, Cantwell, & Mattison, 1980; Loucks, 1987). However, when the situation is less emotionally charged, as when the conflict is imaginary, as part of a problem-solving or role-playing activity, children with SLI are more like their peers. Their solutions to conflict are less varied, but group differences are quite small (Stevens & Bliss, 1995).

Topic maintenance represents another important measure of conversational participation. However, the evidence currently available is quite indirect. Schelletter (1990) found that a group of seven- to nine-year-old children with SLI were less likely to pronominalize a referent more than once relative to five- and seven-year-old normally developing children. One possible explanation for this finding is that the children with SLI changed topics more quickly, and thus new referents were introduced, requiring full noun phrases rather than pronouns.

Turn-taking measures can also provide an indication of conversational participation. Craig and Evans (1989) studied the turn-taking behaviors of children with SLI and both age and MLU controls. The children with SLI exchanged turns smoothly with their interlocutors. Most of their turns were single utterances, but they were temporally adjacent and semantically related to the prior utterance. Relative to younger controls, their utterances were more adjacent; relative to age controls, however, these utterances were less tightly timed to the prior utterance. Age-mates also used a greater number of multi-utterance turns. One indication that the children with SLI might have been less assertive was the finding that these children interrupted the other speaker with relatively low frequency. Fujiki, Brinton, and Sonnenberg (1990) found that during instances of overlapping speech, children with SLI showed a greater tendency to continue their utterance without repair than was seen for age-mates or younger normally developing children matched according to a composite language score.

Conversational repairs have also been examined in a design in which the experimenter makes a clarification request (e.g., "What?") immediately following the child's utterance. Gallagher and Darnton (1978) found that children with SLI modified their original utterances by altering phonetic details or by adding information. However, unlike younger normally developing children with similar MLUs, these children showed very few substitutions of constituents of the type shown in (8).

(8)   *Child*:   It's broken.

        Adult: What?

        Child: The truck is broken.

Similar findings were reported by Stein (1976), but in that study, age controls served as the comparison group.

Brinton, Fujiki, Winkler, and Loeb (1986) examined children's responses to consecutive requests for clarification. That is, one attempt at repair was not sufficient; the experimenter persisted in requesting clarification. The children with SLI recognized the need to clarify in these instances, but their attempts at clarification were less likely than those of age controls to contain the added information necessary. Syntactic limitations might have been a major obstacle for these children. Requests for clarification were also included in a study conducted by Brinton and Fujiki (1982). Children with SLI were responsive but less likely than controls matched for grade level to produce an appropriate response.

If children with SLI have difficulty substituting elements in their original message in response to listener feedback, their ability to paraphrase sentences would be expected to be similarly deficient. Hoar (1977) found this to be the case in a comparison of children with SLI and grade controls. When asked to paraphrase sentences provided to them by the experimenter, the children with SLI were more likely to repeat the original utterance or rely on lexical substitutions. Unlike the control children, the children with SLI produced very few appropriate syntactic reformulations.

Several investigations have examined the production of clarification requests by children with SLI. Fey and Leonard (1984) and Griffin (1979) observed that children with SLI produced such requests as frequently as age controls, though Watson (1977) found lower frequencies on the part of the children with SLI. Gale, Liebergott, and Griffin (1981) reported that children with SLI requested clarification as frequently as did MLU controls, though for their younger subgroup of children with SLI (averaging almost four years of age), requests were more likely to be made through nonverbal means.

Code Switching

By the end of the 1970s, it was established that normally developing children modify their speech style and complexity as a function of the listener's age, language ability, and social status. By the early 1980s, clinical researchers began to ask whether similar speech modifications are made by children with SLI. The difficulties these children seem to have in producing semantically equivalent responses with different forms (e.g., Gallagher & Darnton, 1978) suggest that they might show similar limitations in their ability to adapt their speech to the listener. Yet within certain limits, children with SLI show a clear propensity to make these adjustments. Fey, Leonard, and Wilcox (1981) found that children with SLI produced utterances with fewer morphemes in preverb position and asked fewer questions about internal states when talking to MLU-matched conversational partners than when talking to age-matched partners.

Fey and Leonard (1984) observed variations in the use of internal state questions, total number of questions, and use of commands when speaking with adults, peers, and toddlers. All adaptations were in the same direction as that seen for a group of age controls who interacted with the same types of listeners. However, the children with SLI showed no variation in MLU or preverb length. Examination of the values indicated that limitations in utterance length (and, quite possibly, morphosyntactic complexity) on the part of the children with SLI were probably responsible for the lack of variation. In the condition in which these children showed the highest MLUs (when conversing with adult listeners), their utterances were shorter than those of the age controls in all conditions.

An interesting modification of the multiple listener paradigm was employed by Messick and Newhoff (1979). Children with SLI and MLU controls played a game that required them to request a drink while playing the role of a mother, a father, an adult female, a girl, and a baby. The two groups of speakers were comparable in the number of request variations produced across the different roles. However, on a judgment task in which the children were to associate particular request forms with particular recipients, the children with SLI showed random judgments, whereas the MLU controls' judgments were more in line with those expected of older participants. Requests containing "please," for example, were viewed as more appropriate for adults than for toddlers.

Speech adjustments are also made on the basis of familiarity. Olswang and Carpenter (1978) observed that children with SLI produced more utterances when speaking to their mothers than in a comparable session with an unfamiliar speech-language pathologist. However, no differences were seen on lexical or morphosyntactic measures.

Presuppositions
The message conveyed by speakers is often shaped by what they think the listener already knows about the events being described. Such presuppositions form an important part of conversational behavior, and seem to make their first appearance when children are still limited to single-word utterances. This is seen in young children's tendency to name elements in the situation that constitute new information. For example, when presented with five identical dolls in succession and then a stuffed animal, children are most likely to name the first doll and the stuffed animal. Snyder (1975, 1978) examined this type of presuppositional behavior in children with SLI and younger controls at the one-word level and found that both groups were more likely to communicate about new elements than about already-seen elements. However, the children with SLI communicated by using nonlinguistic means more frequently than did the control children. Rowan, Leonard, Chapman, and Weiss (1983) conducted a similar study but used only objects whose names were in the children's lexicons. The children with SLI performed like the younger normally developing children by providing the names of new elements in the situation.

Presuppositional ability beyond the single-word level has also been studied in children with SLI. Skarakis and Greenfield (1982) found that preschool-age children with SLI were as likely as MLU controls to refer to new or changing elements as opposed to unchanging elements in the situation. However, whereas the MLU controls showed a pattern of omitting old information at lower MLU levels and pronominalizing such information at higher levels, the children with SLI showed a tendency to pronominalize at all levels of MLU.

In one important sense, the new-old information paradigm is not a true test of presuppositional ability because the speaker and listener observe the situation together, and hence the listener is privy to all of the information known to the speaker. In contrast, classical referential communication tasks offer a means of studying speakers' ability to shape their message based on what they can only assume the listener knows and does not know. In these tasks, the speaker describes material for a listener who has no visual access to what is being described. The listener must then select the material that corresponds to the speaker's description. The work of Meline (1978, 1986, 1988) and Meline and Meline (1983) has made use of this kind of task. School-age children with SLI and both MLU and age controls participated in a task requiring them to describe novel geometric shapes to an adult listener. The findings varied somewhat from study to study. In some cases, the descriptions of the children with SLI were interpreted less accurately by the listener than those of the age controls, but more accurately than the descriptions of the MLU controls (Meline, 1978); in other cases, the children with SLI were as effective as age controls in conveying an adequate message, though their messages were not as elaborate as those of either the MLU or age controls (Meline, 1986).

Johnston, Smith, and Box (1988) studied children's ability to describe objects varying in attributes such as size and color to a blindfolded puppet serving as a pretend listener. The messages conveyed by the children with SLI were communicatively adequate, but less mature and more redundant than the messages of mental age controls.

Pragmatics as a Distinct Deficit

A comparison of the literature on the pragmatic abilities of children with SLI and the literature on these children's morphosyntactic abilities reveals some obvious differences. In the area of morphosyntax, one can almost take for granted that children with SLI will look weak relative to same-age peers, certainly through the preschool years and usually beyond. The less certain issue is whether differences will be seen between children with SLI and younger controls matched according to some language measure. Often, differences are seen here as well. On the other hand, placing children with SLI on a continuum with age controls and younger controls is difficult for the area of pragmatics. In a few studies, children with SLI perform below the level seen for MLU controls. In other instances, no differences are found, and in still others, the children with SLI perform at higher levels. Furthermore, a few studies report no differences between children with SLI and normally developing children of the same age. In some of the studies reporting poorer performance by children with SLI than control children, there remains the question of whether the weaker morphosyntactic abilities of the children with SLI were getting in these children's way, restricting their ability to exhibit pragmatic knowledge that they possessed.

Nevertheless, there is also evidence indicating that these children sometimes fail to enter into multiparty conversations. Even extremely limited morphosyntactic and lexical skills should be sufficient to permit the child to make some form of attempt. There are also case studies reporting children with SLI who extended situation-specific phrases to inappropriate contexts, seemingly in the service of some conversational goal. Such instances do not seem to reflect pragmatics as a casualty of a morphosyntactic or lexical problem. Instead, such findings suggest that pragmatics can sometimes be a problem in its own right.

Coordinating Lexical, Morphosyntactic, Phonological, and Pragmatic Elements: A Look at Narratives

Even the simplest sentence cannot be constructed and produced without the coordination of lexical, morphosyntactic, phonological, and pragmatic elements. However, an especially good example of such coordination is seen in narrative use. Narratives require considerable skill in manipulating language, whether they are in the form of telling a fictional story, providing an account of a previous experience, or retelling a story heard from someone else. For example, between- as well as within-sentence syntactic devices are needed for the sake of cohesion, presuppositions must be adjusted on line to take into account information just told to the listener, and the numerous speech acts conveyed as portions of the dialogue must be coordinated and kept subservient to the overarching speech act of telling a story.

Given the complexity of narratives, it is difficult to separate problems of intent or knowledge from problems of execution. On balance, however, the evidence suggests that children with SLI have the greatest problems with the latter. For example, their narratives contain the essential ingredients of a story, organized in an appropriate sequence (Clifford, Reilly, & Wulfeck, 1995). Initiating events are included as well as subsequent major events and their consequences (Graybeal, 1981; Liles, 1985b, 1987). Stories typically contain an ending (Sleight & Prinz, 1985). Furthermore, children with SLI make more explicit reference to characters and events when telling a story based on a movie that hasn't been seen by the listener than when telling the story to a listener who has seen it (Liles, 1985a).

On the other hand, details that make for a more complete, cohesive, and engaging narrative are sometimes missing. Some of these omissions might have a lexical or syntactic basis, such as these children's use of fewer words, propositions, and embedded clauses in their stories (Candler & Hildreth, 1990; Crais, 1988; Clifford, Reilly, & Wulfeck, 1995; Graybeal, 1981; MacLachlan & Chapman, 1988; Newcomer, Barenbaum, & Nodine, 1988; Strong & Shaver, 1991). Problems with cohesion could certainly relate to syntax, such as these children's use of lexical ties when pronominal ties would be expected (e.g., *Then he found some old shoes. The shoes smelled bad*) (Liles, 1985a; see also Craig & Evans, 1993), as well as to their difficulties in repairing story meaning (Liles & Purcell, 1987; Purcell & Liles, 1992). However, the occasional absence of story components, such as the protagonist's internal responses to events (Klecan-Aker & Kelty, 1990; Merritt & Liles, 1987), allows for the possibility that some of the difficulties reflect problems with narratives per se.

Liles, Duffy, Merritt, and Purcell (1995) provided a useful way of characterizing the chief difficulties experienced by children with SLI when producing narratives. These investigators reexamined the data of Liles (1985a), Merritt and Liles (1987), and Purcell and Liles (1992). A factor analysis revealed that the variables employed in the three studies could be attributed to two factors. One of these dealt with the global organization of content, such as how intentions and events were logically related in time or through cause-and-effect relationships. The second factor dealt with the use of linguistic structure in integrating the information expressed in the narrative. Measures such as the frequency and number of subordinated clauses, and the degree and accuracy of intersentential cohesion formed this factor. Discriminate function analyses revealed that the second factor was relatively successful in determining whether a child was a member of the SLI group or the age control group. The first factor, global organization of content, proved unhelpful in determining group membership.

Many of the studies of narrative use of children with SLI have employed age controls only. Findings of group differences in these cases are difficult to interpret. Analyses of the type performed by Liles, Duffy, Merritt, and Purcell (1995) help to determine the relative importance of different variables in separating children with SLI from age controls. However, it is not known if the discriminating variables are diagnostic of SLI or merely of lower narrative ability the likes of which can be seen in younger normally developing children as well. An investigation by Gillam and Johnston (1992) employed younger control children matched according to language test score in addition to age controls. The children with SLI were found to produce narratives that were similar in content to the younger controls but reflected a higher percentage of grammatical errors. These findings lend support to the idea that grammatical deficits contribute significantly to the limitations seen in the narrative ability of children with SLI.

Studies of story comprehension permit an examination of these children's ability to make inferences. Even when children with SLI perform as well as controls on interpreting the literal meaning of a story, they experience greater difficulty when inferences are required (Bishop & Adams, 1992; Crais & Chapman, 1987; Ellis Weismer, 1985).

Problems with inferences are not readily explained by morphosyntactic limitations, and indicate that some of the children's problems with narratives might have a different source. In chapter 12, we will examine one possible reason for these children's difficulties with inferences.

Characterizing the Language Deficit in English

Relative to same-age peers, children with SLI who are acquiring English have limitations in every area of language examined. When differences are not observed, it is usually because of ceiling effects. In principle, this did not have to be the case. Children with SLI might have been found to be clearly average in some language abilities and selectively impaired in others.

Even so, the evidence of deficits across areas of language tells only part of the story. In chapter 2, we looked at five ways of describing the language deficits of children with SLI relative to normally developing children. If we examine each detail of language separately, we find abundant evidence for late emergence and protracted development. For a minority of individuals with SLI, a plateau might also be reached. The nature of the errors observed does not point to anything out of the ordinary; errors generally resemble those seen in younger normally developing children. Unusual phonological errors are more likely in the speech of children with SLI, but, it was noted, these can be found in the literature on normal language development as well.

Of the five ways of characterizing the differences between children with SLI and normally developing children, a profile difference appears to be the most accurate, both at a macro and at a micro level. At a macro level, it seems fair to say that lexical and pragmatic skills tend to be less deficient than morphosyntactic skills, with argument structure and phonology falling somewhere between. At a micro level, we saw, for example, that grammatical morphology is weaker than other areas of morphosyntax, and phonological patterns apply less systematically than the children's phonetic inventories would lead us to expect.

The fact that these children's abilities are uneven across language areas opens up the possibility that different factors operate on different aspects of language, and that these children's language deficits might have several distinct sources. One useful way of narrowing down the list of possible factors is to review the evidence from children with SLI who are acquiring languages other than English. We take up this topic in the next chapter.

Chapter 4
SLI Across Languages

There are two important reasons to study SLI in different languages. The first is that with each language we examine, data useful to the assessment and treatment of children with SLI acquiring that language can be obtained. The greater the number of languages studied, therefore, the greater the number of children who can be served in an informed manner.

A second reason for studying different languages is that one language can have properties useful for testing hypotheses based on another. In chapter 11, we will examine several hypotheses based on children with SLI who were acquiring English. In English, most of these competing hypotheses seem equally plausible. However, data from children with SLI learning other languages say something different. In some cases, the languages studied suggest alternative hypotheses that might not have been considered on the basis of English data alone.

Conducting research on SLI in different languages has all the hazards seen in normal child language research, and a few more. Reliance on native speakers is essential, to verify transcriptions and interpretations of utterances, if not to collect the data in the first place. But who identifies a child as exhibiting SLI? In chapter 1, it was noted that the diagnosis of SLI includes a concern expressed by family members and others that the child's language skills are not what they should be. Such concern implies a culture that views limited language ability as placing a child at social and educational risk. The construct of SLI is questionable for languages spoken in cultures whose members do not hold such a view.

The study of SLI in a particular language is also influenced by the presence in the culture of individuals recognized as experts in the assessment of language ability. Family members and others close to the child will recognize the general problem but may be less than confident, and less than competent in their ability to describe it fully and plan a course of action once it is described. Of course, the culture's experts on language ability—usually professionals devoted to this kind of endeavor—assist not only the child and the family; by serving as the magnet for families with children at risk, they also serve as the primary source of children for investigators of SLI.

The availability of standardized tests of language and other abilities can also benefit the study of SLI in different languages. Professional judgment serves as a necessary first step in selecting children for study, but psychometrically sound instruments are needed for such valuable functions as describing and quantifying the language and nonverbal abilities of the children in order to permit replication of the study, documenting long-term language changes following treatment, selecting control children matched according to some language measure, and making comparisons across languages.

This last function merits special comment. Comparisons of children across languages require some basis for matching. For certain languages, a common metric such as MLU in words can be used. However, the structural characteristics of some languages are so different that a common language measure to be used for matching is not likely to be found. In such cases, standard scores from language tests can be employed.

A study by Lindner and Johnston (1992) serves as a good example. These investigators compared German-speaking and English-speaking children with SLI on their use of grammatical morphology and the gap between grammatical morphology and lexical skills. The two groups were administered a test of language ability standardized in their respective languages. Each child's standard score relative to the norm was then determined, and the children speaking the two languages were matched according to these scores. When the standardized tests in the different languages focus on the same age groups, areas of language (e.g., vocabulary, grammar), and modalities (comprehension, production), matching procedures such as this are most appropriate.

In this chapter, we review evidence of the characteristics of language spoken by children with SLI in a variety of languages. For languages that have been the focus of considerable research, we will begin discussion of the language with a brief sketch of its structure. Where possible, we will review data from less intensely studied languages immediately following discussion of data from a language of the same family.

Work in phonology will not be reviewed here. However, it should be kept in mind that the phonological patterns observed in children with SLI are significantly influenced by the language being acquired, just as is the case with normally developing children. For example, whereas [w] is the most common sustitute for /r/ in English, in Italian the most common substitute is [l] (Bortolini & Leonard, 1991), and in southern Swedish it is [h] (Nettelbladt, 1983). The reasons are twofold. First, substitutes are usually sounds that have phonemic status in the language. It is no coincidence, for example, that coalescence errors such as [fit] for *sweet* occur in English, whereas [xul] for *school* is rarely attested. In terms of the features involved, the two errors should be equally likely. In the first instance, [f] assumes the fricative property of [s] and the labial property of [w]. In the second instance, [x] takes the fricative property of [s] and the velar property of [k]. The difference seems to be that /f/ exists in English, whereas /x/ does not. It is not difficult to find [x] used as a replacement in languages that have this sound in their phonemic inventory—Hebrew, for instance. This factor is relevant in the above example for /r/ because the most typical substitute in English, [w], appears only in limited phonetic contexts in Italian, and is not even reflected in the orthography of the language.

The second reason for the differences in the substitutes seen across languages is that comparable phonemes assume different phonetic values. In Italian, /r/ is a trill and is more anterior than in English; /l/ in Italian is produced in the same place of articulation as /r/. In southern Swedish, /r/ is uvular. The production of [h] in its place, then, represents a reasonable approximation in terms of articulatory location.

Italian

Some Characteristics of Italian Studied in Children with SLI
Italian is a language in which all nouns, verbs, and adjectives are inflected; the only forms not carrying inflections are borrowed words. All inflections are word-final and

Table 4.1
Singular and Plural Noun Forms in Italian

Gender	Singular	Plural	
Masculine	albero	alberi	tree/s
	leone	leoni	lion/s
Feminine	macchina	macchine	car/s
	tigre	tigri	tiger/s

Table 4.2
Adjective-Noun Agreement in Italian

Gender	Singular	Plural	
Masculine	uccello rosso	uccelli rossi	red bird/s
	uccello verde	uccelli verdi	green bird/s
	pettine giallo	pettini gialli	yellow comb/s
	pettine verde	pettini verdi	green comb/s
Feminine	palla rossa	palle rosse	red ball/s
	palla verde	palle verdi	green ball/s
	nave gialla	navi gialle	yellow ship/s
	nave verde	navi verdi	green ship/s

syllabic, involving at least a vowel. Nouns are inflected for number (singular, plural) and gender (masculine, feminine). Table 4.1 provides the noun inflections, with examples. Many masculine nouns end in -o in singular and -i in plural. However, nouns with word-final -e in singular and -i in plural can be either masculine or feminine; cues to their gender lie elsewhere, as we'll see below.

Adjectives agree with the nouns they modify in number and gender, as shown in table 4.2. In many instances, the adjective and noun inflections will be phonologically identical (e.g., *libro rosso*, "red book"; *palla rossa*, "red ball"). However, because some adjectives, like nouns, end in -e in singular and -i in plural, on many occasions the inflections of the adjective and noun will differ phonologically (e.g., *libro grande*, "large book"; *cane piccolo*, "little dog"). Most adjectives follow the nouns they modify, though a few precede them (e.g., *la bella ragazza*, "the beautiful girl").

The article system of Italian is provided in table 4.3. Both definite and indefinite forms are used. Articles precede the noun and agree with it in number and gender. The indefinite plural forms (conveying "some") are partitives constructed from the preposition *di* with the definite article of the appropriate number and gender. It can be seen from table 4.3 that the phonetic characteristics of the following word influence the article form used. For example, most masculine singular and plural nouns are preceded by *il* and *i*, respectively. When the following words begin with *z* or an *s* cluster (or a vowel in the plural), however, the forms *lo* and *gli* are required.

Verbs are inflected for tense, person, and number. The inflections for present tense appear in table 4.4. There are three conjugations (whose infinitive forms end in *-are*, *-ere*, and *-ire*), though there are subclasses within the conjugations that require additional changes in the stem for particular person and number combinations. For exam-

Table 4.3
Definite and Indefinite Articles in Italian

Number	Definite		Indefinite	
	Masculine	Feminine	Masculine	Feminine
Singular	il	la	un	una
	lo[a]		uno[a]	
	l'[b]	l'[b]		un'[b]
Plural	i	le		
	gli[c]			

[a] Used when subsequent word begins with z or s cluster.
[b] Used when subsequent word begins with vowel.
[c] Used when subsequent word begins with vowel, z, or s cluster.

Table 4.4
Present Tense Verb Inflections in Italian

Person	portare "carry"		vedere "see"		dormire "sleep"	
	Sing.	Plu.	Sing.	Plu.	Sing.	Plu.
1	porto	portiamo	vedo	vediamo	dormo	dormiamo
2	porti	portate	vedi	vedete	dormi	dormite
3	porta	portano	vede	vedono	dorme	dormono

ple, for verbs such as *dormire* (sleep) (see table 4.4), there are no changes in the stem. However, for certain verbs of the same conjugation, -isc- is used between the stem and inflection for all but first and second persons plural. Thus, finite forms for *finire* (finish) include *finisco* (I finish), *finisce* (he or she finishes), and *finiamo* (we finish).

Past events are expressed by different tenses depending on several factors. In much of Italy, the two tenses involving inflections on the main verb are restricted to events occurring in the remote past, or actions performed on a habitual basis in the past. Events occurring in the immediate past are expressed with the auxiliaries *avere* (have) or *essere* (be) plus the past participle. Thus *Paula ha dormito bene*—literally, "Paula has slept well"—has the meaning "Paula slept well (last night/last week/at the hotel)."

Italian makes use of pronominal clitics to serve a variety of grammatical functions, such as the reflexive ("I got myself up"), impersonal ("one eats well there"), indirect object, and direct object. Clitics always occupy a slot that is obligatory in the sentence. In the case of a clitic serving as direct or indirect object, it is used when the referent is clear from the physical or conversational context (e.g., *lo* in *Gina vede il ragazzo e poi lo bacia*, "Gina sees the boy and then kisses him"). When a pronominalized referent serving the same grammatical function is highlighted in the context, a full pronoun form rather than a clitic is used (e.g., *No, Gina vede lui, non vede lei*, "No, Gina sees him, she doesn't see her"). Clitics precede finite verbs, whereas full pronouns follow them. Unlike other pronouns, clitics are structurally dependent and cannot stand alone. For example, they cannot be used as a one-word answer to a question. Clitics express person and number, as well as grammatical function. The different clitic forms serving a direct object function are shown in table 4.5.

Table 4.5
Italian Clitics Serving a Direct Object Function

Masculine		Feminine	
Singular	Plural	Singular	Plural
lo	li	la	le

The canonical word order of Italian is subject-verb-object. The subject can be omitted from utterances when the referent is established from physical and conversational context; the richness of the verb morphology usually makes interpretation straightforward. The verb morphology also permits deviations from the canonical word order; pragmatic contexts can be found in which the subject, verb, and direct object appear in all possible arrangements.

Most Italian words are multisyllabic, with primary stress falling on the penultimate syllable. Accordingly, monosyllabic grammatical inflections are usually (word-final) weak syllables that are immediately preceded by a strong syllable constituting part of the stem (e.g., *la ragázza álta prénde la matíta róssa*, "the tall girl takes the red pencil"). For inflections that are multisyllabic, one of the syllables of the inflection will be penultimate and hence will usually receive stress, as in *vediámo*, "we see."

Italian Data

Italian-speaking children with SLI show both universal characteristics of language learning difficulties and characteristics that can be traced to the typology of the language they are acquiring. Lexical development appears to be slow in these children, and the emergence of word combinations is quite late. Probably the most common profile is one in which comprehension is stronger than production, but some children show significant comprehension problems as well (Gulotta, Becciu, Mazzoncini, & Sechi, 1991; Levi, Fabrizi, La Barba, & Stievano, 1991). Differences among children with SLI are apparent in other respects as well. Chilosi, Cipriani, Giorgi, and Pfanner (1993) have developed a system of subtypes that reflects some of the same factors discussed in chapter 1.

For most children with SLI, utterance length is quite restricted, with omissions of obligatory elements in the sentence and variability in the particular elements that are omitted (Chilosi & Cipriani, 1991). Many of these same characteristics are seen across production tasks, including imitation (Levi, 1972; Zardini, Battaini, Vender, & D'Angelo, 1985). Phonology is often quite poor (Bortolini, 1995).

One of the striking features in the productions of Italian-speaking children with SLI is their limited use of function words such as articles and clitics. Using existing normative data as a standard, both Cipriani, Chilosi, Bottari, Pfanner, Poli, and Sarno (1991) and Sabbadini, Volterra, Leonard, and Campagnoli (1987) found that these two types of grammatical morphemes were used to a more limited degree by children with SLI than by normally developing children with approximately the same MLU. In a more recent study, Leonard, Bortolini, Caselli, McGregor, and Sabbadini (1992) reported the same finding in a direct comparison between children with SLI and a group of MLU controls. The children with SLI ranged from age 4;0 to 6;0, with MLU in words ranging from 1.9 to 4.3. The children's mean percentages of use in

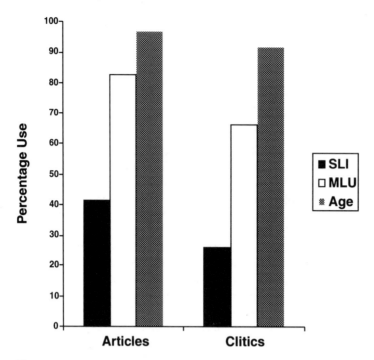

Figure 4.1
Percentage of use of articles and direct object clitics by Italian-speaking children with SLI, younger normally developing children matched according to mean length of utterance (MLU), and normally developing
children matched according to age (Age) studied by Leonard, Bortolini, Caselli, McGregor, and Sabbadini
(1992).

obligatory contexts are provided in figure 4.1. Percentages from a group of age controls are also shown in the figure.

Bottari, Cipriani, and Chilosi (1994) conducted a longitudinal study of one child
with SLI from six to nine years of age and found quite limited correct use of both
articles and clitics throughout the study, with some increase in the final year. For this
child, articles were somewhat more problematic than clitics, in contrast to the findings summarized in figure 4.1.

In each of these studies, the most frequent type of error was omission rather than
substitution of the articles and clitics. When articles are produced appropriately by
children with SLI, their distribution is somewhat similar to that of normally developing children (Leonard, Bortolini, Caselli, & Sabbadini, 1993). Articles that can be used
in a greater number of phonetic contexts (*il, i*) show higher percentages of use in
obligatory contexts than those whose use is restricted to only a few phonetic contexts (*lo, gli*). Similarly, in children with SLI and normally developing children alike,
singular articles such as *la* and *lo* show higher percentages than their plural counterparts *le* and *gli*, respectively. However, children with SLI seem to have special difficulty with *il*.

Close examination of this difficulty in two studies suggests a phonotactic explanation related to the use of a syllable-final consonant when a syllable beginning with
another consonant follows (as it does by necessity in contexts requiring *il*) (Leonard,

Bortolini, Caselli, & Sabbadini, 1993; Leonard, Sabbadini, Volterra, & Leonard, 1988). Interestingly, this difficulty often results in omission of the entire article. In chapter 3, it was noted that phonological avoidance through omission is not especially unusual in children with SLI.

When errors of commission are observed, children with SLI resemble normally developing controls. The substitutions usually seen in both types of children take the form of articles with restricted phonetic contexts being replaced by the more common articles of the same number and gender (*il* replacing *lo*, *i* replacing *gli*). Considering the difficulty that children with SLI experience with *il*, this substitution pattern is noteworthy. Although these substitutions reflect less-than-adult knowledge of the phonetic contexts of *lo* and *gli*, their systematic nature suggests that use of the more common articles was productive and not based on memorized associations between particular articles and particular nouns.

Leonard, Bortolini, Caselli, McGregor, and Sabbadini (1992) also compared the Italian-speaking children with SLI and the MLU controls in their comprehension of articles and clitics. For both types of grammatical morphemes, comprehension was higher for the MLU controls.

Both Leonard, Sabbadini, Volterra, and Leonard (1988) and Leonard, Bortolini, Caselli, McGregor, and Sabbadini (1992) matched groups of Italian-speaking and English-speaking children with SLI in terms of MLU in words and compared the two groups' use of articles in obligatory contexts. In the first of these studies, the Italian-speaking children ranged in age from 4;1 to 6;11, and the English-speaking children ranged from age 3;6 to 6;9. MLUs for the two groups ranged from 2.0 to 3.5 words. In the second study, the Italian- and English-speaking children ranged from age 4;7 to 6;0 and 3;8 to 5;7, respectively. For the Italian-speaking children, MLUs in words ranged from 2.6 to 4.3; for the English-speaking children, the range was 2.7 to 4.2.

In both studies, the Italian- and English-speaking children with SLI were very similar in their percentages of use; for each group, articles were frequently omitted. Similar comparisons for clitics have not been made, owing to the fact that English makes use of full pronouns (e.g., *him*, *her*) both in contexts in which clitics are required in Italian and in contexts requiring full pronouns. However, an inspection of transcripts of spontaneous speech by Leonard, Sabbadini, Volterra, and Leonard (1988) suggested that in contexts comparable with those requiring direct object clitics in Italian, English-speaking children with SLI seemed less likely to omit direct object pronouns.

Not all freestanding grammatical morphemes present Italian-speaking children with SLI with the same degree of difficulty. Cipriani, Chilosi, Bottari, Pfanner, Poli, and Sarno (1991) found that children with SLI were less impaired in their use of prepositions than in their use of articles and clitics. These investigators suggested that prepositions in particular might be learned primarily on semantic grounds, with less reliance on formal aspects of linguistic structure.

To date, there has been little systematic study of the use of copula and auxiliary forms by Italian-speaking children with SLI. The limited data available suggest that the third-person singular present copula form *è* is omitted more frequently by children with SLI than would be expected given their MLUs (Leonard, Sabbadini, Leonard, & Volterra, 1987).

Italian-speaking children with SLI differ from their English-speaking counterparts in the way they treat grammatical inflections. Presumably because nouns, verbs, and adjectives must be inflected in the language, difficulties with inflections take the form

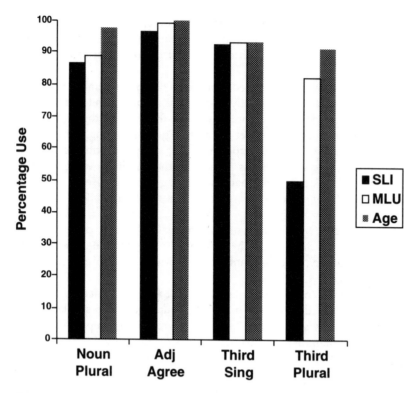

Figure 4.2
Percentage of use of noun plural inflections, adjective agreement inflections, and third-person singular and plural verb inflections by Italian-speaking children with SLI, younger normally developing children matched according to mean length of utterance (MLU), and normally developing children matched according to age (Age) studied by Leonard, Bortolini, Caselli, McGregor, and Sabbadini (1992).

of substitutions rather than use of null forms. Difficulties with the third-person plural verb inflection (e.g., *vedono*), for example, result in the child's use of the third-person singular inflection (e.g., *vede*) in its place.

However, relative to MLU controls, Italian-speaking children with SLI do not appear to have special difficulty with most grammatical inflections, in contrast to their extraordinary problems with forms such as articles and clitics. In a study described earlier, Leonard, Bortolini, Caselli, McGregor, and Sabbadini (1992) included the use of noun plural inflections, adjective agreement inflections, and third-person singular and plural verb inflections in their comparisons of children with SLI, MLU controls, and age controls. The mean percentages in obligatory contexts for the three groups are shown in figure 4.2. Although the age controls showed the highest percentages (as expected), the children with SLI showed percentages that were comparable with those of the MLU controls for three of the four types of inflections. Only the third-person plural inflection produced differences favoring the MLU controls.

For all three groups, the usual error on the third-person plural inflection was the use of the third-person singular. This substitution was not bidirectional; instances of the third-person plural replacing the third-person singular inflection were rare for all three groups. Leonard, Bortolini, Caselli, McGregor, and Sabbadini (1992) observed that third-person plural forms only two syllables in length were produced with accuracy by the children with SLI. These include *fanno* (they do/make), *danno* (they give), and *stanno* (they stay). Because such verbs occur frequently in the language, they might have been memorized forms for the children.

However, other evidence suggests that the limited use of third-person plural was reflecting more than rote memory. For some verbs, phonological details of the stem must change for the third-person plural (as well as the first-person singular). However, when the third-person plural inflection was used with such verbs, the stem was not altered. Because these unaltered forms are not heard in the input language, the children's use of the third-plural inflection in these instances could not have been due to rote learning.

The special difficulty with the third-person plural did not appear to be due to problems with verb inflections in general. In addition to the fact that the third-person singular inflection is used with high percentages by these children, the limited evidence available for other verb inflections does not point to other areas of vulnerability. Bortolini and Leonard (1996) examined the spontaneous speech of several of the children with SLI in the Leonard, Bortolini, Caselli, McGregor, and Sabbadini (1992) study and found very few instances in which first- and second-person inflections were not used appropriately. Accuracy on these inflections was as high for the children with SLI as for their MLU controls. Furthermore, earlier evidence from Leonard, Sabbadini, Leonard, and Volterra (1987) suggests that the difficulty is not due to inherent difficulty in combining the linguistic features of third person and plural. In a group of children with SLI whose use of third-person plural inflections was even lower than that of the children in the Leonard, Bortolini, Caselli, McGregor, and Sabbadini (1992) investigation (averaging 32%), use of the third-person plural copula form *sono*, in the few obligatory contexts identified, was as high as the use of the phonologically identical first-person singular copula form.

Additional information about verb inflection use comes from the work of Bottari, Cipriani, and Chilosi (1995). They asked whether Italian-speaking children with SLI were likely to produce infinitive forms in finite contexts (e.g, *dormire* for *dormo* or *dorme*). Of 46 children with SLI, only 7 showed productions of this type. Furthermore, the child with the greatest use of infinitives in these contexts (6%) showed use of the appropriate finite inflection in 90% of the instances. The remainder (4%) were productions of the wrong finite form.

The similarities between Italian-speaking children with SLI and MLU controls in their use of grammatical inflections appear to hold for comprehension as well. Leonard, Bortolini, Caselli, McGregor, and Sabbadini (1992) found that the two groups did not differ in their comprehension of any of the inflections studied. Comprehension scores for the third-person plural verb inflection were somewhat lower for the children with SLI (65% correct) than for the MLU controls (79%), but this difference was not statistically significant.

Comparisons between Italian-speaking and English-speaking children with SLI have been made for the production of the noun plural inflection and the third-person singular verb inflection in the investigations of both Leonard, Sabbadini, Leonard, and Volterra (1987) (the same children in the Leonard, Sabbadini, Volterra, and Leonard 1988 study, described earlier) and Leonard, Bortolini, Caselli, McGregor, and Sabbadini (1992). Higher percentages of use were seen by the Italian children in both studies, though the difference for the noun plural inflection in Leonard, Bortolini, Caselli, McGregor, and Sabbadini (1992) was not statistically reliable.

Although Italian-speaking children with SLI may resemble MLU controls in their use of many (but not all) inflections, it is also true that children with severe impairments can show problems on more than the third-person plural inflection. Cipriani, Chilosi, Bottari, and Pfanner (1993) described one child with SLI who sometimes produced infinitive forms in place of inflected forms, as in (1).

(1)   *Adult*:   Cosa fai?   "What are you doing?"
       Child: Girare   "To turn around"

Children with such severe deficits not only omit function words frequently, but also occasionally omit main verbs and other obligatory open-class words (e.g., *cane casa*, "dog home," for *il cane va a casa*, "the dog goes home"). According to Cipriani, Chilosi, and Bottari (1995), children with SLI with severe deficits are more likely to omit lexical verbs than are younger normally developing children.

Other Romance Languages

Spanish
Spanish and Italian share many properties, including similar noun, adjective, and verb inflection paradigms, the use of clitics, and the licensing of null subjects. As in Italian, the canonical order is subject-verb-object, but word order can vary according to discourse context. One of the most notable differences is that in some Spanish-speaking regions (e.g., Mexico), past events are commonly described using the simple past—a paradigm with inflections on main verbs—rather than the present perfect, as is often seen in Italian. Certain inflections end in (or consist entirely of) -*s* or -*n* (e.g., *globo-globos*, "balloon-balloons"; *corre-corren*, "he/she runs-they run"). Most Spanish words have penultimate stress.

The available data from children with SLI acquiring Spanish suggests that comprehension skills are often higher than production skills in these children, phonology can be quite limited, and morphosyntactic problems are common (Merino, 1983). Lexical problems and difficulties in formulating the proposition of the utterance have also been observed (Serra-Raventós & Bosch-Galceran, 1992). In most studies, the morphosyntactic abilities of school-age Spanish-speaking children with SLI have been compared with those of age controls. Age controls have shown higher percentages of use of function words such as articles and clitics, with both omissions and substitution errors more abundant in the speech of children with SLI (Bosch-Galceran & Serra-Raventós, 1994; Merino, 1983). Of the inflections studied, adjective inflections have not produced group differences (Merino, 1983). However, noun plural inflections (Merino, 1983) and third-person plural verb inflections (Bosch-Galceran & Serra-Raventós, 1994) seem to be more problematic for children with SLI.

Restrepo (1995) studied a group of somewhat younger Spanish-speaking children with SLI, ranging in age from 5;0 to 7;1. A group of age controls also participated. Both omissions of grammatical morphemes and productions of incorrectly marked morphemes were more frequent in the speech of the children with SLI. The most frequent type of incorrect selection concerned the use of articles that failed to agree with the noun in gender. Verb inflections sometimes failed to agree with the number of the subject. However, tense was marked correctly, and nonfinite forms rarely, if ever, appeared in place of finite forms.

It is clear from these studies that Spanish-speaking children have not mastered many aspects of the grammatical morphology of their language by the time they are in school. On the other hand, because comparisons have been between children with SLI and age controls, it is not yet clear if the grammatical morphemes revealing differences represent areas of special difficulty for children with SLI or merely constitute part of the more general deficit experienced by these children relative to age expectations.

French

French differs from Spanish and Italian in ways that make it a very informative language within which to examine SLI. Its noun, verb, and adjective inflection paradigms are structured as in these other Romance languages, but in spoken form, many of these inflections are homophonous. Clitics are employed, but null subjects are not permitted. Word order can vary according to discourse factors, though the canonical order is subject-verb-object. French differs from Spanish and Italian in its prosodic characteristics. Words often have final- rather than penultimate-syllable stress, and the differences between strong and weak syllables are not large when measured in duration. In fact, fundamental frequency is the major cue to stress in the language.

The study of language disorders in French-speaking children has been an active area of research in recent years, and children with SLI have been among the groups receiving considerable investigative attention. Many of the features of SLI seen in children acquiring other languages are observed in French-speaking children as well (see Chevrie-Muller, 1996; Gérard, 1991). Likewise, individual differences (Le Normand & Chevrie-Muller, 1991) and possible subgroups (Van Hout, 1989) can be found. As has been found in other languages, French-speaking children with SLI during the preschool years are at significant risk for residual spoken language problems and serious reading problems in later years (Billard, Loisel Dufour, Gillet, & Ballanger, 1989).

The distribution of lexical classes reflected in the spontaneous speech of preschoolers with SLI seems to match that seen in MLU controls (Le Normand & Chevrie-Muller, 1989). Nouns and verbs predominate in the speech of both groups, with adjectives and prepositions less common. Function words as a group constitute less than 10% of the lexical types.

In some respects, the use of function words by French-speaking children with SLI resembles the data reported for English-speaking children. Nominative case pronouns are less likely to be used by children with SLI than by MLU controls. However, the greatest difficulty seems to center on the third-person masculine singular form *il* (he) (Plaza & Le Normand, 1996), suggesting that case is not the source of the problem. Because Italian and Spanish allow null subjects, comparisons with these other Romance languages cannot assist in interpreting this finding. However, the phonetic form *il* is a masculine singular article in Italian, and is especially troublesome for children with SLI acquiring that language.

Some of the findings for French differ from those seen for typologically similar languages. A case in point is an investigation of article use in French-speaking children with SLI conducted by Le Normand, Leonard, and McGregor (1993). Children with SLI age 4;0 to 6;0 (MLUs 2.8 to 3.7 words) were found to use articles with percentages in obligatory contexts as high as those seen in a group of MLU controls. The percentages averaged slightly above 90% for each group, and were well above those seen in a group of Italian-speaking children with SLI studied by Leonard, Bortolini, Caselli, McGregor, and Sabbadini (1992). (In both studies the contracted form *l'* was excluded from analysis.)

Le Normand, Leonard, and McGregor (1993) suggested several factors that might have produced these findings. First, the durations of articles in French are more similar to those of adjacent syllables, and thus articles may be more salient than their counterparts in Italian. Second, many French words have a weak syllable-strong syllable pattern, and thus these children might have had practice producing phrase-initial weak syllables, which prepared them for article use once function words entered their repertoire. Of course, articles were not assessed with novel nouns, and thus the possibility remains that the articles produced were learned with their adjacent nouns on a rote basis. Such rote learning might even be promoted in the language, given that in spoken French, the gender of many nouns is made clear only by the singular article.

Methé and Crago (1996) focused on the verb morphology of French-speaking children with SLI, age 6;8 to 8;4. These children's use of verb morphology was compared with that of a group of age controls and a group of younger normally developing children age 4;10 to 5;11. Copula and auxiliary forms were omitted significantly more frequently by the children with SLI than by the children in the other two groups. In addition, the children with SLI were more likely to use indistinct syllables in copula and auxiliary contexts that might have served as fillers rather than as attempts at a specific target. Methé and Crago also reported that subject-verb agreement errors were infrequent in the speech of the children with SLI, as was the use of an infinitive in place of a finite verb form. The latter errors appeared in fewer than 1% of the contexts in which they were possible.

Hebrew

Some Characteristics of Hebrew Studied in Children with SLI

The typology of Hebrew, a Semitic language, is significantly different from other languages discussed in this chapter. All verbs and many nouns and adjectives consist of a root plus a pattern. The root conveys the core meaning, and the pattern modulates the core meaning by conveying notions such as reciprocity and causality. The root is made up of (usually three) discontinuous consonants, and is thus not a well-constructed form by itself. For example, the root for the notion of "wear" is *l-v-sh* (we shall use phonemic notation rather than Hebrew script here). Each pattern (or *binyan*) is formed through the insertion of vowels between the consonants of the root and, in some cases, the addition of syllabic prefixes. For example, the verb form *lovesh* is from the *pa'al* pattern and is used to express the simple transitive (e.g., "he is wearing nice clothes"). The causative sense of making someone wear something (dressing someone) is conveyed through the form *malbish* of the *hif'il* pattern. To express the reflexive act of dressing oneself, a form from the *hitpa'el* pattern, *mitlabesh*, is used. It can be seen that the pronunciation of the second consonant of the root changed from the labial dental fricative *v* in *lovesh* to the labial stop *b* in *malbish* and *mitlabesh*. Such alterations are dictated by phonological rules and do not represent a change in the core meaning of the root.

Within each pattern, verbs are inflected for tense, number, gender, and (in past and future tense) person. These inflections are syllabic suffixes and vocalic infixes that appear between the consonants of the root. For example, *lovesh* is a masculine sin-

Table 4.6
Hebrew Verb Inflections for Three Patterns

	Pa'al Pattern r-x-v "ride"	Pi'el Pattern b-sh-l "cook"	Hitpa'el Pattern l-v-sh "wear"
Present			
Masc. sing.	roxev	mevashel	mitlabesh
Masc. plu.	roxvim	mevashlim	mitlabshim
Fem. sing.	roxevet	mevashelet	mitlabeshet
Fem. plu.	roxvot	mevashlot	mitlabshot
Past			
1 Sing.[a]	raxavti	bishalti	hitlabashti
1 Plur.[a]	raxavnu	bishalnu	hitlabashnu
2 Masc. sing.	raxavta	bishalta	hitlabashta
2 Fem. sing.	raxavt	bishalt	hitlabasht
2 Masc. plu.	raxavtem	bishaltem	hitlabashtem
2 Fem. plu.	raxavten	bishalten	hitlabashten
3 Masc. sing.	raxav	bishel	hitlabesh
3 Fem. sing.	raxva	bishla	hitlabsha
3 Plu.[a]	raxvu	bishlu	hitlabshu

[a] No gender distinction is made.

gular present form. To express masculine plural present, *lovshim* is used, which differs from its singular counterpart in its masculine plural suffix and in having no vowel between the last two consonants of the root. The (third person) masculine singular in past tense is *lavash*, which differs from its present tense counterpart in the particular vowels that appear between the consonants of the root. For some verbs and patterns, both vowels must be considered to determine the specific grammatical function. Examples include the forms for "buy": *kone* (he buys), *kona* (she buys), and *kana* (he bought). Table 4.6 provides the inflected forms for three patterns. A different root is used in each pattern so that a wider range of verbs can be illustrated.

In each pattern, the masculine singular form is considered the basic form in present tense; in past tense, the basic form is third-person masculine singular. These inflected forms are the simplest morphologically because they involve the smallest number of elaborations on the root.

Nouns are marked for number and gender. The masculine and feminine plural suffixes are *-im* and *-ot*, respectively. For many nouns, the distinction between the singular and plural forms is found only in the presence or absence of the plural suffix, as in *kadur-kadurim*, "ball-balls." However, for other nouns, changes elsewhere in the word accompany the plural form (e.g., *sefer-sfarim*, "book-books"; *yalda-yeladot*, "girl-girls").

Usually, singular nouns ending in *a* or *t* are feminine and other singular nouns are masculine. However, exceptions can be found. For example, *kos* (glass) appears to be masculine but is not; thus, it requires the plural suffix *-ot*. Some of the exceptions are complicated because a noun of one gender takes a plural suffix of another gender, yet the plural form is considered to have the same gender as the singular. For example, the plural form for the masculine noun *shulxan* (table) is *shulxanot*, which appears to be feminine but is not.

Adjectives follow the nouns they modify, and must agree with them in number and gender. For example, the adjective "big" must assume a masculine and a feminine form, respectively, in *meil gadol* (big coat) and *xultsa gdola* (big shirt). In exceptional cases such as *shulxanot*, the adjective must agree with the actual gender of the noun, not its surface form. In this case, then, *shulxanot gdolim* is required, rather than *shulxanot gdolot*.

Hebrew differs from the languages already discussed in the way it marks definiteness. The prefix *ha-* is used to mark definite; no overt form is used to indicate indefinite. Unlike definite articles in languages such as English, the definite prefix in Hebrew is affixed not only to common nouns but also to other words that modify the noun. Thus, *ha-kelev ha-gadol* (the big dog) is literally "the-dog the-big."

Definiteness is also marked in accusative case, using the freestanding invariant form *et*. This form appears with both common and proper nouns. For example, "Daddy washes the car" and "Daddy washes Rachel" would be *aba roxets et ha-mexonit* and *aba roxets et Rachel*, respectively.

The canonical word order of Hebrew is subject-verb-object. However, word order is quite variable in conversation. Subjects can be omitted in past tense in first and second person.

Although exceptions are abundant, most words in Hebrew have word-final stress. Consequently, many of the inflections of person, number, and gender carry primary stress. One of the exceptions is the present feminine singular inflection (e.g., *roxetset*, *mevashelet*, *mitlabeshet*), where primary stress is on the penultimate syllable. However, some common verbs take the feminine singular inflection *-a*, which is stressed. Examples include "she wants" (*rotsa*), "she sees" (*ro'a*), "she does/makes" (*osa*), and "she drinks" (*shota*). The frequent occurrence of word-final and penultimate stress means that in multisyllabic words, the full expression of certain patterns will depend on weak syllables located early in the word. This is especially true for the *hitpa'el* pattern.

Hebrew Data

As is true for other languages, children with SLI acquiring Hebrew show the familiar characteristics of slow lexical development and late word combinations. Comprehension is usually superior to production. Phonology and morphosyntax are most often the weakest areas. Pragmatic abilities are adversely affected by the difficulties in other areas of language; when pragmatic behaviors do not depend on age-level language abilities, Hebrew-speaking children with SLI usually resemble normally developing children of the same age (Rom & Bliss, 1981, 1983). A closer look at these children's relative strengths and weaknesses, however, reveals much information about SLI that would be unavailable if study were limited to non-Semitic languages.

These children's difficulties with the verb system vary according to the element of the verb system examined. The inflections coding commonly studied features such as number, gender, person, and tense do not stand out as extraordinarily difficult for Hebrew-speaking children with SLI. Data from two studies support such a conclusion. Rom and Leonard (1990) examined the verb inflections reflected in the spontaneous speech of a group of Hebrew-speaking children with SLI (age 4;4 to 5;3) and a group of younger normally developing children (age 2;4 to 3;3) matched according to mean number of morphemes per utterance (MPU), an utterance length measure devised by Dromi and Berman (1982) for Hebrew. The MPUs ranged from 2.3 to 4.4; corresponding MLUs in words ranged from 2.0 to 3.4. Because the speech samples were small, Rom and Leonard collapsed all present tense inflections (marking number and gender), and all past tense inflections (coding number, gender, and person). The groups were highly similar in the percentages with which they used both present tense and past tense inflections in obligatory contexts. Mean percentages of use averaged over 90% for both types of verb inflections in the speech of the children with SLI.

Dromi, Leonard, and Shteiman (1993) also examined verb inflections for person, number, gender, and tense, but did so using probe items that required the children to produce inflections with a wider range of verbs, some taking less common inflections. Four different patterns were represented in their probes. A group of children with SLI, a group of younger controls matched according to MLU in words, and a group of age controls participated. The children with SLI ranged in age from 4;1 to 5;11 and in MLU in words from 2.0 to 3.1. Percentages of correct use were lower than in the Rom and Leonard (1990) study. For the children with SLI, present and past tense inflections averaged 76% and 56%, respectively. The MLU controls' percentages were no higher; only the age controls produced these inflections with high percentages.

Given that the masculine singular forms in present tense and third-person masculine singular forms in past tense are considered the basic, unmarked forms in Hebrew, it seemed possible that percentages collapsed across number, gender, and person might have obscured real differences among the groups of children. Leonard and Dromi (1994) examined the Dromi, Leonard, and Shteiman (1993) verb inflection data to determine if this were the case. Table 4.7 displays the children's mean percentages of use according to the number, gender, and tense of the inflection. Past tense forms were examined in third person only. The results revealed the same trends in all three groups: (1) percentages for present tense inflections were higher than those for past tense; (2) within present tense, singular inflections had higher percentages than plural inflections, with the feminine plural inflection showing the lowest percentage; and (3) within past tense, masculine singular inflections were used with higher percentages than feminine singular inflections.

Considering the number of inflections that are reflected in the data summarized in table 4.7, one might conclude that the children with SLI were as proficient with verb

Table 4.7
Mean Percentage of Inflection Use in Obligatory Contexts by Hebrew-Speaking Children

	Present				Past[a]		
	Sing.		Plu.		Sing.		Plu.[b]
	Masc.	Fem.	Masc.	Fem.	Masc.	Fem.	
SLI	93	88	87	28	64	50	62
MLU	97	94	95	40	41	27	25
Age	100	99	99	88	87	85	85

[a] Only third-person forms were examined in past tense.
[b] No gender distinction is made for third-person plural in past tense.
Source: Dromi, Leonard, and Shteiman (1993); Leonard and Dromi (1994).

morphology as were the MLU controls. However, inflections for number, gender, person, and tense do not provide a complete picture. Important differences existed between the groups when the overall accuracy of the verb form was examined. The children with SLI were less likely to produce the verbs in an adultlike manner. Many of the differences resided in the weak initial and medial syllables of the verb, an important locus for verb pattern distinctions. In fact, Dromi, Leonard, and Shteiman (1993) found that the children with SLI were significantly more likely than the MLU controls to omit sounds or syllables that resulted in a pattern error. They also exhibited greater difficulty with the phonotactic rules of verb formation.

Yet, the difficulties of the children with SLI nevertheless revealed evidence of creativity. For example, several children produced the past tense forms *bishel*, *bishla*, and *bishlu* as *vishel*, *vishla*, and *vishlu*, respectively. The present tense forms of this verb employ *v* rather than *b* as the consonant variant of the root (see table 4.6). Several of the children seemed to overextend the *pa'al* pattern to other patterns. For example, the forms *bishla* and *bishlu* were produced by some children as *bashla* and *bashlu*, respectively.

The verb production data from Hebrew-speaking children with SLI make it clear that a distinction must be made between verb morphology in the form of inflections for number, gender, and the like, and morphology involving other types of verb formation. Relative to MLU controls, children with SLI seem to have less difficulty with the former than the latter. However, even errors in pattern and phonotactic rule application sometimes provide evidence of use that cannot be the result of rote memory.

Both Rom and Leonard (1990) and Dromi, Leonard, and Shteiman (1993) found no differences between the children with SLI and the MLU controls on noun plural inflections or adjective-agreement inflections. As expected, age controls showed higher percentages than these two groups for each inflection type. The clearest differences were seen for items whose phonological form promoted an inflection that was not in fact the appropriate one. For example, many of the children with SLI and MLU controls produced the plural of the noun *vilon* (curtain) as *vilonim* rather than *vilonot*, probably because the singular form did not end in *a* or *t*. Similarly, the adjective-noun phrase *kos male* (full glass) was produced by many of the children even though the feminine adjective form was required, *kos melea*. This type of use suggests that these children were applying what they knew about Hebrew grammar, even if at times it was not adultlike.

In the studies of Rom and Leonard (1990) and Dromi, Leonard, and Shteiman (1993), the definite accusative case marker *et* was produced with higher percentages in obligatory contexts by the MLU controls than by the children with SLI. The definite prefix *ha-* yielded mixed results. Rom and Leonard observed differences in the use of this form favoring the MLU controls, whereas Dromi, Leonard, and Shteiman (1993) found no differences between groups. The age controls' use of both *ha-* and *et* was above 90%.

There is a parallel between the production findings for children with SLI acquiring Hebrew and those learning Italian. In both instances, nonfinal morphemic material is most vulnerable. In Italian, these are function words such as articles and clitics; in Hebrew, they are prefixal or freestanding definite markers and syllables needed to form the pattern of verbs.

German

Some Characteristics of German Studied in Children with SLI

There are many characteristics of German that can be valuable to the study of SLI. At the top of the list are its heavy reliance on grammatical case, and its variable but rule-governed word order. German also makes significant use of inflections that have a wide range of allomorphs. Collectively, these characteristics pose a significant challenge to children whose grammatical learning abilities are suspect.

In German, nouns are masculine, feminine, or neuter in gender. However, the overt marking of gender appears in the preceding article, demonstrative, or adjective, not on the noun. The noun *Apfel* (apple), for example, is masculine, requiring the definite article *der* if used in nominative case, *den* if used in accusative case, and so on. The word *Kind* (child), in contrast, is neuter, with its own definite article forms required (e.g., *das*). Table 4.8 provides the definite and indefinite articles of German according to gender, case, and number. Adjectives preceding nouns with no accompanying articles have inflections that match those of the definite articles (e.g., *großer, großen, großes, großem* [big] for masculine singular nouns). Slightly different, overlapping sets of inflections are required when adjectives appear with a definite or indefinite article.

There are four noun plural allomorphs: *-(e)n, -s, -e,* and *-er;* the last two have a variant in which an umlaut occurs as well as the affix. In each instance, the prenominal article, demonstrative, or adjective will also be marked for number. Plural can also be expressed with no affix; here, the prenominal form alone, an umlaut alone, or the prenominal form together with an umlaut expresses the number of the noun. Table 4.9 provides a few examples of the plural allomorphs. It should be noted that the choice of allomorph is only loosely linked to the gender of the noun; exceptions are numerous. However, there is one clear connection: feminine nouns whose singular form ends in *e* (e.g., *die Straße,* "the street") will have *-n* as the plural affix (*die Straßen*). Nouns also employ affixes for genitive case of masculine and neuter singular nouns, dative case of plurals, and, in certain expressions, masculine and neuter singular nouns.

Verbs are inflected for person, number, and tense. Polite forms are expressed in second person singular and plural, using inflections that match the third-person plural inflection. Table 4.10 provides the inflections for present and simple past tense for a "weak" verb. Such verbs require changes only in the inflections themselves. "Strong" verbs, on the other hand, require changes in the stem (the vowel, in particular) with different tenses, and differ from weak verbs in the inflectional paradigm. Thus, whereas

Table 4.8
Definite and Indefinite Articles of German

Case	Singular						Plural
	Masculine		Feminine		Neuter		
	Def.	Indef.	Def.	Indef.	Def.	Indef.	Def.
Nomin.	der	ein	die	eine	das	ein	die
Accus.	den	einen	die	eine	das	ein	die
Gen.	des	eines	der	einer	des	eines	der
Dat.	dem	einem	der	einer	dem	einem	den

Table 4.9
Plural Allomorphs of German, Used with Articles in Nominative Case

Singular	Plural	
die Frau	die Frauen	the woman/women
das Auto	die Autos	the car/cars
der Hund	die Hunde	the dog/dogs
die Kuh	die Kühe	the cow/cows
das Buch	die Bücher	the book/books
das Bild	die Bilder	the picture/pictures
der Lehrer	die Lehrer	the teacher/teachers
die Mutter	die Mütter	the mother/mothers
der Apfel	die Äpfel	the apple/apples

Table 4.10
Present and Simple Past Verb Inflections for the Weak Verb *lernen* (to learn)

Person	Present		Past	
	Singular	Plural	Singular	Plural
1	lerne	lernen	lernte	lernten
2	lernst	lernt	lerntest	lerntet
3	lernt	lernen	lernte	lernten

the third-person singular past form for "learn" is *lernte* (see table 4.10), the corresponding form for "find" is *fand*. There are also "mixed" verbs that show changes in the stem but employ the inflectional paradigm of weak verbs.

An inspection of table 4.10 reveals considerable overlap in the phonological forms used for the inflections. The first- and third-person plural inflections in present tense are identical to the inflection used for the infinitive. In past tense, the first- and third-person singular inflections are the same. Because subjects must appear with finite verbs, ambiguity is minimized.

The past tense forms shown in table 4.10 are but one way of describing events in the past. In conversational speech, such events are usually described using the auxiliary plus past participle (e.g., *Chris hat gekocht*, "Chris has cooked"). For most weak verbs, the past participle is formed by adding the prefix *ge-* and the suffix *-t* to the stem. The past participles of strong verbs are formed by adding the prefix *ge-*, the suffix *-en*, and changing a vowel in the stem. Thus, the past participle of the weak verb *lernen* (to learn) is *gelernt*, whereas for the strong verb *finden* (to find), it is *gefunden*. For mixed verbs, the participle employs *ge-* and *-t* as in the weak verbs, but a vowel change is required.

The copula *sein* (be) forms in present tense are distinct for each person in singular (*bin, bist, ist*), but first and third person in plural employ the same form (*sind, seid, sind*). In past tense, first- and third-singular forms are the same (*war, warst, war*), as are first- and third-plural forms (*waren, wart, waren*). All but first- and third-person plural forms in past tense are monosyllabic.

In declarative sentences, the finite verb appears in second position. This "verb-second" rule means that when the sentence begins with an adverbial such as "yesterday" or the direct object through topicalization, the subject is placed after the finite verb, as in (2b) and (2c).

(2) a. die Frau fand die Kinder
         the woman found the children
         "the woman found the children"
     b. gestern fand die Frau die Kinder
         yesterday found the woman the children
         "yesterday the woman found the children"
     c. die Kinder fand die Frau
         the children found the woman
         "the woman found the children"

When auxiliaries are used, these appear in second position and the infinitive or past participle appears in final position, as in (3).

(3) a. Ich will das kochen
         I will that cook
         "I will cook that"
     b. Ich habe das gekocht
         I have that cooked
         "I have cooked that"

In subordinate clauses, the finite verb appears in clause-final position, as in (4).

(4) Peter sagt daß die Frau die Kinder fand
     Peter says that the woman the children found
     "Peter says that the woman found the children"

For sentences with auxiliaries, this placement results in the auxiliary, rather than the infinitive or participle, appearing in final position, as in (5b).

(5) a. Chris kann kochen
         Chris can cook
         "Chris can cook"
     b. sie sagt daß Chris kochen kann
         she says that Chris cook can
         "she says that Chris can cook"

Stress is usually on the initial syllable in German words. Most syllabic inflections appear as word-final weak syllables that immediately follow the strong syllable of the stem, thus conforming to the dominant strong syllable-weak syllable pattern of the language. Grammatical morphemes that fall outside of this pattern are the monosyllabic function words and the prefix of the past participle.

German Data

As we saw in chapter 1, the study of German-speaking children with SLI has a long history. However, since the early 1980s, there has been a notable increase in the number of investigations in this area (e.g., Clahsen, 1989; Grimm, 1983; Kegel, 1981;

Schöler & Moerschel, 1984). These studies reveal considerable heterogeneity among German-speaking children with SLI. However, as has been observed in children with SLI acquiring other languages, comprehension abilities are usually superior to production abilities, and the semantic area is often a relative strength (Grimm, 1993; Grimm & Weinert, 1990). Although phonology can be quite weak in these children, syntax and grammatical morphology—especially inflections involving grammatical agreement—stand out as vulnerable areas (Clahsen, 1991; Grimm, 1993; Schöler, 1985), as we shall see.

Some illustrative data come from an individual with SLI studied by Kerschensteiner and Huber (1975). The errors seem typical of those described by other investigators, though probably more frequent in occurrence. The most remarkable aspect of this case is the individual's age at the time of the study—23 years.

The utterances in (6) were among those provided by Kerschensteiner and Huber. The adult German equivalent appears in parentheses.

(6) a. der sagen wo kommen her
         he say where come from
         (der sagt wo die her kommen)
         (he says where they from come)
         "he says where they come from"
     b. ja ich versteh die Sätze
         yes, I understand the sentences
         (ja ich versteh die Sätze)
         (yes, I understand the sentences)
         "yes, I understand the sentences"
     c. der zuhause n guten Auskommen hat
         he at home a good income has
         (der hat zuhause ein gutes Auskommen)
         (he has at home a good income)
         "he has a good income at home"

Function words such as copula forms and articles were sometimes missing or produced in a reduced phonetic form (n) that obliterated grammatical marking. Verbs often failed to agree with the subject, as in (6a). Word order, too, was problematic. In (6c), the verb *hat* is appropriately finite, and should appear in second position. Its placement in final position would be correct only if this were a subordinate clause.

Subsequent investigations of German-speaking children with SLI have agreed on the observation that word order problems abound in these children. However, they disagree on how such problems should be characterized. Günther (1981) described the sentences of a group of school-age children with SLI as being too tied to a subject-verb-object order, even in sentence imitation tasks (see also Kegel, 1981).

Findings from Grimm's (1983) study of preschool-age children with SLI supported a very different conclusion. These children produced a disproportionate number of sentences with verbs in final position. Similar findings were reported by Kaltenbacher and Kany (1985) in their investigation of preschool-age children with SLI. In both of these studies, the children with SLI differed from a group of normally developing two-year-olds in the degree to which verbs were produced in final position in general, and in the degree to which finite verbs in particular were (inappropriately) produced in this position in main clauses.

Clahsen (1989, 1991) argued that many of the verbs appearing in final position in these last two studies were in fact nonfinite, and that the finite forms used (some incorrectly marked for person or number) usually appeared, appropriately, in second position. This observation led Clahsen to propose that the children with SLI in the Grimm (1983) and Kaltenbacher and Kany (1985) studies were in an early stage of grammatical development in which a morphological paradigm had not fully developed. In essence, when a finite inflection was unavailable, a nonfinite form was used and placed in the customary sentence position for nonfinite forms. Clahsen also suggested that the frequent use of the subject-verb-object order by the children with SLI studied by Günther (1981) was attributable to these children's having achieved a later stage of morphological development, in which a larger set of finite inflections is available, permitting more consistent use of finite verbs in second position.

To test his hypothesis, Clahsen (1989, 1991) examined the spontaneous speech of a group of German-speaking children with SLI ranging in age from 3;2 to 9;6 and in MLU in words from 1.46 to 2.84. He found that most of the children produced verbs in final position, and these were infinitive forms or uninflected verb stems. Auxiliaries and copula forms were usually missing. Occasional instances of a finite inflection or copula were seen (e.g., -t, ist), but these did not always show proper agreement with the subject. According to Clahsen, these findings suggested that the children's problems with word order were due to a more fundamental problem in establishing agreement relations (in this case, subject-verb) in the grammar.

However, one of the children studied by Clahsen produced utterances similar to those described by Grimm (1983). This child produced many finite forms but placed them in final position. Only modals were appropriately positioned in the sentence. Interestingly, these finite forms were correct in their person and number marking, as in (7).

(7)   und jetzt du wieder schreibst
       and now you again throw
       "and now you throw again"

Clahsen concluded that for this child, only modals had the requisite properties to move to second position; the inflections on other verbs, although showing proper agreement, were built up by word formation rules and were thus more like nonfinite forms in structure. For this reason, they remained in final position.

Problems of verb morphology and word order were not the only difficulties forming the basis for Clahsen's (1989, 1991) claim that most of the children had problems with agreement relations. Articles were often absent from these children's utterances, and when they appeared, they frequently assumed the wrong gender and number. Case markings, too, were problematic. Articles were often marked for the wrong case, and if both an article and an adjective appeared in a noun phrase, the case marking appeared on only one of them.

S. Roberts (1995) examined speech samples from eight of the children with SLI studied by Clahsen (1989, 1991) who formed appropriate matches for a group of English-speaking children with SLI (discussed below). She found that when finite verb inflections and copula forms were produced, they were usually correct. For third-person singular verb inflections, accuracy ranged from 87% to 100% (mean = 97%). Bare stems outnumbered both incorrect finite forms and nonfinite forms in third-person singular contexts. For copula forms, the range of accuracy was 86% to 100%

(mean = 96%); omissions constituted the dominant problem. Roberts observed word order patterns to be consistent with those described by Clahsen. Finite forms usually occurred appropriately in second position in main clauses (mean = 92%), whereas correct nonfinite forms almost always appeared in final position (mean = 97%). For nonfinite productions that should have been finite, the verb was usually in final position for most of the children, but two children were more likely to place the verb in second position.

Roberts also identified bare stems whose vowels and consonants permitted them to be considered attempts at finite or nonfinite forms. The same positional pattern was seen; forms that seemed like attempts at finite forms were in second position, whereas forms that appeared to be attempts at nonfinite forms were in final position. Finally, Roberts examined article use and found correct use in obligatory contexts ranging from 30% to 71%. If the correct form was not used, omissions usually occurred. Use of an inappropriately marked article was relatively infrequent (7%).

The more recent publications by Grimm and her colleagues (Grimm, 1993; Grimm & Weinert, 1990) do not seem consistent with the idea that verb-final use is closely linked to limitations in finite verb morphology. These investigators summarized data from preschool-age children with SLI and a group of MLU controls. The examples of utterances that they characterized as typical of their subjects with SLI showed finite verb forms in final position, whereas Clahsen (1989, 1991) observed only one child with such a pattern. The children with SLI studied by Grimm and her colleagues also showed a tendency to produce verbs in final position in their responses on a sentence imitation task, even when the verbs in the sentences to be repeated were located in second position. It is not clear how often the finite morphology of the verbs was retained; however, examples given by Grimm (1993) show that some of the misplaced verbs were produced with finite inflections. Kaltenbacher and Lindner (1990) also presented case studies of childern with SLI who used finite verb forms in final position.

The available evidence is sufficient to conclude that finite verbs appear in final position in the speech of some German-speaking children with SLI. Although such cases suggest word order problems, they also raise the possibility that finite verb morphology might not represent an area of extraordinary weakness in these children. With the exception of the 23-year-old discussed earlier, most of the children with SLI who showed finite verb limitations might have been doing what was expected, given other details of their language, such as their utterance length. Surprisingly, few of the studies comparing German-speaking children with SLI and MLU controls have compared the two groups on their use of finite verb inflections.

A study by Bartke (1994) is a welcome exception. She examined the speech of school-age children with SLI and a group of younger MLU controls and found that the former not only committed more verb placement errors, but also showed less accurate use of three of four verb inflections studied. Furthermore, although the remaining inflection (the second-person singular form -st) was produced accurately by both groups when it was used, it was more likely to be omitted by the children with SLI. This finding seems to confirm that difficulties in both verb morphology and word order are common characteristics in German-speaking children with SLI.

It would appear that the especially serious problems German-speaking children with SLI have with verb morphology do not extend to past participles. Clahsen and Rothweiler (1992) examined the participles used in spontaneous speech by children with SLI (age 3;1 to 7;11, MLU 1.6 to 4.1 words) and MLU controls (age 1;6 to 3;9). The two groups were similar in the degree to which they distinguished participles from infinitives and in the errors they made in participle formation. The prefix *ge-* and the suffixes *-t* and *-en* were omitted to the same degree in the two groups. In addition, both groups were similar in applying the stem of the infinitive to the participle (e.g., *gesingen* for *gesungen*, "sung") and the *-t* suffix to participles requiring *-en* (e.g., *gesungt* or *gesingt* for *gesungen*).

Much remains to be learned about the reduced ability with verb morphology seen in German-speaking children with SLI. The degree to which this difficulty is related to word order problems or more general problems with agreement must be established. Lingering questions notwithstanding, there is little to suggest that these children's limitations reflect a system gone completely awry. Verb morphology may not be up to the standard expected, given these children's utterance lengths, for example, but patterns of use matching those of normally developing children can be seen. For example, Lindner, Stoll, and Täubner (1994) observed that children with SLI, like younger normally developing children, were more likely to use past participles with change-of-state verbs and present tense forms with stative and process verbs.

The use of noun plural allomorphs by German-speaking children with SLI has been the subject of considerable investigation. Clahsen, Rothweiler, Woest, and Marcus (1992) conducted a comprehensive investigation of the use of plural allomorphs by the children with SLI (age 3;1 to 6;11) participating in the Clahsen and Rothweiler (1992) study on participles described earlier. Spontaneous speech samples as well as responses on formal tasks were obtained. In spontaneous speech, the children's correct use of plural allomorphs in obligatory contexts averaged around 79%, more closely approximating normally developing children three years of age. Singular forms were produced in plural contexts by only half of the children; for all children, the most common error was the use of an inappropriate plural allomorph. In most instances this was *-(e)n*; in some instances, *-s*. Examples included *Bäumen* for *Bäume* (trees) and *Vogeln* for *Vögel* (birds). The children's responses on a formal task showed lower levels of accuracy (averaging around 67%). More striking was the children's high degree of singular noun use in place of plural nouns on this task.

Clahsen, Rothweiler, Woest, and Marcus (1992) also examined the children's use of noun compounds. Such compounds are productive in German and require adding a plural allomorph to the initial (non-head) noun, as in *Bild+er+Buch = Bilderbuch* (picture book). It can be seen that the plural allomorphs serve a linking function only; they do not mark number in these instances. An inspection of the children's use of compounds in spontaneous speech revealed that the plural allomorphs *-s* and *-(e)n* were not employed in compounds, even though (in the case of the latter allomorph) it was often required. The other allomorphs were used by the children in compounds. The same pattern of use was seen in the speech of a normally developing child studied by Clahsen, Rothweiler, Woest, and Marcus. Noting that the two allomorphs not used in compounds were the ones most frequently serving as inappropriate substitutes in plural nouns (see above), they took the findings as evidence that children with SLI, like more typical children, are sensitive to the distinction between regular and irregular morphology.

Several other studies have examined the use of noun plural allomorphs by German-speaking children with SLI. These include the investigations of Veit (1986), Holtz

(1988), and, more recently, Schöler and Kürsten (1995). In each of these studies, the children with SLI exhibited greater difficulty in using the appropriate allomorphs than did control children. Although errors were generally similar in the two groups, children with SLI may rely to a greater extent on the allomorph *-(e)n* than do younger normally developing children (Schöler & Kürsten, 1995).

The investigation of Schöler and Kürsten included tasks dealing with other aspects of grammar. Children with SLI and normally developing children age 6 to 12 years were compared on their ability to insert appropriate inflections and articles in texts with such forms missing. The normally developing children reached ceiling on this task by 8 years of age. Even the 12-year-old children with SLI performed well below this level. Both the normally developing children and children with SLI had the least difficulty with adjective inflections, followed by verb inflections, then articles. But the children with SLI differed from the control children in that they frequently inserted inflections or articles that matched the surface form of the inflections of adjacent words in the phrase. For example, the dative case phrase *einem hohen Baum* (a tall tree) was produced as *einem hohem Baum*. This is not a surface matching strategy without a grammatical foundation; the form *hohem* would be appropriate if no article had preceded the adjective.

The children in the Schöler and Kürsten (1995) study also participated in a task requiring them to detect and correct grammatical errors in sentences. The normally developing children reached the highest level of accuracy by age 8;6, a level not achieved by even the oldest of the children with SLI. Furthermore, for those errors that could be detected consistently by both groups of children, attempts at correction were less likely to be successful by the children with SLI.

The syntactic and morphological difficulties of German-speaking children with SLI seem to place a heavy burden on these children's ability to participate in conversations and recount stories (Weinert, Grimm, Delille, & Scholten-Zitzewitz, 1989). For example, Grimm (1987) reported that a group of preschool-age children with SLI was significantly less likely than younger controls to take aspects of their mothers' prior utterances and make some change in one of the constituents, thus contributing new, topic-relevant information to the conversation. Conversational turns of this type require considerable control of syntactic categories (Leonard & Fey, 1991).

Finally, there have been at least two investigations aimed at comparing German-speaking children with SLI and children with SLI acquiring English. Lindner and Johnston (1992) proposed that even when children with SLI have difficulty with their language, German-speaking children should be stronger in grammatical morphology than English-speaking children, and should show a smaller gap between grammatical morphology and vocabulary ability (with higher scores on the latter). The children with SLI in the two languages ranged in age from 4;5 to 6;11 and were matched according to standard scores on tests of language ability. Separate measures of grammatical morphology and lexical use were then obtained from the children. The investigators' expectations were confirmed by the data. Thus, even though the German-speaking children with SLI had grammatical problems relative to their normally developing compatriots, the relative severity of their deficits in grammatical morphology was blunted by the nature of the language they were acquiring.

S. Roberts (1995) reported similar findings in a study that compared eight of Clahsen's (1989, 1991) subjects with SLI and eight English-speaking children with SLI matched according to MLU in words. For example, correct third-person singular verb inflections and copula forms were used with higher percentages in the speech of the German-speaking children with SLI.

Other Germanic Languages

Dutch

SLI in Dutch-speaking children has received increasing attention in recent years (e.g., Bol & Kuiken, 1988; Leemans, 1994). Although the noun and verb inflection paradigms of Dutch are not as elaborate as those in German, Dutch and German share many characteristics, including the verb-second rule.

It appears that children with SLI learning Dutch often exhibit limitations in phonology and morphosyntax, though semantic and pragmatic problems can be seen as well (Beers, 1992; A. Mills, Pulles, & Witten, 1992). The morphosyntactic deficits observed include word order errors, agreement errors, and omission of articles.

However, the few studies conducted to date that compared Dutch-speaking children with SLI and MLU controls have revealed far more similarities than differences. An investigation by de Jong, Fletcher, and Ingham (1994) showed that children with SLI resembled MLU controls in the verb argument structures reflected in their speech. Yet the children with SLI were more likely to use verbs requiring fewer arguments, such as intransitives and verbs with a single, oblique argument (e.g., *op het trottoir,* "on the sidewalk," in *Ik sta op het trottoir,* "I stand on the sidewalk").

Bol and de Jong (1992) found much similarity in the use of auxiliary verbs by children with SLI and MLU controls. These investigators noted that several of the children with SLI omitted aspect auxiliaries (verb forms of *zijn,* "to be," and *hebben,* "to have"), as in (8).

(8) dat Zeppelin daan
    that Zeppelin done
    "Zeppelin done that"

Here, the auxiliary form *heeft* (has) should have appeared in second position (and the correct form of the participle in final position is *gedaan*).

According to Bol and de Jong (1992), the reason for the failure to find the large group differences in auxiliary use documented in the literature on English is that Dutch and English auxiliaries function quite differently. For example, sentence-initial auxiliaries are not required for well-constructed questions in Dutch, nor are auxiliaries needed for tag questions. Furthermore, in many speaking contexts, modal auxiliaries can be used without infinitives following, as seen in the equally grammatical utterances in (9a) and (9b).

(9) a. Ik kan dat doen
        I can that do
        "I can do that"
    b. Ik kan dat
        I can that

Left unexplained, as Bol and de Jong (1992) point out, is why their Dutch-speaking subjects exhibited fewer difficulties relative to MLU controls than seem to occur in children with SLI acquiring German, a language with many of these same auxiliary characteristics. Future research is needed to resolve this apparent discrepancy.

Swedish

Swedish also has a verb-second rule for main clauses, but its word order rules differ from those of German in other respects. When modals or other auxiliaries are used, the accompanying nonfinite verb directly follows the auxiliary rather than being placed at the end of the sentence, as in (10).

(10) a. Du måste köpa mjölk
          You must buy milk
          "You must buy milk"
      b. Du har köpt mjölk
          You have bought milk
          "You have bought milk"

In subordinate clauses, the finite verbs are not located in final position, as in German, and in such clauses, subjects always precede the verb.

Swedish nouns, verbs, and adjectives are inflected, though the inflectional paradigms are not elaborate. The verb inflection -r is used for present tense, and -de or -te is used for past tense; distinctions are not made for person and number. Past events are described by either the simple past or the present perfect; an example of the latter is shown in (10b).

The definite-indefinite distinction in Swedish differs from most other languages. Both the definite and indefinite forms must agree with the gender of the noun. However, the indefinite forms are articles, whereas the definite forms are suffixes attached to the noun. Examples are shown in (11).

(11) a. en docka   dockan
          a doll   the doll
      b. ett tåg   tåget
          a train   the train

Most words in Swedish have stress on the initial syllable, with duration playing a major role in the distinction between strong and weak syllables. An especially salient characteristic of Swedish is its use of grave and acute accents. The distinction between these accent types lies principally in the location and degree of drop in fundamental frequency during the production of the word.

Although children with SLI acquiring Swedish do not constitute a homogeneous group, problems with morphosyntax and phonology seem to be the most common (Hansson, 1992; Hansson & Nettelbladt, 1990; Magnusson, 1983; Magnusson & Nauclér, 1990; Nettelbladt, 1983; Nettelbladt, Sahlén, Ors, & Johannesson, 1989). Pragmatic skills do not appear to be as deficient as other areas of language in these children (Nettelbladt & Hansson, 1990). In most cases, comprehension ability is not as limited as production. However, as a group, these children's comprehension of language is below age level (Magnusson & Nauclér, 1990). Those with comprehension problems and limited lexical ability appear to have the most severe problems in morphosyntax (Sahlén & Nettelbladt, 1991). Even children with SLI not singled out for poor comprehension perform below the level of age controls on grammaticality judgment tasks (Magnusson & Nauclér, 1990).

Hansson and Nettelbladt (1995), Hansson (1997), and Håkansson and Nettelbladt (1996) have compared the morphosyntactic characteristics of Swedish-speaking children with SLI and MLU controls. Spontaneous speech served as the source of data in each study. To the extent that the studies examined the same aspects of grammar, the findings were quite uniform. The children with SLI used indefinite articles with smaller percentages in obligatory contexts than did the MLU controls. Interestingly, this difference was not seen for the definite suffix. Likewise, no differences were observed for noun plural inflections or for adjective agreement inflections. The auxiliary *har* (have), used in the present perfect, was omitted more often by the children with SLI. Taken together, the finite verb forms were produced with lower percentages by the children with SLI than by the MLU controls. However, neither the present nor the past verb inflection revealed differences (in any of the studies) when taken singly, due to considerable variance in the two groups of children. In all studies, word order problems were more frequent in the speech of the children with SLI. These children overused the subject-verb-object order, employing it even when an adverbial or other element was placed in initial position. For example, one child with SLI produced the utterance in (12a) when the order in (12b) was required.

(12)   a.   Sen jag fick en kompis där
            Then I had a friend there
       b.   Sen fik jag en kompis där
            Then had I a friend there
            "Then I had a friend there"

SLI in Other Language Families

Together, Romance, Germanic, and Semitic languages constitute only a small number of the ways in which languages vary. A thorough understanding of how SLI is manifested in language production and comprehension will require examination of children with SLI in many other languages. Fortunately, work is under way in languages from several additional language families. Ljubešić and Kovačević (1992) provide data on children with SLI acquiring (Serbo-)Croatian, a Slavic language. One aspect of grammar examined by these investigators was the distinction between these children's ability to produce nouns marked for plural and nouns marked for "dual," a type of plural reserved for use with the numbers two and (the term "dual" notwithstanding) three and four. Although the school-age children with SLI participating in the study had greater difficulty than a group of age controls on both types of forms, the dual morpheme was easier. This form requires fewer morphophonological alterations of the singular form.

SLI in Hungarian, a language of the Uralic family, has been studied by Vinkler and Pléh (1995). Hungarian has a rich, agglutinative morphology such that a sequence of morphemes appears after the stem of the noun or verb, each morpheme serving a unique grammatical function. For example, for nouns, the stem is followed by a derivational suffix, such as a nominalizer, followed by a suffix that modifies the meaning, such as a plural or possessive marker, followed in turn by a case marker. Relative to other languages, the case system of Hungarian is extensive. Vinkler and Pléh reported a case study of an eight-year-old child with SLI whose scores on a test of language production resembled those of most five-year-olds. In the child's spontaneous speech, articles were often omitted, certain noun or verb suffixes were missing on occasion, and others were overused. Strikingly, the order of the morphemes reflected in the child's multimorphemic words was quite accurate; even if the wrong morpheme was selected, it was properly ordered according to its grammatical function.

Dalalakis (1994) reported a study on Greek-speaking children with SLI. Greek has a rich noun and verb morphology. Noun plurals, for example, differ according to the three genders (masculine, feminine, neuter) of the noun and according to the singular form of the noun in nominative case. Some of the plural forms result in additional syllables in the word; for others, the plural form and singular form have the same number of syllables. In the Dalalakis study, school-age children with SLI and a group of age controls were given a series of comprehension, production, and grammaticality judgment tasks. The children with SLI performed more poorly on all tasks. For certain tasks, the nature of the errors could be examined. For example, on a task requiring the children to produce the plural forms of nonsense nouns, both groups had the most success with feminine nouns and the least success with neuter nouns. Each group produced the singular when the plural was required, but this error was more frequent in the children with SLI. Conversely, both groups overapplied a more common plural suffix, but this tendency was greater among the age controls.

Children with SLI learning Japanese were the participants in an investigation by Fukuda and Fukuda (1994). Several characteristics of Japanese are unlike those seen in other languages serving as the focus of research in SLI. These include the marking of tense on adjectives as well as verbs, and the use of an honorific system to mark the social standing of the speaker relative to the listener. The most frequent means by which honorifics are expressed is through the use of a morpheme affixed between the verb stem and the suffix marking tense. For example, *tabe-mas-u* is the honorific present form for "eat," whereas the corresponding form in the past ("ate") is *tabe-mashi-ta*. It can be seen that both the honorific suffixes and the tense suffixes show a distinction between present and past.

Fukuda and Fukuda (1994) administered a comprehensive battery of grammatical tasks to school-age Japanese-speaking children with SLI and a group of age controls. Differences favoring the age controls were seen for most of the tasks, including the production of tense and honorific forms as well as past tense adjectival forms, the comprehension of passive constructions, and the judgments of case particles placed in appropriate and inappropriate contexts.

Findings from still another language family come from the work of Crago and Allen (1994). These researchers studied SLI in Inuktitut, an Eskimo-Aleut language. Inuktitut is an ergative language (e.g., the subject of an intransitive verb is assigned the same case as the object of a transitive verb). Nouns and verbs are extremely complex morphologically. Verb inflections agree with both subject and object, and distinctions are made for four (rather than the usual three) persons, three (rather than the usual two) numbers, and numerous modalities. Nouns can have eight cases and three numbers. In addition, a multitude of productive noun- and verb-internal morphemes serve as nominalizers, verbalizers, modifiers, and the like.

There are probably more grammatical morphemes in Inuktitut than there are children with SLI who speak the language, testimony both to the richness of Inuktitut morphology and to the relatively small number of speakers of the language. Crago and Allen (1994) examined the spontaneous speech of one child with SLI (age 5;4), along with that of an age control and an MLU control (age 2;1). Although comparisons involving so few children must be made with caution, the three children's utterances appeared quite different. The child with SLI seemed to exhibit word-finding difficulties, and was more likely to omit obligatory verb inflections, produce bare-stem locatives, and produce pronoun forms in contexts where the reference is ordinarily made by means of an inflection. In addition, the child used one particular inflection as a filler, placing it in inappropriate positions in the sentence. Although many of the child's omissions of inflections were seen to a lesser degree in her MLU control, neither of the normally developing children showed evidence of the improper pronoun or inflection productions seen in this child.

SLI Profiles Vary with Type of Language

Children with SLI look first and foremost like speakers of the type of language to which they are exposed and only secondarily like rather poor speakers of that language. The term "type of language" is deliberate; here we do not mean the obvious—that children acquiring German use German words and children acquiring Spanish use Spanish words. Rather, we are saying that children with SLI who hear a language with obligatory noun, verb, and adjective inflections, for example, will use such inflections much more readily than children with SLI whose language permits bare stems and contains only a small number of inflections. Children with SLI whose language permits considerable variation in word order will show greater word order variation than children with SLI acquiring a language with rigid word order. Children with SLI whose language places verbs in one of two sentence positions according to finiteness will place verbs in the same two positions. Children with SLI acquiring a language in which morphemes serving distinct grammatical functions are attached to a noun or verb in a prescribed sequence will show the same sequence in their own productions.

Relative to normally developing peers, children with SLI acquiring each language will look rather weak in language ability. However, the characteristics of language that most sharply distinguish children with SLI from age or MLU controls will not be the same from one type of language to the next.

It would be premature to equate "type of language," as it is used here, with language typology. Italian, for example, has obligatory inflections and relatively free word order, and allows sentences without overt subjects. However, the fact that the inflections are also word-final and syllabic might be important.

Although much needs to be learned about SLI across languages, one conclusion seems safe: if there is a universal feature of SLI, apart from generally slow and poor language learning, it is well hidden. In any given language, children with SLI might show areas of extraordinary weakness. But these areas will vary from language type to language type. If we are to understand SLI, we will need to take this fact into account.

Chapter 5

Evidence from Nonlinguistic Cognitive Tasks

Many children with SLI show weaknesses in areas of functioning that seem to require little or no language ability. Some of these areas are clearly cognitive; children perform relatively poorly in these areas despite achieving age-appropriate scores on standardized nonverbal tests of intelligence.

This fact is not a secret. Empirical findings that point to such deficits have been around for more than 30 years; and for an even longer time, clinicians working with these children have suspected problems in areas that fall outside of language proper. However, it can come as disappointing news to those who hope to study a population in which language is selectively impaired.

For some investigators, findings of nonlinguistic deficits in children with SLI raise the possibility that certain cognitive attainments lie closer to, and perhaps are responsible for, language. By studying children with SLI, then, we may be able to identify the specific cognitive attainments that have this status. In contrast, other investigators hold the view that two higher level systems may be disrupted, not because they are integrally related but because they are both relatively fragile. In this case, the study of the nonlinguistic deficits presumably will shed little light on the nature of the language impairment.

Mental Representation

One of the interpretations of findings that children with SLI have problems in nonlinguistic areas of cognition is that specific language impairment may not be specific at all. Indeed, in previous decades, one of the prevailing theoretical frameworks—based on Piaget—was that language is but one of several interrelated mental representational abilities. An impairment in one area should not occur without some signs of difficulty in other areas.

The structural parallels between some of the representational abilities emphasized in Piaget's work and particular language development milestones are indeed striking. For example, children's ability to employ alternative means to achieve some end, such as using a stick to reach an object otherwise unavailable to them, shares certain structural properties with the use of word forms to have an adult perform an act or reach an object. Children's use of one object (e.g., a computer diskette) to stand for another (e.g., a spaceship) in play bears a certain resemblance to the use of a word to stand for its referent. Relating one pretend object to another (e.g., using a sheet of paper as the landing strip for the diskette representing a spaceship) can be likened to forming a semantic relationship by combining words. Performing reversible operations in conservation tasks shares certain features with the ability to use sentences in the passive.

These possible parallels have not escaped the notice of investigators studying children with SLI.

The study of representational abilities in children with SLI began with Piaget's colleague, Inhelder (1963), and picked up momentum with the publication of articles by Morehead (1972) and Morehead and Ingram (1973), in which these investigators proposed that the grammatical deficits in the children with SLI they examined were attributable to a broader deficit in mental representation. We provide a review of the work on SLI with a distinct Piagetian orientation.

Sensorimotor Abilities

Snyder (1978) studied children with SLI and younger normally developing children whose speech was limited to single-word utterances. She administered tasks that tapped several different abilities seen in the sensorimotor period of development. The two groups performed similarly in most areas. However, the children with SLI were less proficient than the control children in their ability to use alternative means to an end. Snyder also observed that the children with SLI were more likely than the controls to resort to gestures to communicate imperative intent to their interlocutors.

Symbolic Play

The symbolic play of children with SLI has been the object of considerable investigation. Lovell, Hoyle, and Siddall (1968) seem to have been the first to examine symbolic play in this clinical population. Preschool-age children with SLI were divided into older and younger subgroups and their symbolic play compared with that of age controls. The older age controls spent more time in symbolic play than the older children with SLI. The younger age controls differed from the younger children with SLI only in the amount of time they spent in play representing a transitional level between mere practice play and symbolic play.

Morehead (1972) reported a study of five children with SLI who were presented with sets of materials during free play that required varying degrees of symbolic ability. In one condition, for example, the children were first given a doll, a doll bed, and a doll blanket. They were then given a shoe box and a sheet of paper along with a doll. Morehead observed that the children showed only limited symbolic play with the new materials, in contrast to expectations based on their performance on nonverbal intelligence tests.

In a study by Brown, Redmond, Bass, Liebergott, and Swope (1975), children were given specific toys and encouraged to pretend they were performing particular activities. For example, the children were asked to create a birthday party by using an egg carton, a box, straws, a shell, Popsicle sticks, and blocks. Brown et al. reported that a group of preschool-age children with SLI showed less adaptiveness in their use of objects in a pretend manner and less integration of play behaviors around a theme than did a group of age controls.

A different measure of symbolic play was employed by Williams (1978): the number of discrete symbolic play acts performed in a play session. Again, preschool-age children with SLI and age controls participated. The age controls were found to engage in a greater number of symbolic acts than the children with SLI.

Udwin and Yule (1983) also compared age-matched normally developing children and children with SLI in their preschool years. Two symbolic play measures were employed. One was a rating given to the children's spontaneous play based on the degree to which they introduced elements of time, space, and character not immediately given in the perceptual environment. For the other measure, the children were given sets of miniature toys and were then scored on the basis of the number of toys meaningfully used and related to each another. Differences favoring the age controls were seen for both measures.

The second symbolic play measure used by Udwin and Yule (1983) was also employed by Terrell, Schwartz, Prelock, and Messick (1984), but in this investigation preschool-age children with SLI were compared with a group of younger normally developing children showing comparable expressive vocabularies and utterances limited to one word in length. The children with SLI earned higher symbolic play scores.

A symbolic play study employing two control groups—an age-matched group and a group matched according to MLU—was conducted by Terrell and Schwartz (1988). The children with SLI in this study also were preschoolers, though their MLUs (ranging from 1.5 morphemes to almost 2.5) were somewhat higher than those of the children participating in the Terrell, Schwartz, Prelock, and Messick (1984) investigation. The children were given sets of objects that could be readily substituted for other objects in pretend play. Instances of such object substitutions were noted. The findings of this study revealed no differences between the children with SLI and the MLU controls, in contrast to the findings of Terrell, Schwartz, Prelock, and Messick. The greatest degree of symbolic play was seen in the age controls.

The status of the symbolic play abilities of children with SLI relative to those of MLU controls became even less clear from an investigation by Roth and Clark (1987). These researchers used some of the same measures of symbolic play as Terrell, Schwartz, Prelock, and Messick (1984) and Lovell, Hoyle, and Siddall (1968) in comparing five-to-seven-year-old children with SLI and a group of MLU controls. MLUs averaged just under 3.5 morphemes. The MLU controls, rather than the children with SLI, were observed to use higher levels of play.

It can be seen that in each of the studies in which children with SLI were compared with age controls, the children with SLI displayed less developed symbolic play. This unanimity in the findings seems to leave no doubt about whether children with SLI are age-appropriate in their symbolic play. They are not. However, it is far from clear how these children stand relative to MLU controls.

Why have comparisons between children with SLI and MLU controls yielded results that have been all over the map? In one case, children with SLI seemed more mature than MLU controls in symbolic play; in another, equal to MLU controls; and, in still another, poorer than the controls. The problem, of course, might rest with the use of MLU. As we saw in chapter 2, the selection of a language measure for purposes of matching should have a rationale, and it is not clear that the choice of MLU was well motivated. However, the results seem more likely to be attributable to the fact that the functional relationship between symbolic play and language ability, though real, is not so close as to be adequately captured by any single measure of language ability, or of symbolic play, for that matter. This can be seen in the studies that attempt to uncover the precise relationship between symbolic play and language ability, to which we now turn.

The first study examining this relationship provided encouraging results. In the Lovell, Hoyle, and Siddall (1968) investigation reviewed earlier, a relationship was found between the children's MLUs and the amount of time they spent in symbolic play. Folger and Leonard (1978) examined this issue by comparing the proclivity toward symbolic play in two groups of preschool-age children with SLI. One group showed productive use of two-word utterances; the other group was limited to single-word utterances. Half of all the children provided unequivocal evidence of symbolic play. However, the correspondence between ability to engage in symbolic play and ability to use two-word utterances was far from perfect. Approximately one-third of the single-word utterance users engaged in symbolic play, and a similar percentage of multi-word utterance users showed no evidence of this type of ability.

Two longitudinal studies have been conducted to examine the relationship between symbolic play and language ability in children with SLI. Skarakis (1982; see also Skarakis-Doyle & Prutting, 1988) followed three children age 1;10 to 2;7 for a six-month period. At monthly intervals, the children's symbolic play, spontaneous speech, and language comprehension abilities were evaluated. The children's symbolic play abilities, like their language abilities, progressed through the expected stages at later-than-expected ages. However, parallel achievements in the respective areas are not evident in the findings. Each of the three children showed evidence of comprehending multi-word utterances before displaying true symbolic play, and two of the children's MLUs in morphemes were over 2.0 before they exhibited play of this type.

The second longitudinal study was conducted by Shub, Simon, and Braccio (1982). Three two-year-olds with SLI and two normally developing one-year-olds were initially seen when they possessed fewer than five nominal forms in their expressive vocabularies. The children were observed every four months until they achieved an expressive vocabulary of 50 words and showed evidence of multi-word utterances. At this point, the children were seen at 12-month intervals until a total duration of 3 years was reached. Measures of language production, language comprehension, and symbolic play were obtained at each point of observation. Shub et al. reported that for all children, the appearance of object substitutions in play coincided with an emerging ability to name absent objects. Of course, because observations were at least four months apart, simultaneous attainments could not be distinguished from developments occurring two or three months apart, in any order, if they occurred within the same four-month period.

A different approach to studying the relationship between symbolic play and language ability was taken by Panther and Steckol (1981). These investigators employed a treatment paradigm. Preschoolers with SLI were assigned to one of three conditions, in which they were taught to produce words, to use objects in an unconventional (and, from an adult perspective, symbolic) manner, or to do both. The first two groups made gains only in the type of ability for which they received instruction. Only the group in the third condition made significant gains in both areas.

Finally, Capreol (1994) studied six preschoolers with SLI who showed age-appropriate ability in block construction. The children's symbolic play was then observed. Three of the children displayed symbolic play abilities that were comparable with their block construction abilities, whereas the other three children's symbolic play skills were less developed. Capreol noted that the children with the stronger symbolic play also showed a larger lexicon and greater proficiency with word combinations.

Despite positive findings in particular studies, the evidence collectively suggests that the symbolic play and language abilities in children with SLI are related only

mental imagery

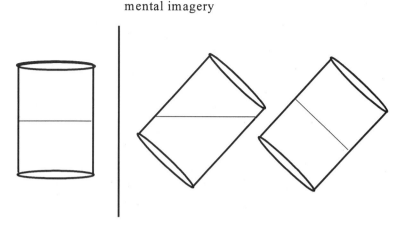

Figure 5.1
Imagery task in which the child must select the direction of water level (from the two options appearing to the right of the vertical line) that reflects the water direction when the container shown on the left is tilted.

within broad limits. These children are likely to have restrictions in symbolic play ability, and in most cases, the poorer the play abilities of these children, the less developed is their language. However, too many exceptions are seen to assume a close relationship between these two types of abilities.

Mental Imagery
Another form of representational ability that has been studied in connection with SLI is mental imagery. Efforts at establishing links between mental imagery and language ability might seem surprising, in that these two types of abilities appear to be quite different. Unlike language, mental imagery is primarily depictive in character. That is, images are generated from stored information about the shape of the physical world. Language does not seem to work this way.

Inhelder (1963) was the first to note the possible relationship between imagery and language ability in a child with SLI, age 9;6. This child experienced considerable difficulty with imagery tasks, such as predicting the direction of the level of water in a tilted glass container. An illustration is provided in figure 5.1. The child seemed unable to dissociate the horizontal direction of the water from the inclined position of the container, predicting the rightmost outcome in the figure. Similar findings were reported in a descriptive study of 17 (mostly school-age) children with SLI conducted by deAjuriaguerra, Jaeggi, Guignard, Kocher, Maquard, Roth, and Schmid (1965). These children had difficulty, for example, in a task that required them to analyze the movement of a projected shadow.

Johnston and Ramstad (1983) administered a battery of tasks to a group of children with SLI 10 to 12 years of age and found that the children's poorest performance occurred on the imagery tasks. For example, the children exhibited difficulty with a task in which they had to feel geometric forms without looking and then, based on visual information, select the shapes that corresponded to them.

This same imagery task was included in a battery of tasks employed by Kamhi (1981), in what may have been the first study of mental imagery to use two comparison groups. The children with SLI averaged about five years of age, and showed a mean MLU in morphemes of almost 5.0. Their performance was compared with those of a group of mental age controls and a group of MLU controls. The children with SLI performed more poorly than the first of these control groups, and significantly better than the second.

The children in the Kamhi (1981) study were about as young as one would want to test children on most conventional imagery tasks. This can be seen from an investigation by Siegel, Lees, Allan, and Bolton (1981). These researchers found that neither three- to five-year-old children with SLI nor a group of age controls was successful in selecting drawings that correctly depicted the water level in a tilted glass.

One of the tasks employed by Kamhi, Catts, Koenig, and Lewis (1984) was the same imagery task used by Kamhi (1981), in which the appropriate shape was to be selected after the children had held a corresponding form in their hand without seeing it. Children with SLI averaging almost six years of age and their age controls participated. Higher performance levels were seen on the part of the age controls.

A different imagery task was used by Camarata, Newhoff, and Rugg (1981). Here, children had to select objects that matched the form as seen from the angle of the experimenter. Thus, the children's ability to imagine the objects in different orientations was under scrutiny. The participants were children with SLI from age 3;6 to 5;6 and MLUs ranging from 1.5 to almost 4.0 morphemes, a group of age controls, and a group of MLU controls. The age controls performed at a higher level than the children with SLI, who in turn outperformed the children with comparable MLUs.

Murphy (1978) employed four different imagery tasks in a comparison between school-age children with SLI and a group of normally developing children matched according to both age and performance on a reasoning task. Surprisingly, the control children scored higher than the children with SLI on only one of the four imagery tasks. This finding may have been the result of the basis for matching; the reasoning task on which the children were matched also seems to have involved imagery. Thus, only normally developing children whose imagery abilities were somewhat depressed or children with SLI whose abilities were approximately age-appropriate might have been included.

Data from Savich (1984) lends support to this interpretation of the Murphy (1978) study. Savich employed five imagery tasks, two of them matching those used by Murphy in a study of school-age children with SLI and age controls. Differences favoring the age controls were found for all five tasks.

Studies of the visual spatial memory of children with SLI might also be construed as involving imagery. In these studies, children with SLI are shown a visual stimulus. The stimulus is then removed and the children must identify its original location after several seconds. Children with SLI are less accurate at identifying the original locations of the stimuli than are age controls (Doehring, 1960; Poppen, Stark, Eisenson, Forrest, & Wertheim, 1969; Wyke & Asso, 1979), even though they can match the stimulus locations with a second, identical set of materials if the original stimuli remain visible.

In all of the tasks discussed thus far, it has been assumed that imagery was invoked, but we cannot be sure. Johnston and Ellis Weismer (1983) took this extra, important step by making use of a mental rotation task. In this task, drawings of a pair of geometric forms were presented and the children were to push one of two buttons corresponding to "same" and "different." The form on the left was always vertical, whereas the form on the right was either vertical or rotated about its center 45°, 90°, or 135°. Response times that increase linearly as a function of degree of rotation are interpreted as evidence that imagery was involved. Johnston and Ellis Weismer found that school-age children with SLI exhibited slower response times than age controls overall, but that for both groups, response times were slower with increasing degree of rotation. Because the slope of the response times as a function of degree of rotation did not reveal group differences, it appeared that the problems of the children with SLI rested with image generation, maintenance, or interpretation rather than image transformation.

The literature on mental imagery can be interpreted with relative ease. Children with SLI do not perform as well on imagery tasks as age controls do. In two studies, MLU controls also participated. These younger children did not do as well as the children with SLI.

Unlike the case for symbolic play, few studies have examined the specific relationship between mental imagery and language ability. The study by Kamhi, Catts, Koenig, and Lewis (1984) examined the relationship between the children's performance on their imagery tasks and vocabulary comprehension. For the children with SLI, a positive relationship was observed. The same calculations could not be performed on the mental age controls, because these children's generally high performance restricted the variability in their scores.

Conservation and Seriation
Other Piagetian abilities examined in children with SLI include conservation and seriation. Two examples appear in figure 5.2. To test conservation of number, children are presented with, for example, two rows of marbles, one consisting of five marbles and the other of seven, but each the same length. The children must then select the row containing the greater number of marbles. To test seriation, children must mentally rearrange the order of items appearing in a stimulus array. For example, children might be asked to select from an array of four randomly ordered lines the one that is the second longest (Siegel, Lees, Allan, & Bolton, 1981).

The evidence from these tasks is mixed. Inhelder (1963) described a child with SLI, age 5;7, who succeeded in a number conservation task similar to the one above. Similarly, Johnston and Ramstad (1983) found that a group of 10- to 12-year-old children with SLI could successfully perform a task of number conservation. However, the task

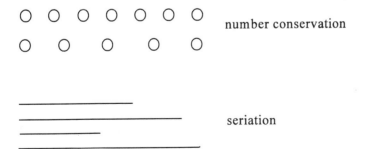

Figure 5.2
Task of number conservation in which the child must ignore spatial arrangement while recognizing differences in number of items, and task of seriation in which lines must be mentally rearranged in terms of length.

employed by Johnston and Ramstad is ordinarily mastered by children at age 6;6, so the fact that the children with SLI in the study had no difficulties is not to say that the children had acquired this ability within the usual time frame. The same children had much greater difficulty with the seriation task employed, resembling typical children more than two years younger.

Conservation and seriation tasks were among those used by Siegel, Lees, Allan, and Bolton (1981) to study preschool-age children with SLI. These children performed significantly more poorly than a group of age controls on these two types of tasks.

Kamhi (1981) included a number conservation task in his study of five-year-old children with SLI. The scores of these children fell between those of the mental age controls and the MLU controls, but the magnitude of the difference did not reach levels of statistical significance.

These findings suggest that abilities such as conservation might not be as poor in children with SLI as other abilities assessed by Piagetian tasks (notably symbolic play and mental imagery). However, they have emerged as problematic often enough in these studies to preclude any declaration that all is well in these areas.

Hypothesis Testing

Imagine the appearance of a new television series dealing with characters from other galaxies. To prepare you for your first viewing, your neighbors' children present you with pairs of pictures, each picture of a pair showing one of the characters. The children point to each picture and say, "He's a Mantorin," "She's a Vaktol," and so on. After a few pictures, you check your recognition abilities by referring to the next lavender creature as a Vaktol (you remember that the last character called a "Vaktol" was lavender), only to learn that you just deeply offended a Mantorin. Color is irrelevant, it turns out, in distinguishing these two species; it's all in the forehead.

Hypothesis testing of this sort would seem to have a bearing on language development. After all, children must hypothesize the meaning and boundaries of new words and syntactic constructions. For this reason, investigators have examined hypothesis testing in children with SLI. If deficits in hypothesis testing are at the root of the problem, these children should do relatively poorly even on nonlinguistic hypothesis testing tasks.

The tasks employed in the literature typically involve distinctions of size, shape, and color, using such down-to-Earth stimulus pairs as a large white circle and a large red circle, a small red circle and a large white circle, and so on. Initially, the experimenter presents several pairs and indicates the member of the pair that corresponds to the rule (e.g., the red one). The child is then given the opportunity to guess the correct item from the next pair, with feedback provided. Additional trials are then presented to determine if the child has formed the correct hypothesis. Depending on the number of dimensions involved at the same time, as well as on other task details, the solution can be rather tricky to discover.

Using a task of this general nature, Hoskins (1979) found differences in the hypothesis-testing abilities of school-age children with SLI and a group of age controls. The children with SLI required a greater number of trials to establish the correct solution, and some children did not arrive at the proper solution at all. Studies by Kamhi, Catts, Koenig, and Lewis (1984) and Kamhi, Nelson, Lee, and Gholson (1985) also included tasks focusing on hypothesis-testing abilities, but differences did not emerge, seemingly due to the difficulty of the items even for the control children.

Nelson, Kamhi, and Apel (1987) compared five- to seven-year-old children with SLI and a group of mental age controls on two types of hypothesis-testing tasks, one in which explicit input was provided during the exposure trials (e.g., "the answer is in the little black circle") and one in which the input was less detailed ("the answer is in this picture"). The explicit input condition resulted in better performance on the part of the children with SLI, though the mental age controls were more successful than the children with SLI in both conditions.

In most studies of hypothesis testing, each pair of pictures is removed as the next pair is presented. Ellis Weismer (1991) wished to determine whether such a procedure might introduce unintended short-term memory demands on the children. She employed a task in which the children could view all previous pairs as they attempted to solve the problem. In other respects, the task resembled that of Nelson, Kamhi, and Apel (1987). Children with SLI averaging around seven years of age participated, along with a group of age controls. There were differences between the two groups' nonverbal intelligence test scores; these were controlled by using intelligence test score as a covariate in the analysis. The age controls performed at a higher level than did the children with SLI. Type of input (explicit, nonexplicit) did not prove to be a relevant factor.

Masterson (1993) examined how school-age children with SLI would perform on the type of hypothesis-testing task used by Ellis Weismer, but added a task that served as a complement. Whereas hypothesis testing involves the discovery of a rule, in the second task the children were given the rule and had to apply it. The children with SLI performed like younger children matched according to vocabulary comprehension test score, and below the level of mental age controls, on the hypothesis-testing task. However, they performed as well as the mental age controls on many of the items on the task of rule application. Their performance dropped off (and thus resembled that of the younger controls) on items in which the rule to be applied was complex, requiring several attributes to be considered simultaneously. According to Masterson, her findings may have reflected a problem centered on rule induction, or on limitations in processing capacity.

Analogical Reasoning

Another cognitive ability examined in children with SLI is analogical reasoning. This type of reasoning is seen in the application of existing knowledge to a new experience, such as applying one's knowledge of the location of the human mouth to determine the location of the mouth on a Vaktol.

Nippold, Erskine, and Freed (1988) examined both verbal and nonverbal analogical reasoning in school-age children with SLI and a group of age controls. Both tasks were of the type "Ear goes with radio as eye goes with ...," the missing item being one of several presented in a multiple-choice format. Differences favoring the controls were seen on both types of tasks. Because the nonverbal intelligence scores of the two groups also differed, Nippold, Erskine, and Freed reanalyzed the data with nonverbal intelligence score as the covariate. The differences disappeared, a somewhat surprising outcome in the case of the verbal task. Masterson, Evans, and Aloia (1993) administered only a verbal analogies task to school-age children with SLI, mental age controls, and a group of younger controls matched according to receptive vocabulary test score. The children with SLI performed like the language test controls and more poorly than the mental age controls.

Kamhi, Gentry, Mauer, and Gholson (1990) studied analogical reasoning by making use of problems such as the farmer's dilemma of how to get a fox, a goose, and corn across the river (without one or two of these being consumed) with only a small boat at his disposal. School-age children with SLI and mental age controls participated. In one condition, the children heard the solution to one dilemma and had to demonstrate an understanding of it by using appropriate toys. In another condition, the experimenter demonstrated the appropriate moves with the toys while describing them verbally. After demonstrating comprehension of the solution with the original set of materials, the children were tested on their ability to apply the same concepts to analogous situations (e.g., transporting a lion, a pony, and a bucket of oats across a precarious bridge). The children with SLI had greater difficulty than the mental age controls on the original task only in the condition in which the description of the solution was not accompanied by demonstration. The two groups were comparable in their ability to apply the solution to the new situations.

The expected interpretation of these findings would be that the children with SLI possessed mental age-appropriate reasoning skills when (and only when) information did not have to rely exclusively on language. However, Kamhi, Gentry, Mauer, and Gholson (1990) argue that the verbal-only condition required the children to process the information more quickly. Accordingly, the results could be as much due to processing speed limitations as to difficulties with language. This issue will be the focus of considerable discussion in chapter 12.

Hierarchical Planning

It is well known that syntax possesses a hierarchical structure. The article *the* in (1), for example, is structurally linked to *Red Sox* via the node marked *E*, and is connected to the preceding word *love* only through a higher-level node, *C*. Failure to recognize this structure could lead to some odd rules of sentence formation.

(1)

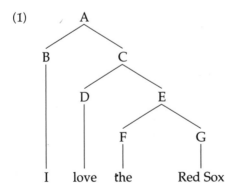

Cromer (1978, 1983) hypothesized that the ability to recognize and process hierarchical structure probably transcends language, and that if children experience difficulties with syntax, these difficulties might be apparent as well on nonlinguistic tasks that require an appreciation for such structure. To test this hypothesis, Cromer adapted a method designed by Greenfield and Schneider (1977). In the Greenfield and

Schneider work, children were shown a mobile made from construction straws and were asked to make one that matched it. The mobile had a symmetrical tree structure, and could be constructed in either a hierarchical or sequential manner.

The structure in (1) is not symmetrical, but it will do for illustration. A hierarchical method of construction might involve the child's connecting two straws to point *A*, then connecting two additional straws to point *C*, and so on. Alternatively, the mobile might be constructed in a sequential manner. For example, a straw could connect points *G* and *E*, another would then connect *E* to *C*, and so on. In the Greenfield and Schneider study (1977), normally developing children under age seven years usually constructed the mobile by using the sequential method; only children age seven and above made use of hierarchical construction.

Cromer (1983) used two different tasks, one in which the children were asked to draw a replica of a two-dimensional symmetrical tree structure, and one in which they were asked to construct a replica of a three-dimensional symmetrical tree structure. For each task, after completing the replica, the child was asked to do it again, using a different method. The first steps in doing so were demonstrated for the child.

The major focus of Cromer's (1983) study was a group of five children with Landau-Kleffner syndrome; these are children with a severe language impairment associated with a convulsive disorder (Landau & Kleffner, 1957). A group of seven school-age children with SLI also participated. The children averaged 12 years of age. A group of 12 age controls and a group of 12 deaf children also participated. All of the age controls and 11 of the 12 deaf children were able to complete at least one of the two tasks in a hierarchical manner. In contrast, only five of the seven children with SLI (and only two of the five children with Landau-Kleffner syndrome) succeeded. They were successful, however, using the developmentally earlier sequential method. Cromer interpreted these findings as indicating that a minority of the children with SLI had a hierarchical planning deficit that might account for both their performance on the two tasks and their syntactic difficulties.

Kamhi, Ward, and Mills (1995) produced results that revealed even less of a relationship between hierarchical planning and grammatical ability. Five- to seven-year-old children with SLI and a group of mental age-matched controls completed four different construction tasks. The control children were more successful in completing the models; however, they were more likely to use a sequential method. The two groups were very similar on all four tasks in their ability to use a hierarchical method. Furthermore, there was no correlation between the children's scores on language measures and their performance on the hierarchical planning tasks.

The Little Paradox

We have seen that many children with SLI have limitations in mental representation and hypothesis testing. Yet, they score at age level on nonverbal tests of intelligence. How can this be? Johnston (1982) has offered a possible solution. She noted that many children with SLI perform relatively well on tasks of visual perception of static figures, shapes, and designs, which are included in commonly administered tests of nonverbal intelligence such as the Leiter International Performance Scale (LIPS) (1979) and the Wechsler Intelligence Scale for Children—Revised, (WISCR-R) (1974). For example, Johnston's item analysis of the LIPS revealed that most of the items in the two-to-eight-year age range are perceptual in nature.

Kamhi, Minor, and Mauer (1990) performed a similar analysis on two additional nonverbal intelligence tests, the Columbia Mental Maturity Scale (CMMS) (Burgemeister, Blum, & Lorge, 1972) and the Test of Nonverbal Intelligence (TONI) (Brown, Sherbenou, & Johnsen, 1982). The first 20 items of the CMMS were perceptual in nature, as were the first 13 items of the TONI. Kamhi, Minor, and Mauer pointed out that if five-year-old children performed correctly only on these first 13 items of the TONI, they would earn a nonverbal score of 128. Identical performance would not yield scores below the conventional cutoff of 85 until the children reached 11 years of age. It is possible, then, that nonverbal tests of intelligence used to select children with SLI are tapping a relatively strong ability in these children. In a real sense, the resulting age-appropriate nonverbal IQ scores might be an overestimation of the children's abilities.

Let's consider an example. One of the children with SLI studied by Leonard, Eyer, Bedore, and Grela (1997), age 5;0, earned a score of 98 on the performance portion of the Wechsler Preschool and Primary Scale of Intelligence-Revised, (WPPSI-R) (1989). The subtests that were her strongest involved identifying the incomplete portion of a drawing (e.g., pointing to the place where the sleeve of a jacket is missing) and drawing copies of geometric designs. The subtest giving her the greatest difficulty (she was below age level on it) was one in which she had to draw the route out of a maze. An inspection of the subtests reveals that the two producing the highest scores could be approached by relying on static visual information. The maze subtest, in contrast, could not be solved without mentally tracing a path from the starting point to the exit. She succeeded only on the early items that involved straight paths or a single detour, solutions that could literally be seen simply by viewing the maze.

This pattern of performance was probably no fluke. Swisher, Plante, and Lowell (1994) administered the Kaufman Assessment Battery for Children (Kaufman & Kaufman, 1983) to a group of children with SLI and a group of age controls. The children with SLI showed overall scores within the norm, but were most distinguishable from the age controls in their relatively poor performance on spatial rotation items.

The Big Paradox

We might ask how it is possible for these children to perform at age level in any cognitive task, given their language problems. Particularly by the age of school entry, language is used in many mental activities. As pointed out by Johnston (1994), "If language symbols are poorly controlled, there should be a cognitive consequence" (p. 109).

It is conceivable that the detrimental effects of language problems on other performance won't show up if the verbal mediation that can be invoked is well within the abilities even of a child with language limitations. For example, the child above might not have known "sleeve" but she surely knew "arm," and might have employed such a word as a guide to her pointing response. However, as children get older, the nonverbal intelligence measures used contain fewer items involving static visual perception items, and for those that are employed, it is not clear that the language that would be helpful in verbal mediation is within the children's grasp.

This state of affairs might lead to the expectation that nonverbal IQ scores of children with SLI will decline over time. In fact, there seems to be evidence of this sort. Tallal, Townsend, Curtiss, and Wulfeck (1991) evaluated the nonverbal IQs of children with SLI at age four years and then five years later, using the LIPS. The children were divided according to whether or not other family members had a history of language problems. Nonverbal IQs for the children with SLI and a positive family history fell from a mean of 108 to 99 over the five-year period. The decline for the children with no such family history was comparable, from 110 to 102.

Tomblin, Freese, and Records (1992) reported that a group of young adults with a history of SLI showed a mean nonverbal IQ of 89.75 on the Wechsler Adult Intelligence Scale—Revised (WAIS-R) (1981). At elementary school age, the same children had a mean nonverbal IQ of 98.50 on the WISC-R. Tomblin et al. pointed out that such a drop is inconsistent with psychometric research showing considerable consistency between WAIS-R and WISC-R scores.

An inspection of the SDs presented in the Tomblin, Freese, and Records (1992) study reveals that several of the young adults scored 80 or lower on the WAIS-R, placing them below the customary boundary used for SLI. Data presented in a longitudinal study by Bishop and Adams (1990) also show declines in nonverbal IQ across time. In this study, some of the children with nonverbal IQs (on the LIPS) in the normal range at four years of age scored below 85 (on the WISC-R) at age eight. In a follow-up study by Aram, Ekelman, and Nation (1984), more than 20% of a group of preschoolers showing age-appropriate scores on the LIPS were found to score below 85 on the nonverbal scale of the WISC-R in their early teens.

Evidence of declining nonverbal IQ with age was presented from a different perspective by Eisenson (1972). Eisenson reported data from an unpublished study by J. Stark in which nonverbal IQs were obtained from three age groups of children: 3;0 to 4;11, 5;0 to 6;11, and 7;0 to 8;11. The children met the usual criteria for SLI except that they were included on the basis of the absence of mental retardation rather than on the basis of a nonverbal IQ of at least 85. Indeed, this feature of the study makes it valuable to answer the question at hand. Two different nonverbal tests were administered to each child, the CMMS and the LIPS. On both tests, mean scores decreased with increasing age. For the CMMS, the means were 100, 87, and 73; for the LIPS, 94, 89, and 79.

Striking results were also obtained by Paul and Cohen (1984). These investigators tested children with severe language disorders at two points in time, separated by at least five years. The five children who showed nonverbal IQs above 85 (rnean = 99) at the outset showed an average nonverbal IQ of 63 upon subsequent testing. Four of the five children showed a decline, and only two scored above 85 at the later testing point.

Of course, many individuals with SLI do not show a decline in nonverbal IQ at later ages. The most obvious conclusion to draw from this fact is that intellectual growth can proceed at a normal rate with recourse only to verbal mediation of an elementary nature. But this might not be the correct conclusion. A preview of some of the data from language intervention studies reviewed in chapter 10 will show why.

The language intervention literature reveals that without treatment, many children with SLI fall further behind their peers across time. With treatment, one of two outcomes can be seen. For some children, there is an acceleration in the rate of language growth so that they begin to show the expected language learning slope, no longer falling further behind their age-mates with each passing year. For other children, there is an actual narrowing of the language ability gap relative to the norm, though sometimes the gap is never completely closed. Evidence for the more favorable of these outcomes takes the form of an increase in standard scores on language measures across time. Among the scores that show increases are verbal IQ scores, because these are most directly affected by the language impairment and its amelioration.

It is difficult to know the extent to which a modest acceleration in language development relative to age-mates can offset the increasingly important role that language plays in all mental activity. According to the line of thinking presented here, if below-age-level standard scores on language measures do not change with increasing age, nonverbal IQs may show a decline (reflecting the drag that poor language begins to have on other cognitive processes). If standard scores on language measures show an increase, on the other hand, the decline in nonverbal IQ will not be seen. An investigation by Stark, Bernstein, Condino, Bender, Tallal, and Catts (1984) provides an example of data of the latter type. The children with SLI in their study were first seen at an average age of 6;6 and again at age 10;3. Although this was not an intervention study, the children were enrolled in language intervention programs during the four-year period between testing. Verbal IQs increased signficantly across this time frame, whereas nonverbal IQs held steady.

The argument presented here cannot be accepted without additional data. The most reasonable interpretation of increases in verbal IQ without corresponding changes in nonverbal IQ is that the latter required no verbal mediation, or a level of verbal mediation that even children with SLI can muster. Converging data are needed. It must be demonstrated that children with SLI who show no gains across time in standard scores on language measures show declines in nonverbal IQ, whereas children with SLI who show real gains on language measures show no decline in nonverbal IQ. Evidence of this type would help us understand how age-appropriate nonverbal IQs can be seen in older children and adults with SLI in spite of the assumption that many intellectual activities during this time of life benefit from the use of language as a mental tool.

Chapter 6

Auditory Processing and Speech Perception

Abilities such as symbolic play and mental imagery are not the only types of non-linguistic abilities that are suspect in children with SLI. Many of these children exhibit deficits on nonlinguistic tasks that implicate lower level processes. Prime among these are processing tasks involving the auditory modality, though visual and tactile factors have been examined as well. This work is reviewed here. Studies employing nonlinguistic stimuli will be covered first, followed by investigations using linguistic material. The latter are, in essence, studies of speech perception. We begin with a brief description of the early work done in the area of auditory processing.

Early Notions of Deficits in Auditory Processing

At least since the introduction of the term "congenital auditory imperception" by Worster-Drought and Allen (1929), the auditory processing skills of children with SLI have been the object of clinical and scholarly attention. The focus on these skills intensified in the 1960s with the publication of several papers in which auditory processing difficulties were posited as a likely cause of language deficits in children with SLI (Benton, 1964; Eisenson, 1966; Hardy, 1965; Masland & Case, 1968; Monsees, 1961). Lowe and Campbell (1965) pursued this idea by comparing children with SLI and age controls on two different tasks. In the first task, the children were tested on their ability to detect the appearance of two identical pure tones presented in close succession. Difficulty with this task results in children reporting only a single tone presentation. In the second task, the children were asked to sequence two pure tones of different frequencies presented at varying interstimulus intervals (ISIs). In the first task, the children with SLI required ISIs that were almost twice as long as those required for control children (36 ms versus 19 ms) before two separate tones were perceived. In the second task, the children with SLI could correctly sequence the tones only when the ISIs exceeded 250 ms, whereas the age controls succeeded with intervals of only 40 ms. Results highly similar to those of the first task were reported in a later study by McCroskey and Kidder (1980) in an investigation of school-age children with SLI and age controls.

Other studies during the 1960s and early 1970s made use of verbal stimuli. For example, Rosenthal (1972) observed that children with SLI had great difficulty in reporting the correct order of the sounds [tʃ] and [ʃ], two consonants that differ in their temporal characteristics. Regardless of ISI, these children never showed the accuracy levels exhibited by control children. The children with SLI also had some difficulty with the sounds [s] and [ʃ], which differ in their spectral properties. However, once ISIs approached 200 ms, the accuracy levels of the children with SLI were similar to those of controls. When [tʃ] and [ʃ] were presented singly, each could be identified accurately by the children with SLI, suggesting that the problem with ordering these two sounds was not a basic problem of discrimination. Other studies of the discrimination and sequencing abilities of children with SLI have employed pairs or strings of words that were part of formal tests or experimental material (Monsees, 1968; J. Stark, 1967; J. Stark, Poppen, & May, 1967; P. Weiner, 1969).

The use of verbal stimuli to assess auditory processing ability has been roundly criticized (e.g., Bloom & Lahey, 1978; Rees, 1981). Rees (1973) diagnosed the problem accurately: "what positive results have been reported could well be interpreted to reveal what is already known—that these children have a language disorder" (p. 308). Although the more glaring problems of interpretation come from studies involving verbal stimuli, even some of the investigations employing nonverbal stimuli are not free of procedural flaws. For example, in the Lowe and Campbell (1965) study, the children's judgments were communicated to the experimenter by using linguistic terms such as *high, low,* and *last*. It is not clear if steps were taken to ensure that the children comprehended these terms. This is especially important because the control group was matched according to age, creating a situation in which the group performing more poorly on the task was also the group with more limited language skills.

Processing of Brief or Rapidly Occurring Stimuli

A significant refinement in auditory processing research occurred with the appearance of the investigations by Tallal and her colleagues. The tasks employed in this work did not require verbal responses (verbal mediation was always possible, as it is in most tasks), and the instructions could be conveyed through nonlinguistic means. The basic task involved presenting one of two stimuli, with the child required to press one of two response panels. Initially, the child was taught to respond to each stimulus by pressing one of the panels for one stimulus, and the other panel for the other stimulus. Once this step was learned, the child heard random sequences of the stimuli and had to press the panel that corresponded to each stimulus when it was detected. In certain experiments, a slightly different method was used. One of the two stimuli was identified as the target stimulus and a panel was pressed only when that stimulus was detected.

Nonverbal Stimuli

Tallal and Piercy (1973a) employed two nonverbal auditory stimuli of 75 ms, composed of frequencies in the speech range distributed in a way that did not correspond to the output of the human vocal tract. School-age children with SLI and a group of age controls participated. After the children were taught to press the panel corresponding to each stimulus, the ISIs between the two stimuli were varied. The age controls performed above the level of chance at all ISIs. The children with SLI, on the other hand, performed at above-chance levels only when ISIs exceeded 300 ms. Identical results were seen when the task was altered so that the children pushed one panel when the two stimuli were the same and the other panel when they were different. Because the same—different task required only discrimination of the stimuli, Tallal and Piercy concluded that the sequencing difficulties exhibited by the children with SLI in the first task were secondary to their problem with discriminating the two stimuli when presented in rapid succession.

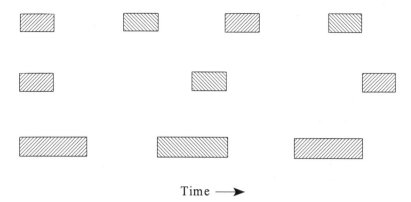

Figure 6.1
Auditory stimuli of varying durations and ISIs. The top row depicts a condition in which both the stimuli and the intervals between them are relatively brief. The middle row shows brief stimuli appearing at longer ISIs; the bottom row shows longer stimuli presented at relatively brief intervals. Performance by children with SLI is poorer in the condition reflected in the top row than in either of the other two conditions.

In a second study involving the same children, Tallal and Piercy (1973b) manipulated the duration of the stimuli in addition to the ISIs. As in the first study, they found that performance of the children with SLI was quite poor when short ISIs were used. However, when the duration of the stimulus was increased, performance improved at shorter ISIs. Thus, it appeared that the total duration of the stimulus pattern was the crucial factor. This is illustrated in figure 6.1.

Tallal and Piercy (1973b) employed the same procedure to determine whether difficulties with short ISIs could be observed by using visual stimuli. Two 75-ms light flashes of different shades of green were presented at ISIs as short as 30 ms. No differences between the children with SLI and the age controls were seen. The difference in performance in the auditory and visual modalities was maintained when the task was altered so that random combinations of the two stimuli were presented in sequences three, four, or five elements in length. The two groups of children performed in similar fashion with the visual stimuli; with the auditory stimuli, the children with SLI matched the performance of the age controls only with the shortest sequence and longest durations.

Tallal (1976) argued that the difficulty with short ISIs seen in children with SLI reflected more than a general immaturity in perceptual skills. She administered the Tallal and Piercy (1973a) task to five groups of normally developing children of different ages. She found that the children with SLI in the Tallal and Piercy (1973a) study performed as well as the oldest normally developing children (age 8;6) when the longest ISIs were used, but performed worse than the youngest children (age 4;6) when the shortest ISIs were employed. Thus, the performance profile of the children with SLI did not match that of normally developing children at any age.

A different group of school-age children with SLI participated in a study by Tallal, Stark, Kallman, and Mellits (1981). The auditory stimuli were those used by Tallal and Piercy (1973a); the visual stimuli were letterlike forms made visible through 75-ms light flashes. The findings for the auditory modality resembled the earlier findings, with age controls outperforming the children with SLI. However, differences between the two groups were also seen for the visual modality. Closer inspection of the data suggested that age played an important role in the findings. The older children with SLI had problems that appeared to be limited to the auditory modality; the younger children with SLI had problems in both modalities.

Problems of tactile perception also have been reported. Tallal, Stark, and Mellits (1985a) observed that school-age children with SLI performed more poorly than age controls in a task in which they had to report whether they felt one or two contacts on the hand (or cheek). In a study reported by Kracke (1975), school-age children with SLI had greater difficulty than age controls in identifying rhythmic sequences presented by tactile means. Difficulties were also observed for sequences presented auditorily (see also Lea, 1975).

A dichotic listening paradigm was used by Haggerty and Stamm (1978). School-age children with SLI and age controls were presented clicks in both ears and had to indicate whether they heard two clicks or one. In the latter case, the ear in which the single click was heard was identified. Half of the pairs were presented with the stimulus to the right ear preceding that to the left, and half of the pairs showed the opposite pattern. The children with SLI required longer ISIs than the age controls before two clicks were reported.

In a study by Hochman, Thal, and Maxon (1977), school-age children with SLI and age controls were presented 100-Hz tones of 20 and 200 ms duration at various intensities, and thresholds for each stimulus duration were obtained. The threshold shift (poorer sensitivity for the 20 ms tone) was not significantly greater for the children with SLI (18 dB on average) than for the control children (16 dB).

If the processing of rapidly occurring stimuli represents an area of difficulty for children with SLI, as seems to be indicated in several studies, it would be important to determine if this problem can be identified at an early age. This is especially true if difficulty with processing rapidly presented material impedes the learning of certain details of language, or at least is correlated with language learning problems due to some more general underlying factor shared by language and this type of processing. Benasich and Tallal (1993) approached this question in a preliminary study of infants from families with a history of SLI. Certainly this strategy does not ensure that the infants tested will in fact exhibit SLI at a later age. However, as we will see in chapter 7, the likelihood of children born into families with such a history having SLI themselves is quite high.

Benasich and Tallal (1993) employed an adaptation of the head turn procedure. The infants were trained to look toward a box (housing toys that moved when a switch was activated) upon hearing a particular sequence of complex tones. Each sequence consisted of two tones. Both tones were 75 ms in duration, but they differed in fundamental frequency. The ISIs between the tones in the sequence were manipulated. The principal measure of interest was the shortest ISI at which the infants produced a looking response. Only three infants (averaging seven months of age) from families with a history of SLI completed the procedures. Their performance was compared with that of a group of 25 infants from families without a history of language problems. The average threshold for the control infants was 69 ms. One of the three infants from families with a history of SLI performed as well as the control infants. The other two showed thresholds of 282 ms and 289 ms, more than 2 SDs greater than the thresholds observed for the controls. It will be important to determine whether these three children's thresholds prove to be predictors of later language and auditory processing abilities.

Verbal Stimuli

The proposal that children with SLI have special difficulties with stimuli of brief duration was first examined with verbal stimuli by Tallal and Piercy (1974). Synthetic sounds of 250 ms—the vowel pair [ɛ] and [æ] and the consonant-vowel pair [ba] and [da]—were presented to the same children who participated in the previous studies of these investigators. No differences were seen between the children with SLI and the age controls in discriminating or sequencing the two vowels. However, fewer than half of the children with SLI could discriminate between [ba] and [da], and only two of those who succeeded in discrimination could sequence the two syllables correctly. The difficulty with [ba] and [da] relative to the vowels was attributed to the fact that in [ba] and [da] only a 43-ms formant transition provided contrastive information. The vowels could be distinguished on the basis of the steady-state formant information that was present for the entire 250 ms of the stimulus.

Convincing evidence that the brevity of the contrastive information played a role in the Tallal and Piercy (1974) findings was offered in the final collaboration of these two investigators (Tallal & Piercy, 1975). Two pairs of stimuli were synthesized. The first pair consisted of the concatenations [ɛɪ] and [æɪ], in which the first vowel was 43 ms in duration, followed by the second of 207 ms. Importantly, these vowels were synthesized as two adjacent steady-state vowels with no formant transitions. The second pair of stimuli represented versions of [ba] and [da] with the formant transitions extended from 43 ms to 95 ms and the steady-state portions of the vowel reduced to 155 ms. An illustration of key aspects of these stimuli is provided in figure 6.2. The children with SLI had difficulty with the vowels, but performed at the level of age controls on [ba] and [da]. This finding appeared to confirm duration as the relevant factor in a sound's difficulty, independent of its transitional or steady-state character. Frumkin and Rapin (1980) and Alexander and Frost (1982) also observed improved performance by school-age children with SLI when the formant transitions in [ba] and [da] syllables were doubled.

Tallal, Stark, Kallman, and Mellits (1980a) studied a new group of school-age children with SLI and age controls. They replicated the Tallal and Piercy (1974) findings for [ba] and [da]. In addition, the children with SLI were found to have great difficulty in a perceptual constancy task in which they had to press one panel for stimuli containing [b] (e.g., [ba], [bi]), and another panel for stimuli containing [d] (e.g., [dæ], [di]). Approximately one-third of the children could perform this task, whereas two-thirds of the age controls were successful. The children who succeeded participated in another task in which words were approximated by presenting the syllables in sequence (e.g., [be]-[bi] "baby") separated by 50 ms (Tallal, Stark, Kallman, & Mellits, 1980b). The children with SLI had greater difficulty than the control children. Tallal et al. interpreted this finding to mean that the added linguistic redundancy available in these stimuli was not sufficient to eliminate the effects of auditory processing deficits in these children.

Tallal, Stark, and Mellits (1985b) attempted to determine whether the same children's discrimination and sequencing performance on [ba] and [da] was related to their language comprehension test scores. Measures of [ba]–[da] performance along with other nonlinguistic and linguistic measures were entered into multiple regression analyses. The results revealed that the three most important variables emerging from the analyses all pertained to discrimination or sequencing of these two syllables.

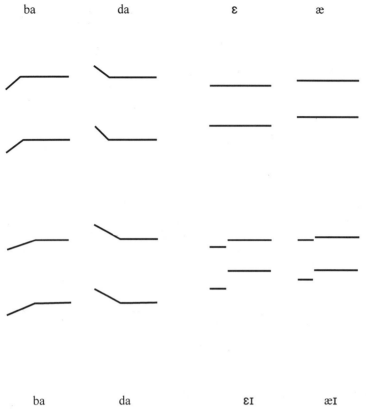

Figure 6.2
The second and third formants of stimulus contrasts presented to children with SLI. Performance on the [ba]–[da] contrast improves when formant transitions are lengthened, as shown on the lower left. However, performance on a vowel contrast drops when vowels are reduced in duration and are followed by another vowel with no formant transitions, as shown on the lower right.

In a related study, Tallal, Stark, and Mellits (1985a) found that discrimination of [ba]–[da] and perception of simultaneous tactile stimulation were two of six rapid perception and production tasks that successfully classified the children into SLI and normally developing groups according to discriminant function analysis.

An expanded list of synthesized syllables was employed in a study by Tallal and Stark (1981). The task required the children to press the panel when they heard one sound of a pair, and to make no response when the other sound was heard. The children with SLI performed below the level of age controls on the pairs [ba]–[da], [da]–[ta], and [da]–[ʃa]. The children with SLI had greater success with [dab]–[dæb], [sa]–[sta], and [ɛ]–[æ]. The finding for [ɛ]–[æ] was interesting in light of the fact that these vowels were only 40 ms in duration. Tallal and Piercy (1975) had used the same vowels and durations in combination with [ɪ] (viz, [ɛɪ] and [æɪ]), and found that the children with SLI had great difficulty. Tallal and Stark suggested that the difficulty experienced by the children with SLI in the earlier study was due to the fact that the contrasting vowels not only were brief but also were followed in rapid succession by other acoustic information.

Henderson (1978) also examined discrimination of the pairs [ba]–[da] and [da]–[ta] by school-age children with SLI and normally developing children of the same age. Differences favoring the age controls were seen for the first of these pairs. In contrast to Tallal and Stark (1981), differences were not seen for [da]–[ta]. However, Henderson noted that the manner in which these two syllables were synthesized permitted them to be distinguished according to degree of first-formant cutback as well as voice onset time.

A just-noticeable-difference task was employed by Elliott and Hammer (1988) to examine fine-grained discrimination according to voice onset time (VOT) and place of articulation. For VOT, an eight-item continuum using 5-ms steps ranging from [ba] to [pa] was employed. For place of articulation, 13 items were used, differing in starting frequency of the second and third formants and the onset bursts. Values ranged from those of [ba] to those of [ga], passing through values appropriate for [da]. School-age children with SLI were studied for a three-year period, beginning at age six. At each age, these children were poorer at detecting differences according to VOT than were age controls. For place of articulation, the children with SLI were poorer only at the youngest age tested. For both kinds of distinctions, performance improved with age.

Elliott, Hammer, and Scholl (1989) explored the question of whether school-age children with SLI and age controls could be successfully classified according to their performance on the same fine-grained discrimination tasks. Using only the results for VOT and place of articulation, they found that more than two-thirds of the six- and seven-year-olds and almost two-thirds of the children age eight years and above could be classified correctly. In a subsequent study, Elliott and Hammer (1993) performed factor analyses to determine how much of the variance in language ability could be explained by the fine-grained auditory discrimination factor. Among six- and seven-year-old children with SLI, 27% of the variance could be accounted for by this factor; for same-age peers, the variance explained was 16%. For children two to four years older, no distinct auditory discrimination factor emerged, even though the children with SLI in this age range demonstrated greater discrimination difficulty than their age controls.

Phonetic identification as well as discrimination were examined by Sussman (1993) in her study of five- to six-year-old children with SLI. The comparison groups were a group of four-year-old controls matched according to standardized language test scores, and a group of age controls. Stimuli were seven syllables whose starting formant frequencies varied on a continuum from [ba] to [da]. In one task, children heard four syllables, the first two identical (the endpoint [ba] syllables), the next two either the same as or different from the preceding syllables. The children touched an "X" placed in front of them whenever they detected a change in the syllables. In the second task, one syllable was presented at a time and the children pointed to a "B" if the syllable sounded like [ba] and to a "D" if it sounded like [da]. Verbal responses also were accepted.

The children with SLI performed as well as the two groups of normally developing children on the discrimination task. However, on the identification task, the children with SLI were less accurate on the endpoint stimuli [ba] and [da] than both control groups, and showed a relatively shallow slope across the continuum. These findings prompted Sussman to propose that the difficulties of children with SLI are not due to a failure to discriminate the auditory stimuli. Rather, the children's problems center on forming a phonological representation, linking the acoustic information to a phonological representation, or preserving the trace long enough to allow this linking to occur.

Identification and discrimination according to VOT were examined by Thibodeau and Sussman (1979). Twenty-one stimuli with [ba] and [pa] as the endpoints were synthesized. The syllables varied in VOT in 10-ms steps. For the identification task, the children pointed to a picture of a man for [pa] and a lamb for [ba]. For the discrimination task, triads were used that contained two identical stimuli and one stimulus that differed by 30 ms. School-age children with SLI and age controls participated (along with other groups of children). No group differences were seen for discrimination. In the identification task, the children with SLI were as accurate as controls in identifying the endpoint syllables; however, they showed boundaries that were not as sharp as the ones emerging from the data of the control children.

R. Stark and Heinz (1996) revisited the question of vowel perception in children with SLI. They examined the contrasts [a]–[i] and [ɛ]–[æ], using both short-duration versions of these vowels (from 5 ms to 40 ms) and long-duration versions (40 ms to 240 ms). Tasks included a discrimination task, in which the children pressed a single switch when they heard any change in the series of sounds being presented, and an identification task in which the children were required to press one of two panels, depending on the sound heard. The discrimination task was used only when performance on the identification task was poor. A temporal ordering task was employed if the children succeeded on the identification task. Here, the children pressed each of the two panels in the sequence that corresponded to the sequence in which the two sounds were perceived.

The children with SLI were school-age; their performance was compared with that of same-age peers. On the identification task, the children with SLI performed as well as the control children on the long-duration contrasts of [a] and [i]. However, they were less accurate on the short-duration versions of these same vowels, and on both long and short versions of [ɛ] and [æ]. The children having difficulties in identification performed well on the same pairs in discrimination. Most of the children who were successful in identification could also perform as well as controls on the temporal ordering task. R. Stark and Heinz (1966) concluded that duration was an important factor in vowel identification, along with formant information; however, in a discrimination task that is stripped of the requirement of forming or maintaining a phonological representation of the stimulus, the role of duration is reduced.

Leonard, McGregor, and Allen (1992) pursued the notion that cues of brief duration are especially difficult for children with SLI if the brief information is adjacent to other material of longer duration. This issue was prompted by the findings of Tallal and Stark (1981) and Tallal and Piercy (1975) that [ɛ] and [æ] could be discriminated even when these were quite brief, yet when they were combined with a vowel of longer duration (creating [ɛɪ] and [æɪ]), the contrast became too difficult. For syllable pairs known to be difficult for children with SLI, this condition also holds; [ba] and [da] differ by a brief formant transition, whereas the longer steady-state portions in these syllables are identical. On the other hand, the syllable pair [dab]–[dæb] was not problematic; here, the contrastive material constitutes the portion of the syllable with the longest duration.

To test this hypothesis, Leonard, McGregor, and Allen (1992) generated the contrasts [dab] versus [dæb], [i] versus [u], [ba] versus [da], [das] versus [daʃ], and [dab]–

[i]–[ba] versus [dab]–[u]–[ba]. The last three pairs were expected to cause trouble for children with SLI because the contrastive portions of the signal were brief in duration relative to adjacent material. The final pair functioned as an important test of the role of relative duration. The medial vowels in these multisyllabic stimuli were 100 ms in duration and were precisely the same tokens used in the contrast [i] versus [u]. (No formant transitions between these vowels and the preceding and following consonants were used.) If the distinction between [i] and [u] was difficult to make only in the multisyllabic cases, it could not be the absolute duration that caused the difficulty. The findings were consistent with these expectations. The children with SLI performed at the level of age controls on [i]–[u] and [dab]–[dæb], and differed from age controls on the multisyllabic pair and the two pairs contrasting in initial or final consonant.

An early study by McReynolds (1966) also provides evidence that children with SLI have difficulty when discriminable material is placed in a larger context. Children with SLI performed as well as age controls in discriminating consonants such as [s] and [ʃ] when each sound appeared in isolation. However, when these consonants were placed in otherwise identical multisyllabic nonsense words (e.g., [həsak], [həʃak]), the children with SLI had significantly greater difficulty than the control children.

Task Effects in Auditory Processing Studies

Overall, the evidence for a deficit in auditory processing seems quite convincing. Nonverbal stimuli that must be processed quickly—either because the stimuli themselves are brief or because stimuli are presented rapidly in succession—are problematic for children with SLI. So, too, are verbal stimuli that must be distinguished on the basis of portions of the signal of relatively brief duration. Because the acquisition of many details of language relies on cues of relatively brief duration, it would be reasonable to expect a relationship between performance on these processing tasks and language ability.

However, there are reasons to believe that if auditory processing tasks provide a window into the language skills of these children, the view is far from perfect. First, of course, many aspects of language do not depend on the ability to process brief acoustic details, and when such brief details are encountered, additional cues are often available. Children with SLI could learn much about language through the redundancies in the acoustic code and those phonetic details possessing greater duration. This partial information could permit the children to hypothesize linguistic units that, in turn, could be used in top-down processing to help decipher other details. If this were not the case, consider how strange the findings of Thal and Barone (1983) would be. These investigators tested children with SLI on their ability to discriminate and sequence words and word strings, as well as a pair of nonverbal stimuli modeled after Tallal and Piercy (1973a). The children had greater difficulty with the latter. Obviously, with the presentation of words, several different perceptual and linguistic operations were set into motion. Not so in the case of 75-ms tones.

A second reason to temper our enthusiasm about how directly auditory processing data reflect language ability is that children improve with practice on these tasks. Using the task of Tallal (1976), Tomblin and Quinn (1983) demonstrated that a group of normally developing school-age children could show successively better performance (with no additional instruction) across four sessions conducted within a one-week period.

Improvement was also seen in a group of school-age children with SLI studied by Robin, Tomblin, Kearney, and Hug (1989). These children, along with age controls, participated in a task in which they listened to a series of square wave tone bursts. One of the ISIs in the series was brief; the children's task was to identify its location. The children with SLI improved significantly on the task with repeated exposure. However, even after a minimum of six sessions, the children with SLI had not reached the performance level seen for the control children after a single exposure to the task. Rate of improvement was slower in the later sessions; however, the possibility of age-appropriate performance with additional practice cannot be ruled out.

Alexander and Frost (1982) gave a group of school-age children with SLI practice on the contrast [ba]–[da] in two conditions, presented alternately. In one condition, the formant transitions were 40 ms; in the other, transitions were initially 80 ms and then reduced in steps of 10 ms. A control group of same-age children with SLI received practice only on the syllables with 40-ms transitions. For all children, eight 30-minute sessions were provided, one per week. Practice with the extended-transition condition resulted in greater gains in discriminating the 40-ms syllables than was seen in the control condition. However, even the children with SLI in the control condition made steady, if modest, improvement.

In two experiments, Merzenich, Jenkins, Johnston, Schreiner, Miller, and Tallal (1996) gave five- to ten-year-old children with SLI extensive practice on (1) non-verbal auditory stimuli containing upward or downward sweeps in frequency covering the same frequency range as consonant–vowel formant transitions of English; and (2) consonant–vowel and vowel–consonant–vowel stimuli contrasting in the consonant (e.g., [ba]–[da], [fa]–[va], [aba]–[ada]). The training procedure involved a progression in which children began with maximally distinct stimuli. The ISIs were 500 ms, the nonverbal stimuli were 60 ms in duration, and the formant transitions in the verbal stimuli were 65 to 70 ms in duration. In addition, in the case of the verbal stimuli, the consonants were differentially amplified 20 dB. As training progressed (through the use of a computer game format) and the children demonstrated success, the durations of the stimuli, ISI, and amplitude were reduced in systematic steps.

After 8 to 16 hours of training (divided into 20-minute sessions) over a 20-day period, the children showed impressive improvement. For example, nonverbal stimuli of 60 ms duration with a starting or ending frequency of 1 kHz required an average ISI of 268 ms early in the study and only 77 ms at the end of the investigation. Similarly, at the end of the study, the majority of the children were successful with verbal stimuli containing formant transitions as brief as 35 ms and ISIs of only 10 ms. In the second of the two experiments, a comparison group of children with SLI also participated. These children played video games instead of the auditory computer games. No gains were seen in this group of children.

Across greater time spans, children with SLI can achieve acceptable performance levels even without deliberate practice, at least with stimuli ordinarily mastered by normally developing children at a younger age. Bernstein and Stark (1985) conducted a four-year follow-up study of the children with SLI who had participated in the investigations of Tallal, Stark, Kallman, and Mellits (1980a, 1980b) and Tallal and Stark (1981). The children were now successful in discriminating and sequencing [ba] and [da], performing as accurately as a group of age controls. These findings indicate that whatever the difficulty with brief duration cues, it does not seem to reflect the permanent state of some perceptual mechanism.

Lincoln, Dickstein, Courchesne, Elmasian, and Tallal (1992) studied the ability of 15- to 20-year-old subjects with SLI to discriminate and report the sequences of two tones. They found that the difficulties were limited to sequencing. Even at ISIs of 20 ms, the subjects could discriminate between the two tones if the sequence was short.

Just as children seem to get better at this type of task with increasing age, so too there is an age below which children cannot perform this task. The youngest age for which this task is appropriate seems to be around 4;6. Unfortunately, many school-age children with SLI have language abilities that approximate those of typical four-year-olds. Unless we know what the performance profiles of four-year-old (and younger) children should look like on this task, the interpretation of the findings for children with SLI can be taken only so far, notwithstanding Tallal's (1976) finding of an uneven performance profile in these children. Of course, logic also leads us to believe that strong task effects are influencing the auditory processing data. If children with SLI really couldn't detect the difference between [ba] and [da] until well into school age, the reviews of language production in chapters 3 and 4 would have been limited to a few paragraphs describing these children's grunts and vowels.

Finally, although processing brief acoustic cues is necessary for language learning, such cues are the most fragile in any less than optimal listening condition, even for individuals whose language abilities are not in doubt. Ludlow, Cudahy, Bassich, and Brown (1983) provide a suitable illustration. These investigators tested the auditory temporal processing abilities of a group of school-age children with attention deficits and found that their performance resembled that of a group of children with SLI. Yet these children had no history of language impairment.

Auditory Processing and Characteristics of Language Production

Even though auditory processing tasks yield data with extra, unwanted baggage, there may still be parallels between the auditory processing profiles and language production profiles of children with SLI. Tallal, Stark, and Curtiss (1976) pursued this question by examining the phonological characteristics of the children participating in the Tallal and Piercy (1973a, 1973b, 1974, 1975) studies. Not suprisingly, the children with SLI exhibited many more phonological errors than did the age controls. More to the point, they had more difficulty with stop consonant and stop consonant cluster productions than with vowels and nasal consonants. As noted by Tallal, Stark, and Curtiss, the former rely on brief acoustic cues; the latter do not.

These same phonological errors were described in greater detail by Stark and Tallal (1979). The greatest number of errors occurred in voicing (e.g., [b] for /p/ in word-initial position), followed by errors in place of articulation. These place errors were usually substitutions of an alveolar for a velar (e.g., [d] for /g/).

Productions of [b] for /d/ and vice versa were not in evidence in the study of Stark and Tallal (1979), nor are they common in the literature on SLI in general, another reminder not to overinterpret the auditory processing findings. It would be difficult to argue that these children couldn't perceive a distinction that they could make in their own speech. Most children with SLI—like younger normally developing children—have the greatest difficulty producing continuant consonants such as /s/,

/ʃ/ and /θ/ (e.g., Leonard, 1982b). Although in many contexts these consonants will have shorter durations relative to adjacent sounds, and thus could pose problems (such as the [das]–[daʃ] contrast studied by Leonard, McGregor, and Allen, 1992), distinctions among them rely on cues that are not as brief as those required to discriminate among stop consonants.

Some rough parallels have been drawn between auditory processing results and data from grammatical tasks. Curtiss and Tallal (1991) found that five- and six-year-old children with SLI not only had greater difficulty than same-age peers in sequencing nonverbal stimuli, but also did more poorly on language test items that depended on word order (e.g., reversible subject–verb–object utterances). Test items that did not depend on order were much easier for the children. In the Leonard, McGregor, and Allen (1992) investigation, the same children with SLI who had great difficulty discriminating stimuli whose contrastive portions were relatively brief were found to make less use of consonantal inflections and function words consisting of weak syllables than did a group of MLU controls.

Auditory Processing and Language Comprehension

Although comprehension problems in children with SLI have sometimes covaried with poor performance on auditory processing tasks (Tallal, 1975), this relationship has only recently received significant attention. Fellbaum, Miller, Curtiss, and Tallal (1995) explored the relationship between auditory processing and the comprehension of a collection of grammatical forms that require discrimination of brief acoustic information. Six-year-old children with SLI and their age controls participated. Auditory processing was examined, using the 75-ms nonverbal stimulus discrimination and sequencing task of Tallal and Piercy (1973a). The grammatical forms tested included those of brief duration (e.g., noun plural -s, regular past -ed, nominative case pronouns *he, she, they*) and those of longer duration (e.g., object pronouns *him, her*, comparative *more*). The children with SLI had greater difficulty than the age controls on both the discrimination and sequencing task and the grammatical forms of brief duration. The remaining forms revealed no differences between the two groups, with one exception: the children with SLI committed a greater number of errors on passives of the progressive form *The clown is being pushed*. Although some of the morphemes in this construction are of relatively brief duration (viz, *is*, *-ed*), the appearance of the more salient *being* and *push* in juxtaposition would seem to rule out an active voice interpretation.[1]

Dollaghan (1995) observed that despite the evidence for difficulties with brief grammatical inflections in children with SLI, inflections had not been included among the stimuli used in perceptual tasks. To examine this issue, Dollaghan compared children's perception of inflected words (e.g., *kissed*) and monomorphemic words with identical consonants in final position (e.g., *taste*). A time-gating procedure was used, in which successively longer portions of the stimulus were presented to the listener. At each interval or gate, the listener guessed the word that had been presented. The dependent measure was the number of gates required before the word was identified correctly. A group of six- to ten-year-old children with SLI participated in this task and were found to require a greater number of gates for the inflected words than for the phonetically similar words.

Of course, in a task of this sort, the increased number of gates required for the inflected words might be due to the fact that the children could guess the bare stem (e.g., *kiss*) when enough phonetic material had appeared. This is not possible in the case of monomorphemic words such as *taste*. For this reason, it is important to determine whether the magnitude of the difference between inflected forms and phonetically similar monomorphemic words is greater in children with SLI than in control children. Dollaghan (1995) obtained data from somewhat older normally developing children that suggested this might be the case. However, a definitive answer to this question must await data from typical children who constitute a better match.

The strongest case made thus far for a meaningful relationship between language comprehension and auditory processing is the investigation of Tallal, Miller, Bedi, Byma, Wang, Nagarajan, Schreiner, Jenkins, and Merzenich (1996). The children participating in this study were the same as the children serving as subjects in the Merzenich, Jenkins, Johnston, Schreiner, Miller, and Tallal (1996) investigation discussed earlier in this chapter. As part of their treatment, the children received extensive practice in listening to stories (and other exercises) in which the verbal material was systematically modified. The duration of the speech was prolonged by 50%, and fast transitional elements were differentially amplified as much as 20 dB. Two experiments employing the above material were conducted. In the second, a control group of children with SLI was added. These children participated in similar activities using unaltered speech material. For the children who listened to the altered speech, gains on standardized measures of language comprehension were extremely large. For some of the children, posttreatment scores fell well within the average range, even though the children's pretreatment scores were quite low. Furthermore, the control children participating in the second experiment made much smaller gains.

Factors of Brevity and Rate Are Implicated in SLI

The conclusion that children with SLI have difficulty processing brief or rapidly presented stimuli seems indisputable. These findings are so consistent and demonstrable across tasks and stimulus variations that it is difficult to imagine that they are not an important piece of the SLI puzzle. At the same time, much work remains to be done to determine whether these difficulties reflect an impairment in some circumscribed mechanism that is devoted to low level auditory processing, or to some more general processing deficit for which brief material is the most problematic. It also remains to be explained how it is possible that brief stimuli and the rapid rate of presentation of stimuli create essentially the same effect. We will return to this issue in chapters 12 and 13, where it will be the object of considerable discussion.

Note

1. The results from the Fellbaum, Miller, Curtiss, and Tallal study were actually more consistent with their hypothesis than they stated. Two other grammatical forms examined by these investigators yielded results that could be interpreted in terms of duration. A difference favoring the control children was seen for the modal auxiliary *will*. This form is not as brief as auxiliary *be* forms, but it is less salient than main verbs (Altenberg, 1987). The comparative *-er* ending did not yield differences between the two groups of children. As a word-final syllable, this form is subject to significant lengthening when it appears in clause-final position (Klatt, 1975).

PART III

Nature and Nurture

Chapter 7

The Genetics and Neurobiology of SLI

The Genetic Study of SLI

It is ironic that for many years, in our search for subtle but powerful causes of SLI, genetic factors were given short shrift. The prevailing assumption was that the basis of the impairment could be found in some pernicious factor within the child, or some prenatal or perinatal disturbance. When the search was extended beyond the child and the delivery room to other members of the family, it was the family's linguistic input to the child that was usually the focus of inquiry. Exceptions to this tendency to ignore genetic factors appear only sparsely in the earlier literature (e.g., Arnold, 1961; T. T. S. Ingram, 1959; Luschinger, 1970).

This oversight is not as incomprehensible as it might seem. As we shall see, many children with SLI come from families in which only the child in question—the proband, in the parlance of genetics—has ever had a deficit in language ability. It is not difficult to find children with SLI whose parents are articulate and well-educated. A few salient cases of this type could easily induce clinicians and researchers to look elsewhere for a cause.

But there are genetic connections, and now that we are aware of them, they seem big enough to trip over. A review of some of the evidence follows.

Familial Aggregation

A first step in the genetic study of SLI was taken when reports of high familial concentration of SLI appeared in the literature. Some of these were case studies. For example, Samples and Lane (1985) described a family in which all six children exhibited SLI. Group studies also revealed high familial concentration. These investigations reported that approximately 30% of the immediate family members of children with SLI also had a history of language problems (e.g., Robinson, 1987; Sonksen, 1979). In these studies, the percentages of affected siblings were on the order of 20% (e.g., Dalby, 1977; Robinson, 1987). The past or contemporaneous language status of the parents and siblings in these investigations was determined through information obtained from parental report and, in some cases, school and clinical records.

Studies of familial aggregation that included a control group began to appear in the mid 1980s. The histories of language problems among family members were determined by questionnaire. The questions were framed in somewhat different ways across investigations. In some studies, spoken language problems were the object of inquiry; in others, language-related skills such as academic performance were also included. The probands were children with SLI who had been identified as having

Table 7.1
Occurrence of Language and Related Difficulties in Families of Children with SLI and in Control Families

	Percentage of Family Members Affected	
Investigators	Proband	Control
Neils and Aram (1986)	20	3
Tallal, Ross, and Curtiss (1989a)	39	19
Tomblin (1989)	23	3

language problems prior to the study. Controls were selected from the same or comparable school districts and socioeconomic class.

Table 7.1 reports the percentage of immediate family members showing a history of language problems in three studies. The studies were similar in the types and ages (four to eight years) of the children serving as the probands.[1]

The clear and significant differences reflected in table 7.1 hold for parents and siblings alike. Across studies, slightly higher percentages of language difficulties are reported for fathers than for mothers. Similarly, percentages are higher for brothers than for sisters. These findings are in line with the male–female ratios for SLI noted in chapter 1.

The study of Tallal, Ross, and Curtiss (1989a) shown in table 7.1 differs from the other two in the higher percentages seen for the families of both the probands and the control children. In this study, a history of having been kept back a grade or of reading problems was grounds for placing a family member into the affected category, whether or not problems in spoken language were reported.

Still higher percentages were seen in a study by van der Lely and Stollwerck (1996) that also took reading problems as evidence for a positive family history. The probands were 12 children with SLI who were selected because their grammatical deficits were more serious than their problems with other areas of language. The family histories of these children were compared with those of 49 control children. Over 75% of the children with SLI had one or more first degree relatives with a history of a language and/or reading problem. For the controls, the percentage was nearly 30%. Both percentages are high, but the gap between the two groups is nevertheless quite large.

Lewis (1992) computed familial concentration in a manner that permitted separation of language problems from reading and other learning problems. For a group of 45 probands with phonological and other language disorders, 15% of other family members showed a history of spoken language difficulties, 4% showed a history of reading problems, 2% showed both spoken language and reading difficulties, and 3% exhibited other types of learning disabilities. For a group of 79 controls, the corresponding percentages ranged from almost zero for combined spoken language and reading problems to 2% for spoken language deficits with no known reading difficulties.[2]

These studies of familial concentration were designed with all the proper safeguards required for research involving questionnaires. Nevertheless, the question remains as to whether the results would be different if direct assessments of the family members' language abilities were possible. Plante, Shenkman, and Clark (1996)

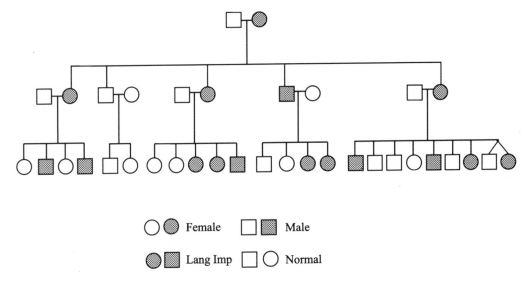

Figure 7.1
Pedigree of a three-generational family containing members diagnosed as language-impaired. Adapted from Gopnik (1994a).

explored this issue. They made use of a test battery developed by Tomblin, Freese, and Records (1992) that successfully differentiated adults with known histories of SLI from adult controls. Plante, Shenkman, and Clark administered this battery to a group of parents of children with SLI and a group of adult controls. They compared the results with those obtained by using a questionnaire. According to the questionnaire data, 38% of the parents of children with SLI had prior spoken language and/or reading problems; no control subjects reported such a history. The test battery data indicated that 63% of the parents of the children with SLI scored at levels comparable with the adults having known histories of SLI in the Tomblin, Freese, and Records investigation, compared with 17% of the adult controls.

Tomblin and Buckwalter (1994) also used the Tomblin, Freese, and Records (1992) test battery to determine the rates of SLI in parents and siblings of probands. They found that 21% of the family members scored at levels consistent with a diagnosis of SLI. As in other studies, fathers were more likely to score poorly than mothers, and brothers were more likely to have low scores than were sisters.

Direct testing was also employed in the case of the large family studied by the independent research teams of Gopnik (Crago & Gopnik, 1994; Gopnik, 1990a; Gopnik & Crago, 1991), Hurst (Hurst, Baraitser, Auger, Graham, & Norell, 1990), and Vargha-Khadem (Vargha-Khadem, Watkins, Alcock, Fletcher, & Passingham, 1995). Sixteen of the members of this three-generation family had been diagnosed as language-disordered. Figure 7.1 provides the pedigree.

Gopnik and Crago (1991) tested 20 of these family members, seven of whom appeared to be acquiring and using language normally. Many of the language tasks administered by Gopnik and Crago revealed differences between the affected and unaffected groups. The language problems exhibited by many of the affected family members were striking.

One of the complicating aspects of research with this family is that some of the affected members would not be considered to exhibit SLI by conventional standards. As noted by Gopnik (1994a) and Crago and Gopnik (1994), two of the affected males had a history of psychiatric problems. Twenty-one of the family members were tested by Vargha-Khadem, Watkins, Alcock, Fletcher, and Passingham (1995) on several nonlinguistic as well as linguistic tasks. One of the tests was a nonverbal test of intelligence. The affected family members' group mean was 86, with a range of 71 to 111. Clearly, some of these individuals' nonverbal IQ scores fell below the traditional cutoff of 85. It also appeared that some of the family members failed a test of oral-motor function. Traumatic events also touched the lives of some of these family members, including physical abuse and the disappearance of a spouse with three of the children (Gopnik, 1994a).

Given the impressive evidence of familial concentration of SLI, it is easy to slip into the view that all children with SLI will have positive family histories for language problems. Yet this is not the case. Tallal, Townsend, Curtiss, and Wulfeck (1991) conducted a questionnaire study of the families of 65 children with SLI and found that for 23 (35%) of the children, there was no evidence of past or present language problems elsewhere in the family. They then examined a subgroup of 23 children with the poorest language skills and found a similar percentage of children (30%) with a negative family history of language problems. Tomblin and Buckwalter (1994) reported an even higher percentage of negative histories (58%) in a study of the families of 26 children with SLI. The percentage of children with SLI showing a negative family history was 40% in an investigation of 53 children by Lahey and Edwards (1995).

Interpreting the Data

What do we make of the dual findings of higher-than-usual familial aggregation of SLI and sizable percentages of families with language problems limited to the proband? Tomblin and Buckwalter (1994) discuss two possible interpretations. One is that there are different causes of SLI, some based on genetic factors, others not. The second possibility is that there is a genetic predisposition in all cases but there is incomplete penetrance. That is, additional factors must be present for SLI to appear.

If there are different causes of SLI, these differences might result in divergent patterns of deficit. In chapter 1, we saw that children with SLI constitute a heterogeneous group; perhaps some of the differences might be attributable to whether or not the deficit has a genetic basis. Tallal, Townsend, Curtiss, and Wulfeck (1991) pursued this possibility by comparing children with SLI who had positive and negative family histories. Those from families with other affected members were more likely to have a record of attention-related behavior problems, and performed more poorly on both standardized academic tests and tests of auditory processing. These families were also lower in socioeconomic status. Lahey and Edwards (1995) found that children with SLI whose problems were limited to language production were more than twice as likely to come from families with other affected members than were children experiencing problems in comprehension as well as production. Byrne, Willerman, and Ashmore (1974) found that children with moderate deficits in language had a greater number of family members with a history of language problems than did children with severe language deficits.

The findings of Lahey and Edwards (1995) and Byrne, Willerman, and Ashmore (1974) highlight the importance of how comparisons are made. Tallal, Townsend, Curtiss, and Wulfeck (1991) first divided children with SLI into those who did and did not have a positive family history. Comparisons of the two groups' language test scores were then made. One of the comparisons made by Lahey and Edwards (1995) was also of this type. In each case, no differences were observed between the two groups. Evidently, quite different results emerge when children with SLI are divided according to language characteristics and then compared according to family history.

Twin studies are an excellent means of examining penetrance. Monozygotic twins are genetically identical, whereas dizygotic twins share 50% of their genes. The difference in the concordance rate (cases in which both twins show the impairment) between monozygotic and dizygotic twins should reflect the contribution of genetic factors to the problem. Furthermore, a concordance rate for monozygotic twins that approaches 100% would suggest complete penetrance. In a study reported by Tomblin and Buckwalter (1994), 82 pairs of twins were identified that contained at least one child meeting the criteria for SLI. Upon testing the second child in the pair, these investigators found a concordance rate of 80% for monozygotic twins and 38% for dizygotic twins. Lewis and Thompson (1992) studied 57 pairs of twins and found concordance rates of 86% and 48% for monozygotic and dizygotic twins, respectively. Monozygotic twins were also more likely to exhibit the same type of difficulty, such as whether or not learning difficulties in school accompanied speech and language problems.

In a study of 61 twins by Bishop (1992a), probands and their co-twins were divided according to whether they continued to exhibit SLI at the time of the investigation or had resolved their language difficulties through time or treatment. For cases in which either the proband or the co-twin had past but not present problems with language (which represented the majority of cases), concordance was more than three times higher for monozygotic than for dizygotic twins. Concordance rate differences were negligible for cases in which both the proband and the co-twin had problems at the time of the investigation. Collapsed across these different combinations, the concordance rates for monozygotic and dizygotic twins were 67% and 32%, respectively. It is also important to point out that another 26% and 23% of the monozygotic and dizygotic co-twins (respectively) showed nonverbal IQs below 85. These percentages are much higher than those seen in the general population and indicate that the conditions that promote SLI may constitute the occasion for other types of disorders as well. This finding suggests that the heterogeneity in disorder types seen in the family studied by Gopnik and her colleagues is not an anomaly. If these low-IQ cases are viewed as part of the same genetic condition that is responsible for language problems only, the penetrance is nearly complete.

A disorder with a genetic basis may constitute a multifactorial trait or a trait attributable to a single gene. Multifactorial traits can be influenced by a combination of environment and genetics, or they can be (even entirely) polygenic—the result of more than one gene. In the case of SLI, certain families may carry several factors that make language learning extremely difficult. In principle, the multiple genes responsible for poor language skills could be the same as those contributing to normal trait variability, rather than constituting a distinct etiology. Methods of distinguishing these two possibilities are being explored (e.g., Gilger, Borecki, Smith, DeFries, & Pennington, 1996).

The methods used to choose between polygenic and single-gene accounts of a disorder are complex and require large numbers of families. Tomblin and Buckwalter (1994) employed a method that involves calculating the rate of language problems in siblings of probands and the rate of language problems in the general population. For the former, they used 24% for male siblings and 6% for female siblings, values that were obtained in their own study of familial concentration. They used 6% and 2% for the male and female figures in the general population. This method compared the relative frequency of language deficits in the siblings of the probands with that predicted by the major modes of inheritance (dominant single gene, recessive single gene, multifactorial pattern). The observed frequency fit the multifactorial model most closely.

Data from the family studied by the Gopnik and Hurst research teams were interpreted by these investigators as reflecting a dominant single gene (Gopnik & Crago, 1991; Hurst, Baraitser, Auger, Graham, & Norell, 1990). Although the definitive work leading to this conclusion has not yet been undertaken, the large number of affected family members gives this viewpoint considerable credence. For example, if one computes the rates of affected siblings in this family in the manner used by Tomblin and Buckwalter (1994), the rate for males ranges from 40% to 58% (depending on which sibling is arbitrarily selected as the proband) and the rate for females ranges from 33% to 43%.

Of course, even if this family's language impairment can be attributed to a single gene, we must be clear about what this means. Pembrey (1992) put it quite well:

> What must be remembered, if this were to be successful, is that it would define a monogenic language disorder, not a "gene for speech" and certainly not "*the* gene for speech." Just because the cause of a watch stopping can be simple, does not mean that the cause of it working is simple.   (p. 53)

There is a final issue pertaining to familial aggregation that bears comment. Tallal (1989, 1991) found that the expected male-to-female ratio for SLI applied only when at least one of the parents had a history of language problems. Furthermore, affected mothers seemed to have a disproportionate influence on the findings of more males than females with SLI. Specifically, affected mothers had three times more male than female children. Approximately half of all children were affected; hence, more male than female children exhibited SLI. Tallal raised the possibility that these findings were related to maternal stress and heightened levels of testosterone, given that these factors have been associated with unusually high rates of male births.

The Neurobiology of SLI

SLI is defined as a language disorder without evidence of frank neurological impairment such as seizure activity or brain lesions. But this does not mean that there is no physical evidence associated with the language problem. Subtle irregularities in brain structure or function might be identified in children with SLI, especially as technological advances are made in neurological testing. These irregularities could represent pathology, or a less typical neurological configuration that makes language learning difficult when certain environmental or biological conditions apply. In a sense, of course, there must be physical evidence of SLI. In the world as we know it, every volitional behavior has a neurophysiological counterpart. And stable behavioral

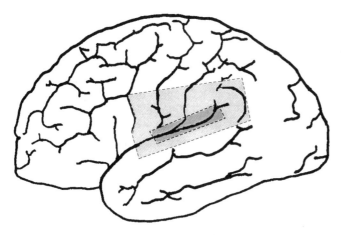

Figure 7.2
Approximate location of the planum temporale (darker shading) and the perisylvian area (lighter shading) of the left hemisphere.

patterns (such as relative language ability) probably should have neuroanatomical correlates as well.

Neuroanatomical Correlates
A valuable lead in the study of neuroanatomical correlates of SLI came from Galaburda, Sherman, Rosen, Aboitiz, and Geschwind (1985). These investigators conducted an autopsy study of four brains of males with a history of developmental dyslexia. Records of three of these individuals indicated earlier problems with phonology and, in one case, other areas of language. One of the key findings in this study was a symmetry of the plana temporale. The planum temporale is located in the upper portion of the temporal lobe of each hemisphere, inside the sylvian fissure. Its approximate location in the left hemisphere is shown in figure 7.2. Typically, the planum temporale of the left hemisphere is larger than that of the right, consistent with the assumption that this area in the left hemisphere is associated with language functioning. In these individuals, however, the two plana were of similar size, the result of a typically sized left planum and an atypically large right planum.

In a series of studies, Plante and her colleagues looked for such left–right hemisphere symmetries in children with SLI, using magnetic resonance imaging (MRI) techniques (Plante, 1991; Plante, Swisher, & Vance, 1989; Plante, Swisher, Vance, & Rapcsak, 1991). Because the shape and location of the planum temporale present obstacles to measurement from MRI scans (see Plante, 1996), these investigators measured a broader area around the sylvian fissure—the perisylvian area—that includes the planum temporale but extends anteriorly to the inferior frontal gyrus and posteriorly to the supramarginal gyrus and angular gyrus. The approximate location of the perisylvian area is also indicated in figure 7.2. Like the plana temporale, this broader area, too, typically shows an asymmetry with the left hemisphere exceeding the right hemisphere in size (Geschwind & Levitsky, 1968).

The first of these studies involved a boy with SLI age 4;9 and his normally developing dizygotic twin (Plante, Swisher, & Vance, 1989). The child with SLI showed symmetry of the left and right perisylvian areas, the result of a larger-than-expected right perisylvian area. A suprising finding was that the co-twin showed an asymmetry in a direction opposite that expected of typically developing children: the right perisylvian area was larger than the left.

Interpretation of these findings was aided by a subsequent study by Plante, Swisher, Vance, and Rapcsak (1991). Measures of MRI scans of the left and right perisylvian areas were obtained from eight boys with SLI, age 4;2 to 9;6. Six of the eight children departed from the usual left-greater-than-right pattern. For three of these six, the right perisylvian area was similar in size to the left because the right side was unusually large. For the other three children, the right perisylvian area was larger than the left, as was found for the normally developing twin in the earlier Plante, Swisher, and Vance (1989) investigation.

Together these findings suggest that atypical left–right perisylvian area configurations consist of a larger-than-usual right perisylvian area that may equal or exceed the left in size. Furthermore, atypical configurations may not be restricted to individuals with SLI. Siblings functioning normally in language may also exhibit this pattern. The latter possibility was explored by Plante (1991). She extended her study to the parents and nontwin siblings of four of the children with SLI who participated in the Plante, Swisher, Vance, and Rapcsak (1991) study. Seven of the eight parents and four of the five siblings showed atypical configurations. Consistent with previous findings of familial concentration of SLI, several of the parents had a history of language learning difficulties, and a few of the siblings exhibited language problems at the time of testing. However, the correspondence between atypical configurations and language status was certainly not perfect. Among the parents or siblings there were instances of atypical configurations with normal language functioning and language problems with typical configurations. Interestingly, two of the four probands showed normal left–right perisylvian area configurations, yet none of the parents and only a single sibling in these families had the same typical configuration.

The work of Plante and her colleagues seems to establish that a larger-than-usual right perisylvian area constitutes a condition that disfavors normal language learning. However, the presence of atypical configurations with no accompanying language problem suggests that a stronger interpretation—atypical configurations causing language disorders—would be too strong. Furthermore, if a larger-than-usual right perisylvian area puts language learning at risk, it is not clear that it represents an abnormal condition, if "abnormal" is defined as "unnatural." In the classic study of Geschwind and Levitsky (1968), nearly one-fourth of the brains studied (presumably from normally functioning individuals) failed to show the more common left-larger-than-right configuration. It would be important to learn whether the typical and atypical patterns form discrete groups, or whether there is a continuum with the typical left-greater-than-right configuration representing the modal pattern.

There is some debate about the origins of the atypical configurations. According to Plante's (1996) interpretation of the literature, asymmetries are established during the third trimester of prenatal development. Autopsy and imaging studies focusing on the perisylvian area in particular show rather stable left–right hemisphere relationships from preschool ages onward. Plante suggests that elevated testosterone levels might be an important factor. Such levels are potentially heritable, and are capable of producing the types of neuroanatomical signs that are so frequent among individuals with SLI. As a test of the viability of this proposal, Plante, Boliek, Binkiewicz, and

Erly (1996) examined children with congenital adrenal hyperplasia, a rare genetic disorder that leads to excessive testosterone production. Both language and MRI testing was conducted on these children, their siblings, and a group of age controls. MRI measures focused on the perisylvian area. The results indicated that language problems as well as the less typical left–right perisylvian area configurations were more frequent among the children with congenital adrenal hyperplasia and their siblings than among the controls.

Locke (1994) takes a different position on the nature of an unusually large right perisylvian area in children with SLI. Citing non-human animal studies, Locke points out that deficits often produce compensatory activity that, in turn, produces compensatory hypertrophy. In the case of SLI, neurodevelopmental delays may promote language learning via less efficient mechanisms in the right hemisphere, which in turn may alter the child's neuroanatomy, resulting in a larger-than-usual right perisylvian area. Such a position does not explain why atypically large right perisylvian areas are also seen in a fair proportion of individuals with no history of neurodevelopmental delays.

Jackson and Plante (1997) turned their attention to the posterior perisylvian region. Neuroanatomical studies have revealed variations in gyral morphology in this area, and studies of individuals with dyslexia showed that they were far more likely than controls to have one or more intermediate gyri between the postcentral sulcus and the supramarginal gyrus (C. M. Leonard, Voeller, Lombardino, Morris, Hynd, Alexander, Anderson, Garofalakis, Honeyman, Mao, Agee, & Staab, 1993). In the C. M. Leonard et al. study, nearly two-thirds of the individuals with dyslexia showed this gyral morphology, whereas over 90% of the controls showed only one gyrus in this region.

Participants in the Jackson and Plante (1997) investigation were ten school-age children with SLI (the probands), their parents, and ten siblings. Fifteen of the parents and four of the siblings exhibited poor language skills. Twenty adults served as control subjects. Both the left and right posterior perisylvian areas were examined for gyral morphology. In the control group, 75% of the hemispheres showed the typical, single gyrus pattern; 23% showed an intermediate gyrus (there were other, less frequent patterns that shall not concern us here). For the 40 family members (including the probands), 58% of the hemispheres showed a single gyrus pattern and 41% showed an intermediate gyrus. The likelihood of the latter pattern did not appear to be related to whether or not the family member had a language problem. For controls and family members alike, the left hemisphere was more likely than the right to reveal the intermediate gyrus pattern.

The morphology of the inferior frontal gyrus (site of Broca's area) in parents of children with SLI was examined by Clark and Plante (1995). The parents were tested for language ability at the time of the study and were divided into those showing and not showing evidence of language limitations. A group of adult controls also participated. MRI scans were analyzed in terms of whether or not an extra sulcus was evident in the vicinity of the inferior frontal gyrus. When all parents of children with SLI were compared with controls, no differences were seen in the tendency toward an extra sulcus. However, an extra sulcus was observed more frequently in the scans of the parents with documented language limitations than in those of the remaining parents.

Jernigan, Hesselink, Sowell, and Tallal (1991) used MRI data to examine the entire cerebral region, dividing it into six zones. Twenty school-age children with SLI and 12 age controls served as the subjects. Several group differences were found; perhaps the most notable pertained to the posterior perisylvian region. When the children in the two groups were divided in terms of whether they showed a larger left than right posterior perisylvian region, a larger right region, or symmetry, group differences were observed that matched those reported by Plante and her colleagues. The control children were more likely to show larger left than right posterior perisylvian areas, whereas the children with SLI were more likely to show a right-larger-than-left pattern. However, when the dependent measure was the actual volume, the differences did not achieve statistical significance.

Another important finding was that the posterior perisylvian regions of the children with SLI showed reduced volume bilaterally, with seemingly larger volume reductions in the left hemisphere (see also M. Cohen, Campbell, & Yaghmai, 1989). This observation is difficult to square with the findings of Plante and her colleagues that the right perisylvian area of children with SLI (and other family members) was atypically large. The precise landmarks used for measurement were different in these studies; however, it is not clear that this factor could have accounted for the discrepancies in the results.

MRIs from the children studied by Jernigan, Hesselink, Sowell, and Tallal (1991) were the source of data in an investigation by Cowell, Jernigan, Denenberg, and Tallal (1994). The corpus callosum (the major nerve-fiber tract connecting the hemispheres) was divided into smaller regions, and the width of each region was measured. No differences were found between the children with SLI and the controls. However, a potentially important group difference emerged when prenatal risk scores for the children were considered. The children with SLI with higher risk scores (due to reports of alcohol use by the mother during pregnancy, high blood pressure, emotional stress, etc.) showed narrower corpus callosum widths. The normally developing controls showed no such relationship between risk scores and corpus callosum measures. Cowell et al. interpreted their findings as an indication that the brains of children at risk for language impairment are especially sensitive to the effects of prenatal environmental factors.

Finally, Trauner, Wulfeck, Tallal, and Hesselink (1995) reported MRI findings in concert with other reports of higher prevalence of neuroanatomical irregularities in children with SLI than in controls. However, of the ten children (out of 34) showing atypical results, six different types of patterns were found, including bilateral ventricular enlargement, right ventricular enlargement, and left ventricular enlargement.

Behavioral Measures of Hemispheric Functioning

Prior to the development of reliable electrophysiological techniques for studying the functional organization of the brain in children with SLI, less direct measures were in use. For example, one measure to assess hemispheric lateralization for language involved the simultaneous presentation of differing digits to the two ears. The subject's task was to report as many of the digits from both ears as possible. The typical result, interpreted as reflecting a left-hemisphere dominance, was the reporting of a larger number of digits that were presented to the right ear than to the left ear. In a study of school-age children with SLI and a group of age controls, Witelson and Rabinovich

(1972) replicated the usual finding for the control children, but found no right-ear advantage for the children with SLI. Similar findings were reported by Sommers and Taylor (1972) and Pettit and Helms (1979), using words as well as digits as stimuli.

There are more recent behavioral procedures to examine lateralization. One of these was employed by H. Cohen, Gelinas, Lassonde, and Geoffroy (1991) to explore the possibility that the speech perception difficulties of children with SLI are attributable to left-hemisphere involvement. Pairs of syllables differing according to either place of articulation or voicing were presented to the children in one ear, and white noise was presented to the other ear. A group of school-age children with SLI showed poorer discrimination of both types of contrasts than did a group of age controls. However, the lateralization findings were similar in the two groups, and were typical of those seen for normally functioning individuals in general. Contrasts of place of articulation were most difficult when presented to the left ear (thus engaging the right hemisphere to a greater extent), whereas contrasts of voicing were more difficult when presented in the right ear (hence, significant left-hemisphere activity).

Neurophysiological Measures

In recent years, methods of measuring the electrical activity of the brain have been refined to the point where such activity can be synchronized to some external event. Specifically, scalp electrodes record the voltage fluctuations produced by large groups of neurons. When repetitions of some stimulus (e.g., a tone) are presented, the voltage fluctuations that are not related to the processing of the stimulus appear random, and those fluctuations that are tied to the stimulus can be extracted. The latter are event-related potentials (ERPs). The peaks of ERPs are defined by their polarity and latency relative to stimulus onset. For example, one well-studied ERP, N400, is especially large in amplitude (as measured in microvolts), negative in polarity (hence, the designation N), and has a latency of 400 ms from the point of the appearance of the target stimulus (see Kutas & Hillyard, 1980). ERPs are often asymmetrical. For example, the N400 in response to closed-class words (e.g., articles, auxiliary *be* forms) is larger in amplitude in the left hemisphere than in the right. Figure 7.3 provides an illustration.

Neville, Coffey, Holcomb, and Tallal (1993) studied the ERPs associated with three different tasks presented to school-age children with SLI and their age controls. During one task, the children monitored a stream of 2,000-Hz tones, listening for instances of a 1,000-Hz tone; upon detection of the latter, the children pressed a button. A second task matched the first except that the stimuli were visual (watching for a small white rectangle presented in a series of larger red squares). The third task required the children to read sentences presented one word at a time. Half of the sentences ended with a semantically appropriate word, the other half ended with a semantically inappropriate word (e.g., "Giraffes have long scissors"). At the end of each sentence, the children pressed one of two buttons to indicate whether or not the sentence made sense. The children with SLI were also administered an auditory processing task developed by Tallal and Piercy (1973a). In this task, sequences of two 75-ms tones were presented at different ISIs and the children were required to press each of two panels in a manner that corresponded to the sequence heard. ERPs were not obtained during this task.

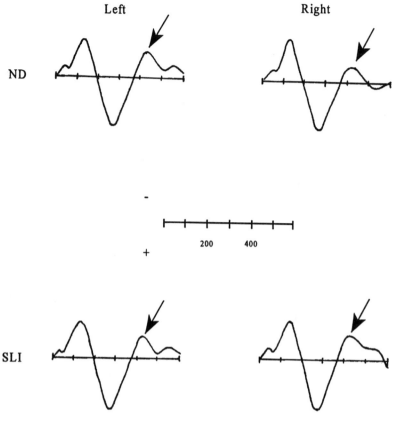

Figure 7.3
The N400 event-related potential in response to closed-class words. In normally developing (ND) indi-
viduals, the amplitude of N400 is larger in the left hemisphere than in the right, as indicated by the arrows.
In children with SLI exhibiting significant deficits in grammatical ability, this asymmetry is not apparent.

During the auditory monitoring tasks, children with SLI did not differ from controls in ERPs. However, when the children with SLI were subdivided according to
their performance on the Tallal and Piercy (1973a) auditory processing task, the
children with the poorer performance showed ERP components that were significantly
reduced in amplitude over anterior regions of the right hemisphere. In addition, one
component, N140, had a longer latency in the same subgroup of children. On the
visual task, the children with SLI as a whole showed lower amplitude for some of the
early ERPs. The children with SLI showed an abnormally large N400 on the sentence
task.

The N400 in response to closed-class words that appeared in the sentence revealed
another type of difference between the children. The typical finding (seen as well for
the children) is a larger (more negative) response over the anterior regions of the left
hemisphere than over the corresponding regions of the right hemisphere. However,
for a subgroup of children with SLI who showed the poorest grammatical skills, this
asymmetry was not seen. Figure 7.3 includes an illustration of this finding. Thus, the
Neville, Coffey, Holcomb, and Tallal (1993) investigation not only produced findings

of differences between normally developing children and children with SLI; it also pointed out that differences among children with SLI in auditory processing and grammatical ability may be linked to neurophysiological differences.

Tomblin, Abbas, Records, and Brenneman (1995) compared school-age children with SLI and age controls on their ERPs in response to a continuous tone that shifted rapidly in frequency from 900 to 1,100 Hz. Only-left hemisphere measures were obtained. The two groups were identical in both amplitude and latency of response. Tomblin et al. did not administer a task such as that used by Tallal and Piercy (1973a), so it is not known if the group with SLI included children with auditory processing problems. Courchesne, Lincoln, Yeung-Courchesne, Elmasian, and Grillon (1989) also found similarities in amplitude and latency between adolescents and adults with SLI and age controls. However, in a study manipulating intensity rather than frequency of tones, Lincoln, Courchesne, Harms, and Allen (1995) found no increase in amplitude of the component N100 with increases in auditory stimulus intensity in a group of school-age children with SLI. Clear increases were seen in a group of age controls. Lincoln et al. took this finding as a possible sign of abnormality in the physiology of the auditory cortex in children with SLI.

Heterogeneity and the Genetic and Neurobiological Data

The evidence reviewed in this chapter gives the strong impression that SLI has a constitutional basis. Children with SLI often are not the first in their families to have language difficulties. The relatively high percentage of atypical neuroanatomical and neurophysiological patterns only reinforces this impression.

The fact that 30% to 60% of children with SLI are the only members of the family with language problems raises the possibility that some children's deficits are not the result of genetic factors. For example, as we saw earlier in this chapter, the traditional distinction between problems in production only and problems in both comprehension and production might prove to be related to whether the language impairment has a genetic basis.

On the other hand, if an assumption of incomplete penetrance is made, genetic predisposition might be a factor in all cases of SLI. The neurobiological findings might be interpreted as consistent with this possibility. For example, larger-than-usual right perisylvian areas have been observed in both affected and unaffected parents and siblings of children with SLI. This configuration might reflect the presence of some genetic factor that can lead to SLI when some other, yet-to-be-identified factor is also operative. Because some children with SLI and some affected family members do not show this configuration, other neurobiological signs reflecting other participating factors will have to be identified before we can be sure that this type of multifactorial account is on the right track.

If some children are genetically predisposed to SLI when other factors are present, what are these other factors? The most obvious candidates—characteristics of the children's linguistic input—are explored in the next chapter.

Notes

1. The criteria used for SLI and the ages studied are critical factors. As we shall see in chapter 9, very different findings emerge from studies in which the probands are two-year-olds with problems limited to language production (Whitehurst, Arnold, Smith, Fischel, Lonigan, & Valdez-Menchaca, 1991). One

important fact surrounding the differences in findings is that many two-year-olds with production delays only show no evidence of language difficulties three years later (Whitehurst, Fischel, Lonigan, Valdez-Menchaca, Arnold, & Smith, 1991).

2. These percentages are based on all family members across three generations. Lewis (1992) also presented corresponding data for immediate family members only, which were approximately twice the above values for both the probands and the controls. However, this particular calculation did not make a distinction between probands with phonological problems only and probands with deficits in phonology and other areas of language. For the extended family calculations, the percentages for the two types of probands were similar, so it seems reasonable to assume that for children with problems in phonology and other areas of language, the percentages would also be considerably higher.

Chapter 8

The Linguistic and Communicative Environment

Casual observation often suggests that children with SLI are quite unremarkable except for their language difficulties. This impression spawned much speculation in the early clinical literature that these children did not enjoy the types of communicative interactions that promoted normal language development. There were two possible problems with the communicative environment. These children might have had little need to talk because other members of the family talked for them. Alternatively, the language that these children heard might have been in some crucial way inadequate.

When research began in earnest, the second possibility was pursued more vigorously, though the first has also received attention. The idea of inadequate input seemed especially plausible, given findings that children living in institutional settings show depressed language levels relative to other abilities. For example, Lamesch (1982) described a French child who was abandoned and placed in an institution. When evaluated some time later, the child displayed appropriate developmental abilities except in the area of language. The child's linguistic environment appeared to be a contributing factor. No person served as the child's primary interactant, and the members of the institution's staff who came into contact with the child devoted their language primarily to the goals of eliciting the child's cooperation during routine activities or discouraging some behavior.

The studies reviewed here vary according to whether the focus is on the input of parents of children with SLI, the input of other adults, or the input of peers. The studies also vary according to whether the comparison dyads include age-matched, MLU-matched, or comprehension score-matched children. The comparison dyad is a critical factor in drawing conclusions from the data. Case studies can provide useful information about the general dynamics of an interaction involving a child with SLI (e.g., Blank, Gessner, & Esposito, 1979), but they do not allow us to interpret the specific developmental appropriateness of the interaction. Of course, "developmental appropriateness" can be defined from different perspectives, and each perspective suggests a different type of comparison dyad. Should parents of children with SLI aim their input at a level commensurate with the children's nonverbal cognitive abilities and world knowledge? Their linguistic abilities? If the latter, should the input be more closely attuned to the children's comprehension abilities or their production abilities?

This line of research also poses difficult questions pertaining to data interpretation. If parents of children with SLI are found to differ from parents of normally developing children in their input, is this the cause or the effect of the child's language impairment? Through the use of some clever research designs, these questions have been addressed.

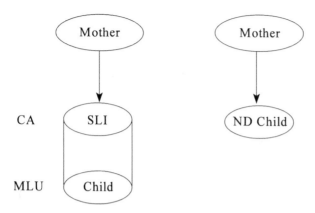

Figure 8.1
Comparison of mothers of children with SLI and mothers of normally developing (ND) children matching the children with SLI in chronological age (CA). Unlike normally developing children, children with SLI show a discrepancy between CA and language ability as reflected in measures such as mean length of utterance (MLU).

The findings of familial aggregation reviewed in chapter 7 complicate the picture further. If parents of children with SLI speak to their children in ways that differ from the speech of parents of typical children, it could be due to their own residual language difficulties. Do we then attribute the children's language problems to the genetic inheritance or to the flawed linguistic input? Fortunately, there are ways to answer this question, too, though only certain studies have the necessary design provisions to do so.

The Communicative Environment of Children with SLI Relative to Age-Matched Controls

Parents as Interactants

The first studies of the communicative environment of children with SLI employed the design illustrated in figure 8.1. Studies of this type examine the language (and other communication) of parents as they interact with their children. The children of one group of parents are developing language normally; the other children exhibit SLI. The two types of children are matched according to age. The figure portrays the fact that the child with SLI shows a gap between age (and various nonlinguistic abilities) and language level (shown in the figure as MLU). Thus, the parents of children with SLI are faced with choosing an input appropriate for the children's age or an input tailored to the children's lower language level. Parents of normally developing children do not have to choose between the two.

Most studies employing this design have reported differences. Wulbert, Inglis, Kriegsmann, and Mills (1975) evaluated the home environments of preschool-age children with SLI and a group of age controls. An inventory was used that included items involving direct observations of the mothers' behavior. The mothers of the children with SLI interacted less with their children than did mothers of the normally developing children, and they were quicker to shout at or threaten their children than to reason with them. However, these mothers also reported that interactions with their children were difficult because the children often rejected their communicative bids.

Siegel, Cunningham, and van der Spuy (1979) studied the interactions between mothers and their three- to five-year-old children. In half the dyads, the children exhibited SLI. During a structured task, the mothers of the children with SLI were more directive, and more likely to interrupt their children's play with a command. There was also evidence that the children with SLI were less responsive than the age-matched children. During play, these children showed less likelihood of responding to their mothers' questions, and if the mother failed to respond to one of their own communicative attempts, they were less likely to reestablish communication. In other respects, the two types of dyads were quite similar. Both groups of mothers showed the same likelihood of initiating interaction, they both facilitated independent play, and they did not differ in the degree to which they responded to their children's attempts to interact.

More similarities than differences were seen in an experiment reported by Cunningham, Siegel, van der Spuy, Clark, and Bow (1985). Mothers of children with SLI resembled mothers of age-matched children in their interactions during free play and structured tasks. However, during the latter activity, the mothers of the children with SLI asked fewer questions. These investigators then conducted two additional experiments. In the second experiment, three types of mother-child dyads participated: (1) mothers of children with SLI showing production deficits but age-appropriate comprehension; (2) mothers of children with SLI exhibiting comprehension as well as production deficits; and (3) mothers of age controls. Cunningham et al. found that the MLUs of the mothers of the children with production problems only were similar to those of the mothers of the control children. In contrast, shorter MLUs were seen in the speech of the mothers of children with both comprehension and production problems. A third experiment was run to identify additional variables of importance. It was found that the greatest discrepancies between the MLUs of mothers of children with SLI and their children's expressive language abilities occurred in dyads in which the child interacted with the mother less frequently, asked fewer questions, and was less responsive to her communicative attempts.

Bondurant, Romeo, and Kretschmer (1983) observed differences between mothers of children with SLI and mothers of age controls on several language measures. The latter group asked more questions, produced a higher number of utterances conveying acceptance, and showed a higher MLU. Laferriere and Cirrin (1984) found no differences in the number of questions asked by mothers of children with SLI and mothers of age controls; however, the mothers of the controls produced a larger number of information-seeking questions, that is, questions for which they did not already know the answer. In a study that included fathers as well as mothers, Stein (1976) also observed greater length and complexity in the utterances produced by parents of age controls.

Eye gaze behavior during parent—child interactions served as the focus of investigation in two studies by Friel-Patti (1976, 1978). In the first study, the mothers of the children with SLI spent greater amounts of time looking at their children than did the mothers of the age controls. However, in the second study, looking time was measured only when mother or child was speaking. On this measure, it was the mothers of the control children who showed a higher degree of eye gaze behavior. In

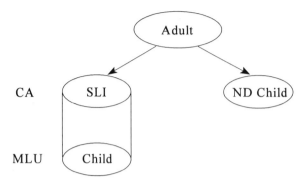

Figure 8.2
Comparison of the input provided by adults to children with SLI and normally developing (ND) children matching the children with SLI in chronological age (CA). Unlike normally developing children, children with SLI show a discrepancy between CA and language ability as reflected in measures such as mean length of utterance (MLU).

both studies, the children with SLI looked less frequently at their mothers than did their age-mates, even while speaking.

Piérart and Harmegnies (1993) examined the prosodic characteristics of mothers of children with SLI. Relative to mothers of a group of age controls, the mothers of children with SLI showed less prosodic variation when speaking to their children. Differences between the mothers were not seen when they interacted with other adults; both groups of mothers showed reduced, but seemingly appropriate, prosodic variation.

Other Adults as Interactants
When parents of children with SLI interact in a manner that differs from the interaction patterns of parents with typical children of the same age, it is difficult to determine who is influencing whom. Parents may interact in the manner they do as a natural response to their children's limited linguistic and communicative repertoire. One way of determining the feasibility of this explanation is to study how adults other than the parents speak to children with SLI. By having the same adults speak to age controls as well, differences in the input might be seen that can be linked to the language abilities of the children. This design is shown in figure 8.2.

This strategy was adopted by Bruck and Ruckenstein (1978) and Fried-Oken (1981) in studies of children with SLI of preschool and kindergarten age. Teachers were the adult interactants in these investigations. Bruck and Ruckenstein noted that when speaking to children with SLI, teachers used fewer clauses per utterance, made more requests for information, repeated more utterances, and asked more naming questions than when they spoke to normally developing children of the same age. Fried-Oken found that questions designed to obtain one-word responses were asked more frequently of children with SLI, whereas requests for explanations were more frequently directed toward same-age control children. In a study involving speech-language pathologists as interactants, Nettelbladt and Hansson (1993) found that requests for clarification were more likely to be issued to children with SLI than to age controls.

The above findings indicate that parents are not alone in behaving differently toward children with SLI. The fact that teachers and clinicians do likewise suggests that these interaction patterns are not the principal cause of the children's language difficulties. Indeed, they are probably a natural consequence of these difficulties. But how much of this is a direct response to obvious signs of language problems demonstrated by the child as opposed to the adults' preconceived notions about what children with language problems require in the way of input?

Newhoff (1977) conducted a study that could address this question. Preschool-age children with SLI and age controls possessing advanced linguistic skills interacted with women they did not know. Each woman was the mother of a young child. The use of age controls with above-average linguistic ability was designed to magnify the differences in the language ability of the two groups of children. The women were found to produce a smaller percentage of complex sentences (those with subordinate clauses) when speaking with the children with SLI.

M. Robinson (1977) used the Newhoff data to examine the women's use of questions when speaking to the two types of children. Differences were found for only one infrequently used question type. The findings of Newhoff and Robinson hint that the differences in input provided to children with SLI and age controls are probably rather small during initial interactions, and widen as adults become more familiar with the children's language abilities.

Peers as Interactants

Even preschoolers seem to be aware that children with SLI communicate in ways that differ from most other children the same age. Hadley and Rice (1991) and Rice, Sell, and Hadley (1991) studied children with SLI and same-age normally developing peers in a preschool setting. The normally developing children were more likely to direct their interactions toward their normally developing peers than toward the children with SLI. In addition, conversational attempts by children with SLI were more likely to be ignored by peers than were the attempts of normally developing children.

The Communicative Environment of Children with SLI Relative to Children Matched on Measures of Language

Even if the speech directed toward children with SLI is a direct and natural response to these children's apparent failure to converse in a typical manner, we cannot easily conclude that this speech plays no role in the children's problems with language. It is easy to see how a vicious cycle of nonoptimal interactions could be set in motion once the adult or peer perceives the child's limitations in language. The modifications made by the adult or peer might be an overcompensation or in some other way abnormal, which might in turn aggravate the child's language learning difficulties.

This argument would be harder to make if the product of these modifications resembled the way in which adults spoke to younger normally developing children with language skills roughly comparable with those of the children with SLI. It is well known that adults speak to younger children in a way that differs in many respects from their speech to older children. It is likely that this tuning according to the child's age actually benefits rather than impedes the child by providing a more accessible (to comprehension) and attainable (in production) model. Several studies of this type have been conducted.

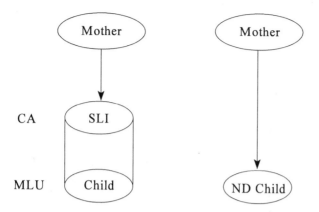

Figure 8.3
Comparison of mothers of children with SLI and mothers of normally developing (ND) children matching the children with SLI in mean length of utterance (MLU). Unlike normally developing children, children with SLI show a discrepancy between chronological age (CA) and language ability as reflected in measures such as MLU.

Parents as Interactants

In most studies of this type, the design shown in figure 8.3 is used. As in the design appearing in figure 8.1, the language of mothers of children with SLI and of mothers of normally developing children is compared. However, in this case, the children are matched according to a language measure such as MLU rather than age. In these studies, the children with SLI have been of preschool age, with MLUs ranging from well below 2.0 to almost 4.5 morphemes. The children serving as MLU controls, of course, have been considerably younger.

The results of these investigations have been mixed. However, one consistent finding has emerged. Mothers of children with SLI appear to make less use of recasts than do mothers of MLU controls (Conti-Ramsden, 1990; Nelson; Welsh, Camarata, Butkovsky, & Camarata, 1995). Recasts are responses to the child's utterances that not only render the utterance grammatical but also convert it to a particular morpho-syntactic construction. For example, the adult's utterances in (1) are recasts in the form of *wh-* questions.

(1) *Child:* Get ice cream.
     Adult: Where can we get some ice cream?
     Child: Kitchen. But now it all gone.
     Adult: Why is it all gone?

The line between a recast and an expansion of the child's utterance is sometimes blurred. For example, recasts designed to add auxiliary verbs, as in (2), expand the child's utterances by necessity.

(2) *Child:* Jenny not going.
     Adult: Jenny's not going?

A compelling aspect of the findings for recasts is that the same studies reporting differences for this measure also report that the two groups of mothers were similar in other types of semantic contingency, such as continuations of the topic reflected in the child's prior utterance (Conti-Ramsden, 1990; Nelson, Welsh, Camarata, Butkovsky, & Camarata, 1995). An example of a continuation appears in (3).

(3) *Child:*   Sammy go eat dinner.
     Adult: He's really hungry.

This suggests that both groups of mothers were responsive to what the children had to say; they differed in the frequency with which their semantically related utterances represented a structural enhancement of the child's own utterances.

Other differences reported in these studies have included mothers of control children producing fewer unintelligible utterances (Millet & Newhoff, 1978) and mothers of controls acknowledging their children's utterances more frequently (Conti-Ramsden & Friel-Patti, 1983). Moseley (1990) found just the opposite for acknowledgments; these were more frequent in the speech of mothers of children with SLI. In a study in which children were matched for vocabulary comprehension test scores, Johnson and Sutter (1985) observed that mothers of children with SLI were more likely than mothers of control children to refer to time events by using temporal

In other respects, the mothers appear to be similar in their interactions. They do not appear to differ in their use of self-repetitions, imitations, or semantic extensions (Macpherson & Weber-Olson, 1980); the number of conversational turns and communicative acts, and responsiveness to topic changes (Messick & Prelock, 1981); their use of requests, directives, assertions, and regulating devices (Conti-Ramsden & Friel-Patti, 1983); the form and level of their initiations, and the adequacy of their responses to their children's questions and comments (Conti-Ramsden & Friel-Patti, 1984); their tendency to assist children in answering questions asked by others (Hill & Clark, 1984); their use of their own names when referring to themselves (e.g., *Mummy needs to get up now*) (Conti-Ramsden, 1989); and their use of requests for action, information, or clarification (Moseley, 1990). Father–child dyads were examined in a study by Silverman and Newhoff (1979), using the set of measures employed by Millet and Newhoff (1978). In this case, no significant differences were found. Lasky and Klopp (1982) also failed to find differences in a study in which children with SLI and normally developing children were matched on language test scores. The mothers of the children in the two groups were similar in their use of self-repetitions, expansions, and acknowledgments, among other measures.

The mixed results can be explained in part by the fact that so many different measures have been employed. Unfortunately, though, there are a few contradictions. It can be noted, with some dismay, that acknowledgments have produced group differences in both directions, as well as no difference.

Studies by Hoffer and Bliss (1990), Schodorf and Edwards (1983), and Cross and her colleagues (Cross, 1981; Cross, Nienhuys, & Kirkman, 1985) have reported the greatest number of differences. Hoffer and Bliss found that mothers of children with SLI were generally less responsive to their children's utterances, and were more likely to ignore or shift the topic of these utterances than were mothers of MLU controls. Schodorf and Edwards examined both mother–child and father–child dyads. Mothers and fathers differed very little; the main differences occurred between the parents of the children with SLI and the parents of the MLU control children. They found that

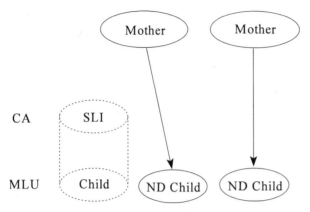

Figure 8.4
Comparison of mothers of children with SLI when interacting with their normally developing (ND) children matching the children with SLI in mean length of utterance (MLU), and mothers of (only) normally developing children matching the children with SLI in MLU. Unlike normally developing children, children with SLI show a discrepancy between chronological age (CA) and language ability as reflected in measures such as MLU.

the parents of the children with SLI used fewer total words, expansions, models, verbal routines, intelligible utterances, and grammatically complete sentences.

In the investigation of Cross (1981) and her colleagues, mothers of children with SLI and mothers of MLU controls differed on 27 of the 70 measures examined. The differences concerned discourse features, syntax, and amount of speech. For example, the mothers of the control children produced a greater number of partial and complete expansions of their children's utterances, asked a larger number of wh- questions, spoke more slowly, and produced fewer unintelligible utterances than did the mothers of the children with SLI.

Cross, Nienhuys, and Kirkman (1985) then made the deft move of examining the speech of mothers of children with SLI who also had younger normally developing children matching the MLU control children of their first study in age and language skills. The speech of these mothers when speaking with the younger siblings was then compared with the data obtained from the mothers of the MLU controls. This design is illustrated in figure 8.4. Only 11 of the 70 measures proved statistically significant, most of them concerning the amount and intelligibility of the mothers' speech. Thus, as Cross and her colleagues noted, some features of mothers' speech seemed to depend more on the communicative proficiency of the children than on characteristics inherent in the mothers. On the other hand, their investigations may have uncovered several fundamental differences between mothers of children with SLI and mothers of more typical children.

Conti-Ramsden and Dykins (1991) continued the line of research initiated by Cross and her colleagues, this time adding within-speaker comparisons. Families were recruited in which children with SLI had younger siblings showing the same MLU. Interactions between the mother and each child were then examined and compared. A group of mothers with normally developing children also participated to provide an additional basis of comparison. The MLUs of their children matched those of the children with SLI. An illustration of this design is provided in figure 8.5. On most

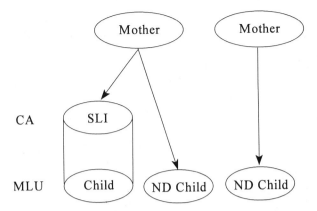

Figure 8.5
(1) Comparison of mothers of children with SLI when interacting with their children with SLI and their normally developing (ND) children matching the children with SLI in mean length of utterance (MLU); and (2) comparison of these same mothers and mothers of (only) normally developing children matching the children with SLI in MLU. Unlike normally developing children, children with SLI show a discrepancy between chronological age (CA) and language ability as reflected in measures such as MLU.

conversational measures, there was greater similarity between mothers speaking to their children with SLI and the same mothers speaking to their younger children than between these mothers (in either interaction) and the control mothers. Though this finding might suggest that mothers of children with SLI can differ more generally from other mothers in their interactive patterns, the fact that the younger siblings were developing language normally suggests that these patterns cannot be assumed to be a detriment to language learning.

In a few respects, the mothers of the children with SLI did not interact in the same way with their two children. Most notably, they used a larger number of utterances per conversational turn when speaking to their younger normally developing children. Conti-Ramsden and Dykins (1991) pointed out that these differences might have been due to differences in the two children. For example, the mothers might have produced fewer utterances per turn when interacting with their children with SLI because these children exhibited poorer attention. Alternatively, these children might have been less intelligible, making it difficult for the mothers to elaborate on the children's utterances.

More recently, Conti-Ramsden, Hutcheson, and Grove (1995) reported an additional difference in the input of the children. Mothers were more likely to provide recasts of the spontaneous utterances of the younger siblings of the children with SLI and the other MLU-matched controls than of the spontaneous utterances of the children with SLI. When recasts were provided to children with SLI, they most often followed the child's response to a question. According to Conti-Ramsden et al., this difference could be important because (in agreement with Nelson, Welsh, Camarata, Butkovsky, & Camarata, 1995) if children hear recasts of utterances of their own choosing, they are more likely to benefit from the additional information contained in them. It is interesting to note that in the Conti-Ramsden et al. study, data from fathers were added to the analysis and the same results emerged. The findings from this study add to those of Conti-Ramsden (1990) and Nelson, Welsh, Camarata, Butkovsky, and

Camarata (1995) in suggesting that recasts figure importantly in distinguishing the linguistic environments of children with SLI and younger normally developing children.

There is one important factor left uncontrolled in studies that employ MLU (or some other language-based) matching only. The children with SLI in these studies display nonlinguistic abilities that exceed those of the younger MLU matches. The design employed by Cunningham, Siegel, van der Spuy, Clark, and Bow (1985), discussed earlier, addressed this question by recruiting children with SLI who differed in language comprehension abilities. We saw that the production abilities of the children seemed to exert a larger effect on the mothers' input. However, there remains the question of whether parents' adjusting their speech to their children's language level is at the expense of the children's more advanced conceptual abilities.

Grimm (1991) approached this issue by examining mothers' speech in terms of conceptual complexity. For example, utterances with a naming function were scored as low in complexity; utterances that expressed a causal relationship were scored higher. Comparisons were made between mothers of children with SLI, mothers of age controls, and mothers of MLU controls in interaction with their children. The utterances of the mothers of children with SLI were more similar in conceptual complexity to those produced by the mothers of the MLU controls. This was not simply a necessary consequence of the reduced length and structural complexity of the mothers' speech; the mothers of the MLU controls produced utterances that were slightly in advance of their children's utterances in conceptual complexity, whereas the utterances of the mothers of children with SLI were on the same conceptual footing as those of their children.

The joint influence of mother and child was captured in a study of maternal book reading by Evans and Schmidt (1991). One mother and her child with SLI and one mother and her MLU control participated. The mother of the child with SLI asked more questions than the other mother, and her child with SLI was more likely to err or ignore the question than was the MLU control. The mother–control child dyad showed greater synchrony, with both mother and child making more statements and acknowledging comments of their communicative partner. There was no evidence that the mother of the child with SLI was misgauging her child's language learning style or ability. For example, the mothers were equal in their ability to predict their child's expressive vocabulary.

The joint contributions of mother and child were also examined in a longitudinal study by Grimm (1984, 1986, 1987). At the outset of the study, mothers of pre-school-age children with SLI were similar to mothers of MLU controls in their use of imitations, corrections, and recasts. However, one year later, the mothers of the children with SLI produced fewer recasts than did mothers of controls when their children were at a comparable point in grammatical development. Grimm proposed that this reduction in recasts relative to controls might have been related to the fact that the children with SLI were less likely than controls to respond to their mothers' prior utterance with a structurally related utterance of their own.

Horsborough, Cross, and Ball (1985) explored the role of the child in mothers' interactive behaviors by comparing the speech of mothers of children with autism, mothers of children with SLI, and mothers of normally developing children. The three groups of children were similar in MLU and language comprehension test scores.

However, an earlier study had found marked differences between the two clinical groups of children in their use of pragmatic functions, favoring the children with SLI (Ball, Cross, & Horsborough, 1981). The mothers of the normally developing children and mothers of the children with SLI differed on 14 of 72 measures. For example, the mothers of the children with SLI used fewer idioms, repeated part of their own prior utterance more often, and imitated the child's utterance more frequently. Somewhat surprisingly, the mothers of the children with autism closely resembled the mothers of the children with SLI. Differences were seen on only four measures, about the same number that would be expected by chance, given the number of comparisons involved. According to Cross, Nienhuys, and Kirkman (1985), these similarities make sense only if one assumes that the mothers' speech was influenced more by factors such as the children's utterance length than by the children's communicative and social skills.

Peterson and Sherrod (1982) compared the speech of mothers of children with SLI with that of mothers of children with Down syndrome and mothers of younger normally developing children. The children were matched according to MLU. Differences among the mothers were seen for 6 of the 23 measures employed. The mothers of the children with SLI were found to use fewer semantically related utterances and less verbal approval than the mothers of the normally developing child~~~~. ~~~~~~~ ~~ were intermediate and did not differ from the values for either of the other two groups of mothers.

Other Adults as Interactants
Nakamura and Newhoff (1982) examined the type of input that speech-language pathologists provide to children with SLI and MLU controls—the design represented in figure 8.6. No differences were seen for the six main adult-language measures employed. One of these, semantically related responses, was analyzed in greater detail. The speech-language pathologists used more expansions with the children with SLI and more imitations with the younger control children. Of course, speech-language pathologists have greater familiarity with language impairments than do

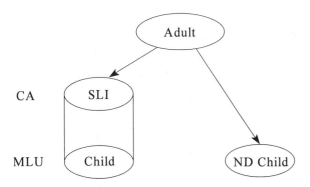

Figure 8.6
Comparison of adults interacting with children with SLI and normally developing (ND) children matching the children with SLI in mean length of utterance (MLU). Unlike normally developing children, children with SLI show a discrepancy between chronological age (CA) and language ability as reflected in measures such as MLU.

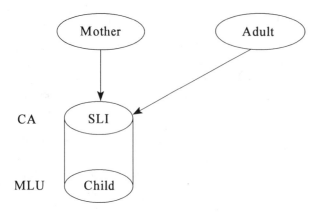

Figure 8.7
Comparison of mothers of children with SLI and other adults interacting with the same children. Unlike normally developing children, children with SLI show a discrepancy between chronological age (CA) and language ability as reflected in measures such as MLU.

most adults, so it is possible that the findings are not representative of the speech directed toward children with SLI by adults other than their parents.

A different experimental approach was taken by van Kleeck and Carpenter (1978) to explore adults' input to children with SLI. Four preschool-age children with SLI participated. All of the children's speech was limited to single-word utterances. However, two of the children possessed language comprehension abilities that approached age level, whereas the other two showed poor comprehension ability. Each of the adults recruited for the study interacted with one child from each comprehension group. When interacting with the low-comprehension children, the adults restricted their lexical diversity and relied more on nonverbal cues. However, on other measures, such as self-repetition and requests for confirmation, there were no differences. The investigators concluded that variables in addition to language comprehension level probably influence adults' interaction patterns.

Different Adults Speaking to the Same Child with SLI
A nice complement to the research designs already discussed is a design in which comparisons are made between mothers and other adults speaking to the same children with SLI. This design is shown in figure 8.7. In the studies of this type that have been conducted, speech-language pathologists have served as the other adults, and thus the proviso noted above concerning familiarity with language impairment must be borne in mind. Giattino, Pollack, and Silliman (1978) examined the use of requests aimed at children with SLI by these two types of adult speakers. No differences were seen in the overall percentages of requests produced. However, the mothers tended to request actions most often, whereas speech-language pathologists were more apt to request verbal information. Olswang and Carpenter (1978) also employed this type of design, though children with cognitive deficits and their mothers participated along with mothers and their children with SLI. Mothers and speech-language pathologists were similar on most of the measures selected for study. However, questions about materials in the environment were more frequent in

the mothers' speech, whereas comments about these materials were more likely to be produced by the speech-language pathologists.

Peers and Siblings as Interactants

In two studies by Newhoff and her colleagues, the input provided by normally developing five-year-old children served as the focus (MacKenzie, Newhoff, & Marinkovich, 1981; Marinkovich, Newhoff, & MacKenzie, 1980). These children interacted with age-matched children with SLI, normally developing children of the same age, and a group of younger normally developing children whose MLUs approximated those of the children with SLI. The five-year-olds' speech was more sophisticated when speaking to their normally developing age-mates than when interacting with the other two groups. For example, they were more likely to issue later-developing requests when talking to the age controls.

If children with SLI are perceived as being less adept in language, simplified input might not be the only result. Wellen and Broen (1982) explored the question of whether children with SLI were more likely to be interrupted by their siblings than were other children. Younger siblings of preschool age who were either normally developing or exhibited SLI were read stories and asked questions with their older, normally developing siblings present. The older siblings' interruptions were noted. Children with SLI were interrupted as frequently as a group of normally developing children who were more than one year younger. Furthermore, the nature of the interruptions differed. Interruptions of the younger siblings with normal language development were typically rephrasings of the questions or prompts. Interruptions of the children with SLI were usually the answers to the questions. This difference might have been related to the fact that the younger normally developing siblings displayed language abilities exceeding those of the children with SLI despite their younger age.

Adult Input as a Predictor of Language Use

Another way to evaluate the input received by children with SLI is to determine whether the characteristics of adult input that predict later language use by normally developing children also serve as predictors for the language of children with SLI. One of the best candidates for this kind of analysis is the relationship between information-seeking questions (e.g., "Do you like this?" "What do you want to play?") by the parent and later use of auxiliaries by the child. The connection between the two seems to be attributable to the fact that auxiliaries frequently appear in such questions. Hoff-Ginsberg (1986) found that the use of such questions by mothers of normally developing 26-month-olds predicted the auxiliary use of their children four months later. Yoder (1989) observed the same result in a study of mothers of preschool-age children with SLI; the mothers' use of information-seeking questions predicted their children's use of auxiliaries when the children were 12 months older.

Adult Input as a Determinant of SLI

A great deal of the evidence on the linguistic environment of children with SLI can be interpreted as reflecting the natural consequence of speakers adjusting their speech to accommodate the needs of a less proficient interlocutor. This is most clearly true for studies comparing the input provided to children with SLI and age controls. Parents, other adults, and peers seem to make the expected adjustments.

The evidence from studies employing MLU controls is less clear-cut. Fewer differences in input are seen in these investigations. Some of the differences that are observed can probably be interpreted as within-family differences that have no bearing on how well a child will acquire language. That is, younger siblings seem to be receiving much the same input, and they are progressing adequately in language development. A few of the remaining differences could be related directly to the poor language abilities of the children with SLI. In earlier chapters, we saw that children with SLI often lag behind MLU controls in semantic and especially morphosyntactic abilities. These same differences should hold between children with SLI and MLU controls who participate in studies that focus on the children's coconversationalists. If differences in the adult interlocutors' input are seen, they may be a natural result of the differences between the children.

The most reliable difference between the speech directed toward children with SLI and MLU controls is the use of recasts. Of the differences observed, the lower frequency of recasts by parents of children with SLI seems to be the least likely to be attributable to the children themselves. If, as just noted, children with SLI are weaker than MLU controls in morphosyntactic abilities, parents of these children would be expected to make morphosyntactic modifications more, rather than less, frequently. An additional reason to place importance on the findings for recasts will be seen in chapter 10; increasing the frequency of recasts appears to facilitate the grammatical learning of children with SLI.

Converging evidence for the effects of recasting could be obtained through the use of a design in which the input of mothers of children with SLI is compared with the input of mothers of age controls and mothers of MLU controls. Among mothers of typically developing children, the use of recasting declines with the child's age, presumably because older children's language provides evidence of semantic and morphosyntactic structures not seen in younger children. Nelson, Welsh, Camarata, Butkovsky, and Camarata (1995) have proposed that mothers of children with SLI may calibrate their responses to the children's age or nonlinguistic cognitive abilities rather than to their weaker language abilities. If this is true, differences might be seen between mothers of children with SLI and mothers of MLU controls (as have already been reported), but not between mothers of children with SLI and mothers of children matched according to age.

There are many details of these children's input that remain unexplored. In the period in which most studies of the linguistic environment of children with SLI were conducted, studies of the speech directed toward young normally developing children were still focused on ways in which the input varies as a function of age. Learnability problems and possible solutions to these problems based on notions of semantic, morphosyntactic, and prosodic bootstrapping had not yet been discussed. Consequently, it is not known whether the input provided to children with SLI offers the same cues for detecting clause boundaries, hypothesizing grammatical categories, and narrowing the range of possible meanings of a new word. Although there is little in the literature to suggest that these cues are less frequent or reliable in these children's input, this work remains to be done.

PART IV

Clinical Issues

Chapter 9
Problems of Differential Diagnosis

In this chapter, we single out three clinical issues that represent thorny problems for professionals serving children with SLI: (1) how SLI might be recognized in children at an early age; (2) the degree to which the nonverbal IQ criterion for SLI should influence clinical decisions; and (3) whether deficits in reading should be considered part of the clinical diagnosis of SLI. These are not the only problems for which solutions are needed, but they are among the most important. Decisions about language treatment, school placement, and provision of assistance in areas that go beyond language depend on how they are handled.

The Early Identification of Children with SLI

There is an old joke about a mute child who, one day at the dinner table, blurts out, "This food is terrible!" The parents, astonished that the child talks, ask why he hadn't spoken before. "Everything was fine until now," is his response. The joke is amusing, at least the first time it is heard, but it hits a little close to home if one is grappling with the problem of identifying SLI at a young age.

Identifying SLI at an early age would be a lot easier if all normal children began to speak at about the same time. Unfortunately, some children destined to be typical language users are surprisingly late in reaching the early milestones of first words and first word combinations. There is nothing obvious that distinguishes these late talkers from other typical children. They seem no more likely to be gifted in areas outside of language (a few famous cases notwithstanding), and there is no evidence that they reach an advanced age before they are served their first bad meal.

The option of waiting until a child reaches three or four years of age before seeking assistance seems too risky. In chapter 1, we saw that high percentages of preschoolers with SLI continue to experience problems through childhood and beyond. If these children could be identified even earlier and their language development facilitated, their outcomes might be considerably better.

The clinical task of early identification of SLI received a significant boost in the late 1980s when several independent research teams began to examine this question. This work was aided by the development of two psychometrically strong yet highly practical vocabulary checklists, the MacArthur Communicative Development Inventories (Fenson, Dale, Reznick, Thal, Bates, Hartung, Pethick, & Reilly, 1993) and the Language Development Survey (Rescorla, 1989). The checklists contain words that have been documented in children's speech at early ages; parents respond by indicating which of the words their child comprehends and/or produces.

The Outcomes and Characteristics of Late Talkers

Most studies on the early identification of SLI have been longitudinal in nature. Children who at age two years show slow development of language based on formal tests and vocabulary checklists are then followed through ages three years and beyond. The criteria used to identify these "late talkers" (the name given these children at age two) have varied somewhat from investigator to investigator. Rescorla (1989) regarded a child as a late talker if at age two the child produced fewer than 50 words or had no two-word combinations. Thal and Bates (1988) extended the late talker category down to 18 months if the child scored below the tenth percentile on the MacArthur Communicative Development Inventories and produced no two-word combinations. Paul (1991a, 1991b) treated children in the 18- to 23-month range as late talkers if they had fewer than ten intelligible words. More recently, Thal, Oroz, and McCaw (1995) used their original (tenth percentile) measure as their initial boundary for late talkers and then subdivided children into those who had ten or more words and those who had fewer than ten words. Several studies examined children whose delays were limited to language production; others looked at children with and without accompanying problems in comprehension.

The outcomes for two-year-olds with delays in production only seem to be more favorable. Thal and her colleagues (Thal & Bates, 1988; Thal & Tobias, 1992; Thal, Tobias, & Morrison, 1991) found that six of ten two-year-old late talkers performed within normal limits one year later. All six had age-appropriate comprehension abilities at the younger age. The four children who remained below age level had comprehension as well as production difficulties. The delays in comprehension were accompanied by limitations in the use of recognitory gestures (e.g., pretending to drink from an empty cup). The children who caught up to peers in language development had made active use of gestures to assist communication. Quite clearly, the use of gestures instead of words and phrases by this group did not stunt their language growth by reducing their incentive to verbalize. In a subsequent study, Thal and Tobias (1994) found that two-year-old expressively delayed children's use of single gestures and gesture sequences (e.g., pretending to pour juice and then pretending to give a drink to a teddy bear) was similar to that of age controls and more advanced than that of younger children matched for expressive vocabulary.

Nevertheless, as a group, children whose language problems center on production are at risk. Rescorla and Schwartz (1990) found that out of 25 two-year-olds with expressive delays, one year later 52% showed MLUs that were at least 1.5 SDs below the mean, and 64% showed scores on grammatical measures that were just as low. Rescorla and Goosens (1992) later reported that the symbolic play behaviors of the children with expressive delays at age two were less developed than those of age controls. Similarly, the children with delays showed more limited phonological abilities (Rescorla & Bernstein Ratner, 1996). The most notable difference was seen for syllable shapes; the children with expressive delays were much more likely to rely on open syllables (e.g., syllables consisting of a single vowel or a consonant–vowel sequence). A study of mother–child interactions when the children were 24 months of age did not reveal clear differences between the mother-child pairs (Rescorla & Fechnay, 1996). Mothers of the children with expressive delays resembled mothers of control children in degree of synchrony (e.g., maintaining or elaborating the partner's topic) and use of social cues (e.g., pointing or nodding while speaking). Similarly, the children with delays interacted like the control children with the (expected) exception of producing fewer intelligible utterances and relying more heavily on gestural communication.

It is not known whether the children who made the greatest gains in language by age three differed from the other children with delays in some subtle way, in any of these abilities, at age 24 months. Stoel-Gammon (1989) studied two children with expressive delays who, like the Rescorla and Bernstein Ratner (1996) subjects, showed a reduced number of syllable shapes. Both of these children approached normal limits in language ability shortly after 24 months.

Manhardt, Hansen, and Rescorla (1995) tested the same children on narrative tasks at ages six, seven, and eight years. At ages six and eight, these children performed below the level of a group of age controls on several of the measures employed, including number of story grammar components expressed, clear character references, and plot-advancing events. These differences are notable in part because no distinction was made between those children who had made significant gains in grammar by age three and those who had not.

Paul (1991a, 1991b, 1993) traced the development of 37 children with language production delays and found that at age three nearly 60% scored below the tenth percentile on tests of phonology and/or syntax. At age four, this figure ranged from 47% to 57%, depending on the measure used (see Paul & Smith, 1993). The children with persisting problems in syntax at age four also showed limitations in narrative skill and exhibited limited lexical diversity (Paul & Smith, 1993). The limitations in narrative ability were more subtle by first grade; by second grade, the children resembled normally developing children on the narrative measures employed (Paul, Hernandez, Herron, & Johnson, 1995). Factors accompanying the children's earlier delays at two years included poorer socialization skills (Paul, Spangle-Looney, & Dahm, 1991), reduced tendency to establish or maintain joint attention (Paul & Shiffer, 1991), and a limited repertoire of syllable structures (Paul & Jennings, 1992). Although the children were initially selected on the basis of their expressive delays, as a group their language comprehension scores were slightly lower than those of a group of age controls. However, these differences evaporated by age three (Paul, Spangle-Looney, & Dahm, 1991). The nature of the mothers' interactions with the children at age two did not appear atypical, considering the children's limited production skills (Paul & Elwood, 1991).

Fischel, Whitehurst, Caulfield, and DeBaryshe (1989) found that after five months, 39% of their 26 subjects with expressive delays showed no improvement in their standard scores on a test of expressive vocabulary, and 26% showed small improvement without reaching normal limits. The best predictors of progress were a relatively small degree of nonlinguistic vocalizations and a relatively high degree of word usage (Whitehurst, Smith, Fischel, Arnold, & Lonigan, 1991).

Whitehurst, Fischel, Arnold, and Lonigan (1992) studied changes in a group of children with expressive delays from age 2;4 to 2;10 on two tests of expressive vocabulary. At age 2;10, 59% of the children had not yet reached the normal range on one of the measures and 43% were still below age level on the other. At age 3;8, only 12% and 16% of the children showed problems on the two tests. These values dropped to below 10% for each test when the children were assessed at age 5;5. Whitehurst, Fischel, Lonigan, Valdez-Menchaca, DeBaryshe, and Caulfield (1988) found that parents of children with early expressive delays appear to provide their children with input commensurate with that heard by younger normally developing children with comparable production skills, suggesting that the children's limited output affects parents' input rather than vice versa.

Given the fact that late talkers identified at around two years of age vary in their outcomes, it might be the case that their development prior to age two provides useful prognostic information. Ellis Weismer, Murray-Branch, and Miller (1994) had a rare opportunity to view this pattern of development. These investigators were following a group of normally developing children in a longitudinal study, beginning when the children were 13 to 14 months of age. (In fact, these children were serving as controls in a longitudinal study of children with Down syndrome.) By around 19 months of age, 4 of the 23 children were clearly behind the others in the number of words used; this gap remained at 25 months of age. By three years of age, one of the late talkers scored well within the normal range in vocabulary and MLU, and two others showed borderline MLU but age-appropriate vocabulary. The fourth child remained significantly below age level in both areas. During the course of the study, individual children scored below age level on select measures (e.g., language comprehension, symbolic play); however, these deficits were not stable across observations and did not predict the children's outcomes. Family history appeared to be the only factor with predictive value; two of the four late talkers—the child who remained low and one of the two who attained only borderline skills—had immediate family members with a history of language problems.

However, the evidence of familial concentration for late talkers seems to hold primarily for those children with comprehension as well as production limitations. Whitehurst, Arnold, Smith, Fischel, Lonigan, and Valdez-Menchaca (1991) found that children with early expressive delays were no different from control children in the number of family members with a history of language problems. Paul and Unkefer (1995) reported similar results for their late talkers with production delays only. However, for the children with delays in comprehension as well as production, the rate of family members with a history of language problems was approximately five times higher.

Getting a Late Start: The Neurophysiological Correlates

Most late talkers—even those with poor outcomes— begin to speak considerably before they reach three years of age. Nevertheless, recent findings suggest that language processing at later-than-average ages is associated with configurations of brain activity that are unlike those seen during language processing at more typical ages. Mills, Thal, Di Iulio, Castaneda, Coffey-Corina, and Neville (1995) examined the event-related potentials (ERPs) of late talkers age 28 to 30 months and a group of age controls. The ERPs were recorded while the children listened to words that they already comprehended. In this type of task, ERPs show a positive deflection at 100 ms, referred to as P100 (see chapter 7). The control children showed the expected left-greater-than-right hemisphere asymmetry for the P100 amplitude. The late talkers did not show such an asymmetry. Most of the late talkers were first identified at approximately 24 months of age, and a few of these children had reached normal limits in language ability by the time they participated in the ERP study. Mills et al. found that these particular children's ERP data conformed to the asymmetrical pattern seen for the control children. Interestingly, these investigators had ERP data from one of these children at 24 months of age, and discovered that the child showed the left-greater-than-right asymmetry at the younger age as well.

Late Talkers Are at Risk for SLI

Late talkers vary in their outcomes to the point where we cannot be certain if any given child will exhibit SLI by age three or four. However, as a group, these children are at much greater risk than children who hit the early milestones of language development on schedule. Most of the evidence suggests that from one-fourth to one-half of late talkers at age two are at risk for SLI by school age. For the children with the poorest outcomes, vocabulary skills often approximate typical levels by age three or four, but syntactic and narrative skills reach marginal levels at best. Age-appropriate language comprehension and symbolic play abilities at age two may be taken to be good prognostic signs, but good outcomes have also been observed for some children who seemed delayed in these areas as well at two years.

The work of Whitehurst and his colleagues seems to be at variance with this general picture. The outcomes of the children they studied were considerably better. The fact that they selected late talkers with age-level comprehension skills does not seem to be the critical factor, as children with similar profiles in other investigations are reported to have had poorer outcomes. One factor that might explain the discrepancies is that Whitehurst and his research team focused on vocabulary abilities as the outcome measure. Judging from other studies, vocabulary seems to be a relative strength in late talkers.

A factor that might reduce the precision of almost all of the studies of late talkers is the choice of intelligence test used to document that these children do not have more general developmental deficits. For this young age, the most frequently used tests are the Bayley Scales of Infant Development (Bayley, 1969) and the Stanford-Binet Intelligence Scale (Terman & Merrill, 1960). Both of these tests include a large number of verbal items for the developmental period of two years (see Paul, Spangle-Looney, & Dahm, 1991). Thus, although the children who qualify as subjects in these studies are late in talking, they are probably not among the most seriously language delayed. The latter are probably excluded from consideration because their poor language skills would prevent them from providing correct responses on the verbal items of these tests.

It is also worth asking whether we are barking up the wrong tree. An assumption behind this work is that children clearly meeting the criteria for SLI were in fact slow in acquiring their early lexicon. In principle, it is possible that children with significant deficits in language ability at, say, age five years reached the early language milestones on schedule but then stopped developing at a normal rate. If this were true, even the most careful study of late talkers would not allow us to identify those who will be diagnosed with SLI in a few years' time. However, the pattern of development just described does not, in fact, portray the development of children with SLI. From all indications, children with SLI acquire not only their first words but also their first word combinations at significantly later ages than normally developing children (Trauner, Wulfeck, Tallal, & Hesselink, 1995).

Even if fully half of all late-talking two-year-olds have significant language problems when they reach school age, the fact that the remaining half do not indicates that a diagnosis of SLI before age three is not currently feasible. Until we know more, late talkers can only be regarded as children at risk for a language disorder. This does not mean that clinical and other professional services are inappropriate; on the contrary, they might lead to an acceleration of the child's language development (as we shall see in the next chapter). However, the variability in the rate of language development among to-be-normal children and the lack of consistency in the factors associated with good and poor outcomes have thus far precluded the establishment of a basis for differential diagnosis.

Heterogeneity and the Nonverbal IQ Criterion

In chapter 1, we discussed efforts to classify children with SLI into subtypes. These efforts were a natural response to the fact that children with SLI constitute a heterogeneous group. Continued clinical research may yet produce a classification system that holds up empirically.

Some have argued that whereas we require a high standard of empirical support before we are willing to recognize subtypes within the broad category of SLI, we stop at the nonverbal IQ border of 85 without exercising the same degree of scientific rigor. In effect, we are setting a boundary for at least one subtype of language impairment when we declare that children must score 85 or above.

Of course, 85 falls 1 SD below the mean; therefore, scores at this level and higher can be defended as "normal." But we know from chapter 5 that children achieving this score often perform below their age-mates on other kinds of nonlinguistic activities, such as symbolic play and mental imagery. So it is a bit of an act of faith to accept a score of 85 on a nonverbal IQ test as the best estimate of nonverbal intelligence.

Many clinical researchers are not so gullible. They treat a score of 85 as a reasonable gauge of nonverbal intelligence, certainly not as the only word on the matter. For most of these researchers, a nonverbal IQ of 85 also serves the important function of keeping the heterogeneity down to a dull roar. Replicability demands specification of the characteristics of the participants, and without reliable subtypes, the 85 cutoff is better than most.

Note, however, that if replicability were the sole aim, an upper as well as a lower boundary might be specified, say, 85–115, or even (to reduce heterogeneity further) 90–110. But a range such as 80–100 is not seen in the literature (see Camarata & Swisher, 1990). Clearly, then, the need to reduce heterogeneity is only one of the functions of the 85 cutoff. We also want to indicate that the participants are safely clear of the boundaries associated with mental retardation. From the standpoint of diagnostic categories, the 70–84 range constitutes a type of no-man's-land, the "borderline range" of intelligence. A diagnosis of mild mental retardation requires an IQ of 55–69 (as well as limitations in adaptive behavior, which do not concern us here).

There are psychometric reasons for taking the 85 cutoff as something less than the gold standard. Every test has a standard error of measurement such that the child's "true" score may fall above or below the observed score. Thus, some children might have an observed score of 85 but a true score of 80. For other children, the opposite might occur; these children would be excluded from the category of SLI (see Lahey, 1990).

On theoretical grounds, the use of 85 as the standard cutoff also has problems. In chapter 5 we noted that language deficits probably limit the degree to which children can call on verbal reasoning to assist them in solving problems that are, on the surface, nonverbal in nature. If this is the case, nonverbal IQ scores might decline with age as nonverbal problems become more conceptual and less perceptual in nature.

Does it make sense to have children declassified because their scores dip below 85 when the reason for the decline was the language problem itself?

The 85 cutoff is also difficult to defend in the face of empirical evidence. Unlike the case of severe mental retardation, where organic factors are often identifiable, the factors associated with borderline levels of intelligence (indeed, even mild mental retardation) are not clearly distinct from those responsible for the variability within the normal IQ range (see Tomblin, 1991).

An investigation by Fey, Long, and Cleave (1994) provides another empirical reason to treat the cutoff of 85 as somewhat arbitrary. They analyzed the data from an earlier treatment study (Fey, Cleave, Long, & Hughes, 1993) to determine whether the gains in grammatical skills made by the children varied according to whether their nonverbal IQs were above or below 85. The findings indicated that the children with IQs between 70 and 84 made gains that were comparable with, if not greater than, the gains made by the children meeting the criteria of SLI. There are other factors that distinguished the two groups. Most notably, the gaps between mental age and language test scores were somewhat smaller in the group with lower IQ scores. Future research might determine whether gains in treatment in children with similarly narrow gaps are any different if IQs are 85 or above. In the absence of any such evidence, the Fey, Long, and Cleave (1994) findings should be interpreted as evidence against making a sharp distinction on the basis of nonverbal IQ score.

The argument put forth here is not that SLI knows no bounds. Distinctions are probably needed. For example, future research might reveal that children whose nonverbal IQ scores at preschool are high (or whose average scores are maintained throughout childhood) have a disorder that is different from those children whose nonverbal IQ scores begin at average levels and then decline. However, once empirically determined boundaries are set as the result of this research, there is no assurance that children with nonverbal IQ scores just above and just below 85 will be in separate groups. In the meantime, 85 should be treated as a provisional boundary that has, at least, statistics on its side.

Children with SLI and Children with Specific Reading Disabilities

For some time, scholars and practitioners have speculated that the basis of specific reading disabilities (dyslexia) is linguistic in nature (e.g., Orton, 1937). However, it is only in the past 20 years that this view has become the prevailing one (e.g., Catts, 1989; Kamhi & Catts, 1989; Perfetti, 1985; Snyder & Downey, 1991; Stanovich, 1988; Vellutino, 1979). Not surprisingly, this shift has prompted researchers to consider whether children with reading disabilities might have had spoken language problems at a younger age, or whether such problems are still present in a subtle form when the reading deficit is identified.

The prevalence figures certainly invite this kind of thinking. Reading disabilities are observed in 4% to 5% of the population (e.g., Silva, McGee, & Williams, 1985; Yule & Rutter, 1976). These figures are slightly lower than the prevalence figures for SLI at age five years (Tomblin, 1996a, b). However, as discussed in chapter 1, a small percentage of children with SLI at age five will no longer have language problems several years later. And, of course, reading disabilities are not diagnosed until school age. If difficulties with spoken language are seen in children with reading disabilities and, conversely, problems with reading are seen in children with SLI, the distinction between these two groups becomes blurred. Here we consider whether SLI and reading disorders should, in fact, be viewed as spoken and written manifestations of the same impairment.

Reading Deficits in Children with SLI

In chapter 1, it was observed that many children with SLI have long-standing problems. Residual deficits in spoken language are sometimes seen, and academic difficulties are frequent. Reading is an area especially at risk. Most of the studies that have followed up children with SLI into school age (retrospectively or prospectively) have reported atypically high proportions of reading and other learning problems (e.g., Aram, Ekelman, & Nation, 1984; Aram & Nation, 1980; Catts, 1991; Hall & Tomblin, 1978; R. King, Jones, & Lasky, 1982; Padget, 1988; Stark, Bernstein, Condino, Bender, Tallal, & Catts, 1984; Strominger & Bashir, 1977; Tallal, Curtiss, & Kaplan, 1988).[1] There are at least two exceptions. Richman, Stevenson, and Graham (1982) and Silva, McGee, and Williams (1985) found that children identified as language-disordered without regard to accompanying factors were at risk for later reading problems. However, in each of these studies, a subgroup of children with age-appropriate nonverbal IQs was identified, and these children did not show later deficits in reading.

Deficits in spoken language comprehension in particular would seem to portend difficulties in reading because such deficits are more likely to reflect limitations in language knowledge. Investigations by Tallal, Curtiss, and Kaplan (1988) and Wilson and Risucci (1988) provided data consistent with this assumption. Placement in special classes for learning disabilities is also predicted by poor comprehension scores (Rissman, Curtiss, & Tallal, 1990). However, Bishop and Adams (1990) found that a language production measure, MLU, served as a better predictor of reading ability in their study.

The area of spoken language that is affected also matters. Children with problems in vocabulary and/or syntax along with phonology are more likely to experience reading problems than those with phonological problems only (e.g., Bishop & Adams, 1990; Catts, 1993; Levi, Capozzi, Fabrizi, & Sechi, 1982; Lewis & Freebairn, 1992). But this does not mean that phonology plays no role. Phonological awareness has emerged as an especially good predictor of reading ability. An example of a phonological awareness task is seen when the child is provided a word and asked to repeat it back after deleting the initial sound or syllable (e.g., responding with "all" after hearing "tall," "son" upon hearing "person"). Blending tasks are also used to assess phonological awareness, as when the child must say *keep* in response to the string /k/-/i/-/p/. Scores on phonological awareness measures serve as a good predictor of reading ability in children with SLI as well as in children in general (e.g., Catts, 1991; Magnusson & Nauclér, 1990).

Given that children with problems limited to phonology are not at special risk for reading difficulties, the findings that phonological awareness predicts reading ability is somewhat perplexing. However, this apparent discrepancy probably says more about the heterogeneity of children with phonological problems. For some children with phonological problems, the difficulty does not reside in the ability to analyze and synthesize sounds and syllables on a conscious level (Catts, Swank, McIntosh, & Stewart, 1990). However, when children with phonological problems are found to show deficits in phonological awareness, reading deficits are also seen (Bird, Bishop, & Freeman, 1995).

Phonological awareness is not the only predictor of reading ability in children with SLI. Rapid naming ability also has predictive power (Catts, 1991). Furthermore, it appears that the best predictors vary with the particular reading measure under study. Catts (1993) has reported that printed word recognition seems to be predicted by phonological awareness and rapid naming, whereas reading comprehension is best predicted by measures of spoken language comprehension and production, and, to a lesser extent, rapid naming.

We would be closer to answering the question of whether SLI and specific reading disabilities boil down to the same problem if all children with SLI were found to have deficits in reading. The findings of Richman, Stevenson, and Graham (1982) and Silva, McGee, and Williams (1985) noted above seem to put a damper on this prospect. However, SLI was diagnosed in the preschool years in these cases, so it is possible that by the time the children reached the age at which reading ability was assessed, they were no longer below age level in spoken language skills. The findings of Bishop and her colleagues nicely illustrate this eventuality (Bishop & Adams, 1990; Bishop & Edmundson, 1987a). In this work, children with spoken language deficits at age four years who achieved normal limits by age 5;6 were not at high risk for reading difficulties. Only those children whose spoken language problems persisted were more likely to have reading problems.

The manner in which SLI is defined can also be a factor. Magnusson and Nauclér (1990) singled out four children from a group of 37 children with language disorders who were found to perform as well as or better than most age controls on tests of reading and spelling at the end of first grade. However, these children had also performed well during preschool on tests of phonological awareness, and on tests of spoken language comprehension and production. According to these investigators, these children's original clinical diagnosis was based either on phonology alone or on early delays in other areas of language that were not in evidence by preschool.

For children whose language delays continue through age four or five, possible signs of reading difficulties can be identified quite early. Gillam and Johnston (1985) found that children with SLI averaging just under five years of age performed more poorly than age controls on a task of print awareness that employed high frequency environmental print (e.g., soft drink and potato chip names) and various degrees of contextual support (label still on package, label without package). The two groups of children did not differ in their experience with these items, suggesting that even before reading instruction, the children with SLI were at a disadvantage.

However, there are case studies of children with SLI who seem to be undaunted—and perhaps even aided—by reading. Weeks (1974) described one such child:

> After Leslie started to learn to read, there were certain situations in which it appeared that Leslie analyzed written language, stored it more efficiently for reproduction—both written and spoken reproduction—than she did spoken language. For example, after reading a word, she seemed to pronounce it more accurately than after hearing it pronounced carefully. She also remembered how to spell it after seeing it written, whereas she didn't necessarily remember how to pronounce a word after hearing it. Leslie was a good speller, usually spelling

correctly words she had seen. By 5;0 she could hold her own in a game of Scrabble with other family members. She also followed written directions more easily than verbal directions. Her teachers at school considered this to be a factor in placing her in the first grade class where many instructions are written, rather than putting her in kindergarten where instructions are all verbal.   (p. 140)

Early on, this child showed most of the hallmark features of SLI, such as omission of grammatical inflections and function words, use of imprecise lexical items, and phonological limitations. However, by age four years she scored at age level on most measures of spoken language. Therefore, this might represent another case of a language problem resolved before school age; thus the apparent lack of difficulty with reading requires no explanation. And yet, this is complicated by the fact that informal reading instruction began before this child was four years old. Therefore, the possibility remains that reading helped her to catch up in spoken language.

Spoken Language Abilities of Children with Specific Reading Disabilities
The archetypal case of specific reading disability is a child who appears typical in all respects until faced with the task of learning how to read. Then, once the reading difficulties are in plain sight, additional probing reveals problems in spoken language as well.

Snyder and Downey (1991) provide an example of the range of problems in spoken language that might be observed. Children with reading disabilities and age controls were tested on phonological awareness tasks, sentence completion tasks requiring the child to supply grammatically obligated material, and discourse tasks that required the child to remember details of a story and make inferences. The children with reading disabilities performed below the level of the control children on all of these tasks. Slow and less accurate responses in naming tasks are also well documented in the reading disability literature (e.g., Wolf, 1982; Wolff, Michel, & Ovrut, 1990).

Evidence of questionable spoken language ability crops up even when attempts are made to select children with reading disabilities and no deficits in spoken language. Stark and Tallal (1988) selected a group of children with reading disabilities who scored no lower than six months below age level on standardized tests of language production and comprehension. However, these children proved to be significantly lower than age controls on seven of the nine production and comprehension measures employed.

One possible explanation for the seemingly late discovery of spoken language problems in these children is that language learning may have been adequate in these children until they reached reading age. At that point, their spoken language skills failed to keep pace with those of their peers because their reading problems did not permit them to benefit from all of the language instruction that takes place in the written mode. Consequently, the spoken language skills that are assessed in the school years—less frequently occurring vocabulary words, figurative language such as idioms and metaphors and the like—did not develop sufficiently in these children.

This seems like an easy hypothesis to test. If spoken language skills do not begin to suffer until reading age, the gap in spoken language ability between children with reading disabilities and children with normal reading ability should widen with age.

At the outset of reading instruction, differences should be small or nonexistent, given the simple lexical and syntactic levels reflected in early primers.

Even if the pace of language development is hobbled by reading difficulties, there is plenty of evidence to suggest that this factor is not the only one involved. It appears that for many children, deficits in spoken language predate the commencement of reading instruction.

The earliest evidence of this sort came from retrospective studies (e.g., T. T. S. Ingram, Mason, & Blackburn, 1970; Rutter & Yule, 1975). More recently, prospective studies have supplied the relevant data. Scarborough (1990, 1991) studied 34 children from families with a history of reading deficits, along with a group of 44 children serving as age controls. The children were first seen at age 2;6 and were followed until age 8;0. As expected, a high percentage of the children with positive family histories were found to have reading problems as eight-year-olds (65%, compared with 5% of the controls). At ages 2;6, 3;0, 3;6, and 4;0, the children who later had reading difficulties showed lower syntactic abilities than the children who later became normal readers. Over 75% of the children were correctly classified as reading disabled or age-appropriate in reading based on the syntactic measures obtained from ages 2;6 to 4;0. At ages 3;0 and 5;0 the poor readers' scores on naming and phonological awareness tasks also were lower. However, Scarborough found that the early syntactic abilities accounted for some unique variance in later reading scores.

Auditory and Visual Processing
In chapter 6, we saw that many children with SLI have problems on tasks requiring them to process auditory and, in some cases, visual stimuli presented at rapid rates. Some work along the same lines has focused on children with specific reading disabilities. Tallal (1980) examined the auditory processing abilities of children with specific reading disabilities using the tasks developed by Tallal and Piercy (1973a, 1973b). These children were successful in discriminating and noting the temporal order of two complex tones when the ISI was over 400 ms. However, they made significantly more errors than age controls when the interval was shorter. Somewhat similar results were observed by Stark and Tallal (1988). In this study, visual as well as auditory stimuli were used. The children with reading disabilities showed greater difficulty than did a group of age controls on the more complex auditory and visual tasks. Mody (1993) obtained results that called into question a purely auditory interpretation of observed perceptual problems. She studied a group of poor readers who also performed poorly with [ba]–[da] stimuli as well as a group of good readers who had no difficulty with these stimuli. In spite of these differences, the two groups performed similarly with nonverbal stimuli composed of sine waves with durations and frequency trajectories that matched those of the center frequencies of the second and third formants of [ba] and [da].

Comparisons of Children with SLI and Children with Specific Reading Disabilities
Direct comparisons between children with SLI and children with specific reading disabilities were made by Kamhi and Catts (1986). The latter group had no history of language impairment. A group of age controls also participated. Tasks included nonsense word and sentence repetition tasks, phonological awareness tasks, a grammaticality judgment task, and a sentence division task (e.g., repeating back "the man" when given the sentence "the man came to my house"). The age controls performed

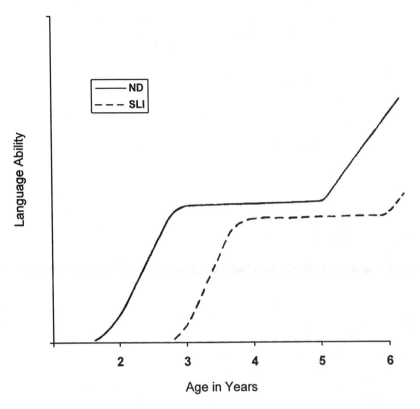

Figure 9.1
Language abilities of children with SLI appear to catch up to those of normally developing (ND) children, only to lag behind again when new types of language skills must be acquired. Adapted from Scarborough and Dobrich (1990).

better than the other two groups; the groups with SLI and reading deficits were similar on most measures. The children with SLI had greater difficulty with word and sentence repetition. On the remaining tasks, the two groups did not differ.

Additional similarities were seen in a subsequent study by Kamhi, Catts, Mauer, Apel, and Gentry (1988). Children with reading disabilities outperformed children with SLI on a task requiring the repetition of multisyllabic nonsense words. However, the two groups performed comparably—and significantly worse than age controls—on several other tasks. These included rapid naming tasks, tasks requiring the repetition of single monosyllabic nonsense words, and tasks of mental imagery. As discussed in chapter 5, mental imagery tasks are often problematic for children with SLI.

One Disorder or Two?
At every turn, one is faced with the similarities in the findings for the two groups of children. Phonological awareness, speech perception, syntactic proficiency, and rapid naming loom as major factors in both groups. Even mental imagery shows up as a weakness in each case.

The overlap can be illustrated in another way. Scarborough (1990, 1991) selected her subjects on the basis of a positive family history of reading problems. However, based on an inspection of these children's early grammatical data, they would have been considered (and indeed were) children with SLI. In turn, many of the children with SLI studied by Catts (1993) and Tallal, Curtiss, and Kaplan (1988) joined the ranks of children with specific reading disabilities.

This is more than semantics. For children with both spoken language deficits and reading disabilities, there is no way to argue that one and not the other diagnostic category is the more appropriate. This view seems to be widely accepted. In fact, one common method of recruiting school-age children with SLI is to administer spoken language tests to children enrolled in classes designated for children with learning (including reading) disabilities (e.g., Crais & Chapman, 1987).

Although many children can be placed in both of these diagnostic categories, we are not yet in a position to treat the two categories as synonomous except for the mode—spoken or written—to which they apply. At least two types of patterns require explication. First, there are children who exhibit spoken language deficits as preschoolers and then appear to catch up to peers by school entry. As pointed out by Scarborough and Dobrich (1990), this may represent an illusory recovery; the differences may reappear at a later time when other kinds of language skills are expected to be attained. This is illustrated in figure 9.1.

If the later-appearing deficiencies include spoken as well as written language skills, the assumption of a single diagnostic category can be maintained. But what if these deficiencies are limited to reading? In that case, we would not know if there was true recovery of the spoken language deficit, with the later problem in reading attributable to a separate source, or whether the recovery was illusory, with written, not spoken, language the most vulnerable area when the problem resurfaced.

The second pattern that confounds our use of SLI and specific reading disability as synonyms is the case of a reading disability with no reported history of problems in spoken language. We now know that children with this pattern are the exception and not the rule; however, studies of different subtypes of reading disabilities have not yet identified the source of such a pattern. If this pattern is brought about by non-linguistic factors, then it will be appropriate to treat SLI and specific reading disability as the same phenomenon only if this pattern is defined out of the diagnostic category. On the other hand, this pattern may yet have a basis in language. If so, our conceptualization of the interaction among components of language might have to be revised, for it is not yet clear how the same fundamental problem can be manifested late and only in the written mode in some cases, and early and more pervasively in others.

Note

1. Although reading is emphasized here, there is evidence that academic skills such as mathematics may also be below age level in children with SLI (e.g., Padget, 1988). Some of these problems seem to have an early beginning. Fazio (1994) studied four- and five-year-old children with SLI and found that they understood basic rules of counting (e.g., the last number reached when counting objects constitutes the number of objects present) but had more difficulty than age controls in recalling the sequence of number words. Problems were still apparent two years later (Fazio, 1996). We shall take up this line of work again in our discussion of limited processing capacity in chapter 12.

Chapter 10

The Nature and Efficacy of Treatment

This chapter is concerned with language treatment, defined as deliberate efforts by professionals and family members to help children with SLI strengthen their language abilities (Fey, 1986). The terms "intervention" and "therapy" are also used frequently in referring to this process. One of the noteworthy things about treatment is that there doesn't seem to be a theory of SLI according to which it is contraindicated. Even the accounts of SLI that characterize the disorder in the gloomiest of ways—as involving some type of permanent absence or loss—recognize treatment as a way of helping children develop compensatory strategies. The perceived contribution of treatment goes uphill from there.

Treatment approaches vary widely, both in the procedures employed and in the areas of language receiving the greatest emphasis. Because children with SLI have serious language learning problems, it would be reasonable to assume that the treatment procedures employed would look very different from the kind of language stimulation that goes on in ordinary circumstances. However, in many cases, this is not true. In an insightful paper that still has currency, Muma (1971) pointed out that many of the techniques that parents and other adults employ naturally with their young normally developing children are, in effect, teaching procedures that can be transferred to clinical practice. Modifications are made in these procedures, most of them having to do with increasing the frequency with which the language targets are presented to the children and ensuring that the targets appear in linguistically unambiguous contexts.

Someday SLI may be understood well enough to develop procedures for its prevention. Until then, treatment is the most sensible course of action. A substantial amount of research has been conducted to discover which types of treatment are effective, and for which children. Much of the chapter will be devoted to this topic.

Does Language Treatment Work?

To determine whether treatment leads to gains in language ability in children with SLI, certain safeguards must be put into place. Even children with SLI show maturation in language, albeit at a slow pace. Accordingly, it is not sufficient to demonstrate gains in language following treatment; it must be shown that these gains are not likely to be the result of maturation or some other, uncontrolled factor. The studies reviewed in this section meet this criterion in one of several ways. The most common types of provisions used in these studies are (1) the use of a no-treatment control group; (2) the use of a comparison group receiving treatment unrelated to the linguistic forms of interest; (3) the use of a multiple-baseline design, in which the child's progress on untreated as well as treated forms is monitored; (4) the use of a nonsense form whose acquisition could be attributed only to experimental treatment; and (5) the use of statistical estimation as a means of determining the amount of gain that could be expected by maturation alone.

There are other designs, but they present interpretive problems, especially where the teaching of language is concerned. For example, for some behaviors, treatment effects can be demonstrated by providing the child with treatment, temporarily suspending treatment, and then reinstituting treatment. A plateau in response frequency or accuracy is seen during the period of no treatment, flanked by periods of increase when treatment is provided. This type of design seems to work only during the early phases of language treatment. Once the child reaches a certain level of ability, the child's use of the treated form escapes experimental control. At this point, periods of withdrawal and periods of treatment are not sharply distinguishable in terms of the child's accuracy levels (e.g., Hargrove, Holmberg, & Zeigler, 1986).

Most treatment studies have examined language production. This could be a function of the fact that almost all children with SLI display deficits in production; only a subset of these children also experience clear problems in comprehension. The emphasis on production in the treatment literature does not mean that the needs of children with comprehension problems are not being met. Many treatment approaches that focus on production are more than sentence planning and execution activities; they also highlight the grammatical, semantic, or pragmatic function of the material to be acquired. Thus, in gaining practice in the production of the target forms, the children are also learning the appropriate contexts in which they are used.

The particular details of language serving as the focus of treatment cover a wide range. There has been a steady stream of treatment studies dealing with morphosyntax since the late 1960s. The earliest of these studies were steeped in the tradition of operant conditioning (e.g., Gray & Fygetakis, 1968). Most of the studies of the 1970s continued the procedures of operant conditioning as a means of teaching morphosyntax, but the forms chosen as targets were more in keeping with then-current linguistic theory (e.g., Hegde, Noll, & Pecora, 1979; Leonard, 1974), a trend that continued through the 1980s (e.g., Connell, 1986b; Thal & Goldenberg, 1981). This combination of operant principles and principles of generative grammar was not as theoretically schizophrenic as it might seem; most of the operant teaching procedures employed can be characterized in current terms as providing the child with both an idealized input and negative evidence. The latter was sometimes overt (e.g., telling the child "No, I think you forgot something") or implied by virtue of positive feedback for well-constructed utterances, and the absence of such feedback for ill-constructed attempts. In the 1990s, morphosyntax remains a common target in treatment studies (e.g., Camarata, Nelson, & Camarata, 1994; Fey, Cleave, Long, & Hughes, 1993), although, as we shall see, the methods of teaching have changed considerably.

The procedures used to strengthen children's phonological abilities show the same historical trends observed for morphosyntax. Early procedures emphasized motor execution with the aim of improving the phonetic accuracy of individual speech sounds. Soon these procedures were placed within an operant framework. Gradually the goals became more linguistically driven, such as the teaching of distinctive feature classes (e.g., [+ continuant] sounds) or the elimination of phonological processes (e.g., velar assimilation); however, the operant framework persisted. In the 1990s, the procedures take on the appearance of being more natural and less drill-like. Furthermore, investigators have reached new levels of sophistication in selecting phonological targets and appropriate controls. We shall not review the treatment studies on phonology here. However, the efficacy of treatment in this area seems clear. Of 63 studies of this type reviewed by Sommers, Logsdon, and Wright (1992), only one failed to report significant gains in phonological abilities.

In the 1980s, the semantic abilities of children with SLI began to serve as the focus of treatment studies. Some investigations were concerned with teaching semantic relations in early word combinations, such as action–object (e.g., *push car*) and possession (e.g., *mommy shoe*) (e.g., Connell, 1986a; Leonard, 1984; Olswang & Coggins, 1984). However, most studies dealt with the lexical abilities of children with SLI. Investigations of younger children with SLI were concerned with teaching new words (e.g., Leonard, Schwartz, Chapman, Rowan, Prelock, Terrell, Weiss, & Messick, 1982; Rice, Buhr, & Nemeth, 1990; R. Schwartz, 1988). Investigations of older children with SLI, in contrast, focused on methods designed to help these children overcome problems of word-finding (see chapter 3) (e.g., McGregor, 1994; McGregor & Leonard, 1989; Wing, 1990; Wright, 1993).

The pragmatic abilities of children with SLI have received relatively little investigative attention in the treatment literature. However, a few studies have appeared since the mid-1980s. These have been concerned with topic initiations (Bedrosian & Willis, 1987), requests for additional information (Dollaghan & Kaston, 1986), and the use of language for purposes of pretending (Skarakis-Doyle & Woodall, 1988).

Approaches to Treatment

Imitation-Based Approaches During the early years of treatment research, the dominant approach employed was that of elicited imitation. In this approach, the experimenter produces the exact sentence or phrase required of the child and the child is asked to repeat it. It is recognized that children's imitations of meaningful sentences are often filtered through their linguistic systems; asking children to repeat adultlike structures is therefore no guarantee that the response will be of the desired form. Two general provisions are made to increase the likelihood of success. First, the children's attention is usually drawn to the detail in the utterance serving as the target of interest. Second, during the early phases of the treatment program, the target structure of interest may be presented in smaller units; only later are the children asked to imitate the structure in longer units. Articles, for example may first be presented in isolated noun phrases (e.g., *the car*), inverted auxiliaries may first be presented in short yes–no questions (e.g., *Can she see?*), and so on.

In many studies, pictures or enactments with toys are presented along with the sentences to be imitated, in an effort to reduce the drill-like appearance of the activity. Gradually, the children are encouraged to make use of the target structure without the benefit of first hearing the experimenter produce it. This is typically handled by ensuring that in the early stages of treatment, a question (or statement) is presented along with the sentence to be imitated. Then, when the imitation component is dropped, the question is retained as a prompt for the child to produce the target sentence. In some cases, the child's reliance on imitation is reduced in small steps. This can occur in at least two ways. First, the experimenter may produce the sentence to be imitated and then insert additional linguistic material prior to the child's imitation, to enforce a slight delay in the child's response. Second, the experimenter may proceed from providing the entire sentence for the child to imitate to providing successively smaller portions of it (Fygetakis & Gray, 1970; Gray & Fygetakis, 1968; Gray & Ryan, 1973).

A few studies have employed a variation of the typical imitation approach. Here, children are asked to respond to pictures and questions, and are provided with the sentence to imitate only if their original utterance lacks the target structure (e.g., Hester & Hendrickson, 1977). Finally, imitation has been incorporated into comprehensive treatment programs that include a variety of techniques. For example, Warren and Kaiser (1986) assisted children with SLI in the use of two-to-five-word utterances (e.g., *Girl eat cookie*) in a program that included pointing to pictures (e.g., "Show me 'girl eat cookie'"), describing pictures (e.g., "Tell me about this picture"), performing the requested action (e.g., "Susan, eat cookie"), imitating the target sentence (e.g., "Say 'girl eat cookie'"), and describing the actions of others (e.g., "What's Jane doing?").

Although some investigators have employed imitation-based approaches to teach new lexical items (Olswang, Bain, Dunn, & Cooper, 1983; Whitehurst, Fischel, Caulfield, DeBaryshe, & Valdez-Menchaca, 1989; Whitehurst, Novak, & Zorn, 1972), most studies making use of imitation have been concerned with facilitating children's morphosyntactic abilities. Targets have included (1) telegraphic utterances reflecting basic sentence relations such as subject + verb, verb + object, and subject + verb + object (Gottsleben, Tyack, & Buschini, 1974; Warren & Kaiser, 1986; Whitehurst et al., 1989; Zwitman & Sonderman, 1979); (2) the progressive *-ing* inflection (Connell, 1980b; Gottsleben et al., 1974); (3) noun plural *-s* (Gottsleben et al., 1974); (4) possessive *'s* (Connell, 1980b; Hegde, Noll, & Pecora, 1979); (5) nominative case pronouns (Connell, 1980a; Gottsleben et al., 1974; Hegde & Gierut, 1979); (6) copula *be* forms (Hegde, 1980; Hegde et al., 1979); (7) auxiliary *be* forms (Hegde, 1980; Hegde & Gierut, 1979; Hegde et al., 1979; Hester & Hendrickson, 1977); (8) copula and auxiliary inversion (Mulac & Tomlinson, 1977); and (9) invented suffixes (e.g., use of *-am* with English-speaking children to mark a notion such as "broken") (Connell & Stone, 1992).

In each of these studies, the children demonstrated gains exceeding those that could be attributed to maturation. That is, these gains were greater than those seen by children receiving treatment on unrelated structures, or those made by the same children on untreated structures.

Modeling Approaches   Modeling is another type of approach that is frequently employed in treatment research. There are two versions of modeling reflected in the literature. In each, the child observes someone (the model) produce examples of utterances containing the linguistic form serving as the focus of treatment. The child is instructed that the model will be talking in a special way, but is not asked to imitate the modeled utterances. At this point, the two versions of modeling differ. In one version, the child observes only; in the other, the child is asked to take turns with the model producing new examples of the target form once the observation period is over. Within each of these types, there are variations. Pictures or enactments with toys frequently accompany the modeling, and the model may vary from a third person in the session to a puppet manipulated by the experimenter to a computer whose digitized utterances are preprogrammed. In some studies, a few of the utterances produced by the model are (intentionally) incorrect, and are corrected by the experimenter to assist the child in determining the desired form. The rationale behind the modeling approaches is that the problem-solving undertaken by the child in an effort to discover what form the utterance is expected to take is an effective means of learning. In a sense, this is a type of imitation, but what is imitated is a rule for combining, inserting, or sequencing morphemes, not the particular utterances produced by the model.

Although morphosyntactic forms represent the majority of targets selected for treatment in studies on modeling, targets of a pragmatic or semantic nature have also been employed. Treatment has dealt with (1) the auxiliary plus negative form *don't* (Leonard, 1975); (2) auxiliary *be* forms (Ellis Weismer & Murray-Branch, 1989; Leonard, 1975); (3) auxiliary *be* inversion (Ellis Weismer & Murray-Branch, 1989); (4) *wh*-word plus auxiliary *be* and *do* inversion (Wilcox & Leonard, 1978); (5) nominative case pronouns (Courtright & Courtright, 1976; Ellis Weismer & Murray-Branch, 1989); (6) invented suffixes (Connell & Stone, 1992; Roseberry & Connell, 1991); (7) miniature languages using novel words and suffixes but paralleling rules found in natural languages (Johnston, Blatchley, & Olness, 1990); (8) topic initiations (Bedrosian & Willis, 1987); and (9) descriptions of events during pretend play (e.g., *It's milk* while the child pours imaginary liquid from a toy pitcher) (Skarakis-Doyle & Woodall, 1988). Each of the above studies showed evidence of gains attributable to treatment effects.

One grammatical structure has produced mixed results. Courtright and Courtright (1979) taught the unusual structure noun + *means* + *to* + verb + *ing* (e.g., "Girl means to running") to a group of children with SLI. Utterances of this type were taught in response to pictures and questions of the form "To what means the (e.g., girl) doing?" Courtright and Courtright reported success using a modeling procedure to teach such forms. However, Connell, Gardner-Gletty, Dejewski, and Parks-Reinick (1981) were unable to replicate this finding. Subtle, as yet unknown procedural differences may have led to the discrepant findings, but the relationship between the children's existing grammars (which weren't specified) and the unusual structure being taught is likely to have been a major factor. For example, even if the dialects of the children allowed "mean" to be used in the sense of "intend" (where "I mean to go to the party" has the meaning of "I intend to go to the party"), thus permitting the infinitival complementizer *to*, the grammars of these children probably required the following nonfinite verb to be in bare-stem form ("to run"). Unlike the case of learning a structure that forms part of a novel language, many of the children were probably learning a structure that seemed to fly in the face of what they already knew about English.

Focused Stimulation   A treatment approach that bears a close resemblance to modeling is "focused stimulation" (Leonard, 1981). There are many variations of focused stimulation; they share the characteristic of providing children with concentrated exposure to particular semantic or morphosyntactic forms. Unlike modeling, where the child is told explicitly to attend to the special way the model is speaking, focused stimulation relies primarily on the high frequency of presentation of the target forms, as well as the unambiguous contexts in which these forms are used. The target forms can be embedded in stories or in simple descriptions of play activities. In some cases, following periods of exposure, the children are asked direct questions whose answers obligate the use of the target form; in other cases, the nonlinguistic events are simply arranged to increase the likelihood of an utterance containing the target form, should the child offer one.

One example of a focused stimulation task is seen in a study by Rice, Buhr, and Nemeth (1990). The goal of this investigation was to examine the lexical learning abilities of children with SLI. Children viewed animated television programs with voice-over narration. The story narration contained multiple instances of each target word. For example, the word *trudge* appeared in sentences such as "He takes his viola and trudges down the road" and "Billy keeps trudging down the road." Following the viewing of the television programs, the children's comprehension of the target words was tested. Results indicated that the children with SLI assigned to the viewing conditions showed greater comprehension of the target words than a group of children with SLI serving as controls.

Culatta and Horn (1982) employed a variation of focused stimulation designed to teach children with SLI the use of grammatical forms such as copula *is* and *am*, and the modal auxiliary *will*. Naturalistic situations such as cleaning house and repairing appliances were used. The experimenter described the situation, introduced appropriate props, and produced the target forms in utterances pertaining to the situation. The child's use of each target form was evoked by means of arranging events that had a high likelihood of promoting an utterance requiring the target form. For example, to promote use of copula *is*, the experimenter asked the child to assist in placing items on a display shelf and proceeded to hand the child an item that was too large for the shelf. As treatment progressed, the experimenter's speech contained a smaller percentage of utterances that made use of the target form. Employing a multiple-baseline design, Culatta and Horn demonstrated that gains made in the target forms exceeded those seen for forms whose inclusion in treatment was postponed.

The forms taught successfully by means of focused stimulation are varied. They include (1) verb complexity (Fey, Cleave, Long, & Hughes, 1993; Lee, Koenigsknecht, & Mulhern, 1975); (2) *wh-* questions (Brooks & Benjamin, 1989; Lee et al., 1975); (3) copula forms (Culatta & Horn, 1982); (4) auxiliary forms (Brooks & Benjamin, 1989; Culatta & Horn, 1982); (5) new lexical items (Leonard, Schwartz, Allen, Swanson, & Loeb, 1989; Leonard, Schwartz, Chapman, Rowan, Prelock, Terrell, Weiss, & Messick, 1982; McGregor & Leonard, 1989; Olswang, Bain, Rosendahl, Oblak, & Smith, 1986; Rice, Buhr, & Nemeth, 1990; R. Schwartz, 1988; R. Schwartz, Leonard, Messick, & Chapman, 1987; Wing, 1990); and (6) novel suffixes (Swisher, Restrepo, Plante, & Lowell, 1995; Swisher & Snow, 1994).

Milieu Teaching   The past 20 years or so have seen the evolution of an approach to treatment that makes significant use of highly natural settings. This type of approach has been termed the "milieu approach," to make explicit its reliance on the child's surroundings to set the occasion for language teaching (Kaiser, Yoder, & Keetz, 1992). Often treatment adopting this approach is conducted in a play setting. The experimenter arranges the setting to increase the likelihood that the child will make some attempt at communication. However, it is the child who chooses the specific activity to be performed at any given point. When the child indicates interest in an activity or object, the experimenter shows attention and, as necessary, provides the child with increasingly specific cues for production of the target. Natural contingencies are applied to the child's production of the target form. For example, if the target is two-word utterances, and the child says *want truck*, the experimenter will give the child the requested object.

There is overlap between milieu teaching and variations of focused stimulation that employ natural settings such as play. The difference in these cases is one of emphasis. In milieu teaching, the emphasis is on obtaining communicative attempts and responding to them with natural contingencies. In focused stimulation, the emphasis is on providing the child with a large number of examples of the target form.

Although the literature on milieu teaching is growing rapidly, most studies have been concerned with children exhibiting global developmental deficits. One study that did examine children with SLI was the investigation of Warren, McQuarter, and Rogers-Warren (1984). These investigators used the milieu teaching approach to help children with SLI increase their total number of utterances, and the number of utterances produced that were not obligated by a prior question.

Conversational Recasting   Another treatment approach that has attracted considerable attention in recent years is conversational recasting. In this approach, the experimenter and child participate in play activities and the experimenter responds to utterances produced by the child in a manner that serves as a relevant conversational turn and contains some linguistic form serving as the focus of treatment. For example, if the target form is the modal *can*, the experimenter might respond to a child's utterance such as *Spiderman go on roof* with an utterance such as *Yeah, Spiderman can jump up on the roof*; a child's utterance such as *We open the box now?* can be recast as *Sure, we can open the box now*.

Camarata and Nelson (1992), Camarata, Nelson, and Camarata (1994), and Nelson, Camarata, Welsh, Butkovsky, and Camarata (1996) have provided evidence that conversational recasting is effective for teaching a range of structures. Some of the structures taught successfully to children with SLI include modal auxiliaries, regular past inflections, passive constructions, and relative clauses.

Connell (1980b) had less success with a method that resembles conversational recasting in certain respects. The targets in this case were the progressive *-ing* and possessive *'s*. To teach the former, the experimenter asked the children to describe pictures of ongoing actions. When the children produced a telegraphic utterance such as *Boy comb*, the experimenter provided an expansion containing *-ing* (*The boy is combing*). This procedure did not result in gains in the target forms. The discrepancy between Connell's finding and those of other investigators might be found in the fact that in Connell's procedure the child is merely responding to a question rather than initiating a conversational turn. Under such circumstances, children may be less receptive to the new information contained in the experimenter's follow-up utterance.

Expansion Approaches   Expansion is a well-known process first documented in the literature on normal language development. It was reported that mothers often responded to their young children's telegraphic utterances (e.g., *Carrie mommy*) with a more complete and grammatical rendition (*Yes, Carrie is sitting on mommy's lap*). Studies employing expansion as a treatment approach with children with SLI are rare. The method used by Connell (1980b) in the study cited above might be considered expansion, though its focus on specific narrow targets (*-ing*, *'s*) seems to justify our considering it a special case of recasting.

An adaptation of expansion can be seen in an investigation by R. Schwartz, Chapman, Terrell, Prelock, and Rowan (1985). In this variant of expansion, the experimenter, rather than the child, initiated the interchange. For example, in teaching one child locative word combinations, the experimenter arranged interchanges such as that shown in (1).

(1) *Experimenter:* What's this?
     Child: Block.
     Experimenter: What's the block in?
     Child: Truck.
     Experimenter: The block is in the truck.

Schwartz et al. found that this type of approach was effective in increasing the children's use of early word combinations.

Other Approaches   The treatment approaches reviewed above are among the most common, but this list is not exhaustive. For example, there are approaches in which the child hears the target form and is asked to respond by pointing to corresponding pictures. Production attempts are not requested. Paluszek and Feintuch (1979) found such an approach successful in teaching plural -s. Following treatment, both production and comprehension of this form were seen. Several other studies used similar procedures, though only comprehension was assessed following treatment. Examples of such studies are those of Bouillion (1973), Boyd (1980), and Foley, Schwartz, and Shamow (1976). In each case, real gains in comprehension were observed.

Some studies have examined the efficacy of approaches that cover a wide range of language abilities, often using a variety of teaching techniques. Cooper, Moodley, and Reynell (1978, 1979) employed an approach with activities designed to facilitate such skills as attending to directions, understanding the properties of objects, and associating objects with their names. Similarly broad approaches are reflected in the studies of Evesham (1977) and Weller (1979). In several cases, the broad approach to treatment was a preschool or elementary school educational experience designed for children with SLI. These experiences provided the children with situationally appropriate language stimulation in a group context and, in some studies, individual attention focused on specific target forms. The investigations of Ripley (1986), Rice, Sell, and Hadley (1990), Haynes and Naidoo (1991), and Rice and Hadley (1995) were of this type. In all of the studies involving broad-based intervention, genuine gains in language ability were observed.

It can be seen, then, that many different treatment approaches have been successful, at least when success is measured in terms of language gains exceeding those that can be expected by maturation alone. But, along with the encouraging results, there are a few unsettling findings. One of these was reviewed earlier: Connell (1980b) found no gains in children's use of grammatical morphemes following a type of recasting procedure. Perhaps the most disturbing finding emerged from an investigation by Whitehurst, Fischel, Arnold, and Lonigan (1992). These investigators studied the outcomes of children who had exhibited limitations in expressive language skills at age 28 months. After a period of treatment from age 28 to age 34 months, the children were followed for another two and a half years. During this period, many of the parents sought treatment for their children in the community. However, Whitehurst et al. found that the children who did not receive treatment fared as well as those who did. The type of treatment received by the latter children varied considerably. Furthermore, most of the children had already achieved age-appropriate levels of language ability by 44 months of age. Nevertheless, this finding serves as a call for further study, especially given the advantage shown for treatment groups over control groups in so many investigations covering a variety of treatment approaches.

It is clear that some of the investigative attention in the future should be devoted to some fairly basic issues. We still know surprisingly little about the learning curves of children with SLI using each treatment approach, and how these curves interact with the children's level of severity at the outset of treatment. Studies that have examined long-term outcomes of children with SLI typically record whether each child received treatment, and if it was received, how much. However, in these studies, it has not been possible to document the particular type of treatment approach that was used with each child. From these investigations, it seems that initial severity level is a better predictor of a child's later abilities than is the amount of time enrolled in treatment (Aram & Nation, 1980; Bishop & Edmundson, 1987a). In fact, no correlation has been found between duration of treatment and posttreatment language ability.

Although predicting children's success with any given procedure is extremely difficult, probably the most difficult case is one in which the children show no evidence of producing the language target that is to be taught. Even if the children give evidence of comprehending the target, the absence of the target from their speech provides no indication of their readiness to acquire the target in production. Some of these children may be on the verge of acquiring the target on their own; others may be months away from learning it even with assistance (Fey, 1986; Leonard, 1981).

Investigators are beginning to explore this issue, employing the notion of dynamic assessment (e.g., Olswang, Bain, & Johnson, 1992). In a study designed to assist children with SLI in the use of two-word utterances, Bain and Olswang (1995) evaluated the children prior to treatment, using a protocol that examined the amount of support the children needed to produce utterances of this type. Six degrees of support were used, ranging from explicitly asking the child to imitate (e.g., "Tell me, 'dog walk'") to simply calling the child's attention to the stimulus materials (e.g., "Oh, look at this"). The degree of support needed to elicit a two-word utterance during the evaluation phase served as a good predictor of the children's gains during treatment.

Comparisons of Treatment Approaches
It seems doubtful that any single treatment approach can be ideal for all children with SLI and for all structures of language that might be taught. A meta-analysis by Nye, Foster, and Seaman (1987) seems to support this assumption. These investigators found a large mean effect size for treatment collapsed across type of approach. However, although the mean effect size for particular approaches was quite high (e.g., modeling), no one approach differed significantly from any other.

Although there is no single best treatment approach, it is reasonable to expect certain approaches to be more effective than others for some structures and for some types of children with SLI. Following this assumption, many investigators have conducted comparison studies of two approaches in terms of their efficacy in helping children with SLI learn particular structures. In some studies, children with SLI have been divided into ability groups, in case the efficacy of a particular approach is limited to children operating at a particular level.

Several investigators have compared imitation-based approaches with other types of approaches, such as milieu teaching (Cole & Dale, 1986), focused stimulation (Friedman & Friedman, 1980; Hughes & Carpenter, 1989), and general language stimulation (Dukes, 1974). In most of these studies, specific grammatical forms were the targets. Gains were seen when using imitation but, with one exception (Hughes &

Carpenter, 1989), these gains were no greater than those seen for the other approach employed, at least when the children with SLI participating were taken as a single group, regardless of ability.

Considerable attention has been devoted to comparisons between imitation-based approaches and modeling approaches. Two studies by Courtright and Courtright (1976, 1979) provided evidence that modeling was more effective than imitation in the teaching of grammatical forms. In the first study, the version of modeling used permitted the child to produce examples of the target form during the treatment phase; in the second study, this was not the case. Connell (1987) found just the opposite in a study designed to teach children the use of an invented suffix. The children with SLI made greater gains using an imitation approach than a modeling approach. In the particular modeling approach employed by Connell, the children were not permitted to produce the form until posttesting. Interestingly, a group of normally developing children of the same age as the children with SLI also participated in the Connell study, and these children were more successful with modeling than with imitation.

Connell and Stone (1992) also conducted a comparison between imitation and modeling using invented suffixes. In their study, children with SLI, age controls, and younger controls matched according to language comprehension test score participated. The three groups' acquisition of the target forms in comprehension showed no advantage of one treatment approach over the other. However, in production, the children with SLI showed greater gains with the imitation approach. The two groups of normally developing children showed no such differential effects. According to Connell and Stone, this pattern of findings suggests that children with SLI have greater difficulty accessing new rules for production than acquiring them in the first place. As in Connell's earlier study (1987), the modeling approach used required no production from the children. Thus, even though the children with SLI could acquire the meanings of invented suffixes through this approach, the approach did not provide them with the practice needed to ensure access of these suffixes for production; only the imitation approach provided such practice.

Imitation has also been pitted against conversational recasting. Camarata and Nelson (1992) employed a design in which children with SLI learned one or two target forms through an imitation approach and an equal number of forms through conversational recasting. The forms taught via conversational recasting were produced spontaneously by the children after fewer presentations. Camarata, Nelson, and Camarata (1994) and Nelson, Camarata, Welsh, Butkovsky, and Camarata (1996) replicated this finding, and observed an additional difference between the two approaches: conversational recasting led to a greater number of spontaneous productions of the target than did imitation.

The apparent superiority of conversational recasting over imitation is surprising in at least one sense. Earlier, it was noted that imitation may be better than approaches that provide children only with an opportunity to hear the target forms because the latter approaches do not permit the children to practice accessing the newly learned target forms for production. Yet, in recasting, the children are not asked to produce the target form. Shouldn't imitation result in greater production gains than recasting?

One important asset of the recasting approach might be its use of the child's original utterance as the platform from which to introduce the target form. When children describe some event, they presumably have an interest in talking about it. A recast of their utterances, then, may provide the children not only with the target form but also with a highly and personally relevant context in which the form can be used. In contrast, imitation approaches make use of contexts chosen by the experimenter. Practice in accessing the form is provided, but the contexts in which such practice is acquired may be somewhat remote from the children's usual communicative contexts.

Ellis Weismer and Murray-Branch (1989) compared two types of modeling, one like that used by Connell and his colleagues, in which production responses were not requested, and one in which the child was given intermittent opportunities to produce examples of the target form. Both versions produced gains, but the version that permitted production by the child resulted in a slightly more stable pattern of acquisition.

This picture of production practice facilitating gains in treatment receives support from an early investigation by Barbeito (1972). In that study, children with SLI participated in one of two focused stimulation conditions. In one, the children simply heard instances of the target form; in the other, the children were requested to produce it. Gains were greater for the condition involving production.

Other useful comparative studies can be identified in the literature. Some compare general language stimulation approaches (e.g., Weller, 1979), whereas others compare approaches involving a combination of teaching techniques with a very specific focus, such as facilitating children's acquisition of new lexical items (Olswang, Bain, Rosendahl, Oblak, & Smith, 1986), two-word combinations (Olswang & Coggins, 1984), and word-finding skills (e.g., McGregor & Leonard, 1989; Wing, 1990). Conclusions must be guarded, given the diversity of approaches and target forms in these studies. However, it seems fair to say that the most successful approaches were those that encouraged production and provided multiple yet natural cues for the desired response.

Comparative studies have also been conducted to determine whether parents can, with instruction, engage in language-facilitating activities that lead to the types of gains in their children's language that are seen in treatment provided by clinicians. A nice example of a study of this type is an investigation by Fey, Cleave, Long, and Hughes (1993). These investigators taught parents of children with SLI to employ an approach with their children that incorporated focused stimulation, recasting, and questions to assist the children in learning grammatical forms. A comparable group of children with SLI participated in sessions employing the same procedures but with a clinician providing the treatment. The children in both treatment programs made greater gains than children in a control group. However, the most consistent changes were seen in the children working with the clinician. Fey et al. concluded that with careful monitoring by professionals, parents can serve as effective intervention agents.

Thus far, we have discussed comparisons of treatment approaches taking children with SLI as a single group. However, it is important to consider whether certain types of children with SLI respond more favorably to one approach and other children with SLI respond better to another approach. Friedman and Friedman (1980) compared an imitation-based approach and a focused stimulation approach in a study designed to teach children with SLI the use of particular grammatical forms. The gains were comparable for the children in the two treatment conditions. However, when Friedman and Friedman examined the children's gains as a function of their initial ability level, it became clear that the imitation-based approach was more effective for the children with lower abilities, whereas the focused stimulation approach was superior for the children with more advanced abilities.

Cole and Dale (1986) also examined the role of initial ability in the relative efficacy of treatment approaches. Children with SLI were assigned to either an imitation-based approach or an approach representing a version of milieu teaching. A set of syntactic, semantic, and pragmatic skills constituted the targets. Both approaches resulted in demonstrable improvement. However, Cole and Dale found no evidence for the kind of aptitude-by-treatment interaction seen by Friedman and Friedman. There are several possible reasons for the discrepancy between the findings of the two studies, not the least of which is the fact that the specific approaches being compared in the two studies were not the same. As noted by Cole and Dale, Friedman and Friedman did not randomly assign children to the treatment conditions, whereas Cole and Dale did. Further, the children in the Cole and Dale study showed lower nonverbal IQs; in fact, some of the children in each treatment condition did not achieve the customary level (IQ of 85 or higher) necessary for the diagnostic category of SLI.

This last observation is not trivial. Yoder, Kaiser, and Alpert (1991) suggested that the children with the higher nonverbal IQs in the Cole and Dale (1986) work were the most comparable to the lower ability children in the Friedman and Friedman (1980) investigation, and therefore might have responded more successfully to the imitation-based approach. Their own comparative study of imitation-based and milieu teaching approaches provides some support for this possibility. The average IQ of the children in the Yoder et al. study was 69. However, a few children exhibited IQs that were sufficiently high to fall into the SLI category. These children approximated the ability levels of the lower ability group in the Friedman and Friedman study, and these children, too, responded more favorably to the imitation-based approach.

It is fair to conclude that we have not reached a point of knowing which approaches are the most effective for teaching particular target forms. Similarly, it is not yet clear which children benefit most from particular treatment approaches. Although the work conducted to date constitutes a healthy beginning, there are many gaps to fill. Until the evidence is more definitive, there is the consolation that educated guesses about the best treatment approach for a child probably won't be far off the mark; as we have seen, most treatment approaches result in demonstrable gains.

What Is Learned?

But maybe the above conclusions are too optimistic. Findings of real gains can take many forms. Some of these may constitute little more than tricks the children have learned to perform in the treatment setting. We need evidence that the abilities resulting from treatment represent real language that is maintained and used in natural circumstances. The relevant evidence is reviewed here.

Use of Newly Learned Structures in New Utterances

It is safe to conclude that the gains seen as a result of treatment reflect more than rote learning. Even in studies with a narrow focus, in which children are taught only one or two target structures, the probes employed ensure that the children use the structures in new utterances. In the study by Wilcox and Leonard (1978), for example, children with SLI were taught to ask questions such as *What does the boy push?* and subsequently produced similar questions not included in treatment, such as *What does the baby want?*

However, there may be exceptions to this type of finding. Studies designed to teach a nonsense suffix to children in only one or two sessions seem to produce lower levels of generalization (Connell, 1987; Connell & Stone, 1992; Roseberry & Connell, 1991; Swisher, Restrepo, Plante, & Lowell, 1995; Swisher & Snow, 1994). In each of these studies, fewer than half of the children with SLI showed an ability to apply the suffix to new words. Yet, this could be a function of the limited amount of training. Age controls also participated and, in each of these studies, a few of them likewise failed to apply the suffix to new lexical items. In some studies, the proportion of age controls failing to generalize approached one-third (Swisher & Snow, 1994).

Use of Newly Learned Structures in Spontaneous Speech

Many studies have assessed the children's ability to employ newly learned structures in untrained utterances by using tasks that resemble those used in treatment. However, a number of investigators have made the requirement stiffer: the children must use the structure in new sentences during conversational speech. Most studies employing this type of measure have reported positive results (e.g., Camarata, Nelson, & Camarata, 1994; Connell, 1986a, 1986b; Olswang, Bain, Rosendahl, Oblak, & Smith, 1986; Warren & Kaiser, 1986; Warren, McQuarter, & Rogers-Warren, 1984). An investigation reported by Hughes and Carpenter (1989) appears to be an exception. In that study, children with SLI showed evidence of gains in the use of copula *is* and *are* on structured probes after treatment, but generalization to spontaneous speech was minimal.

Maintenance of Learned Structures

Gains made during treatment are not very meaningful if they are short-lived. Therefore, some investigators have made efforts to determine whether the abilities reflected immediately after treatment are maintained. The results suggest that they are. For example, studies by Dollaghan and Kaston (1986), McGregor (1994), McGregor and Leonard (1989), Warren, McQuarter, and Rogers-Warren (1984), and Wright (1993) all provided evidence that levels of ability approximately one month after treatment ended were much higher than levels seen prior to treatment.

Broader Gains

Many investigations report gains in grammatical structures other than the ones treated. In some cases, the untreated structures closely resemble those treated. For example, findings showing improvement in copula forms after treatment on phonetically identical auxiliary *be* forms or vice versa are abundant in the literature (e.g., Gray & Fygetakis, 1968; Hegde, 1980; Hegde, Noll, & Pecora, 1979; Leonard, 1974). This type of extension seems to be based on two factors, phonetic similarity and grammatical similarity, operating in tandem. Neither of these factors is sufficient by itself. For example, treatment on auxiliary *is* does not seem to lead to immediate improvement on auxiliary *are* (e.g., Leonard, 1974), nor does treatment on noun plural *-s* result in gains on the third-person singular *-s* inflection (e.g., Gottsleben,

Tyack, & Buschini, 1974). Later in this chapter we shall have more to say about such findings.

In other studies, the language gains exhibited by children with SLI following treatment are broad-based. That is, the children provided evidence of increases in aspects of language that included vocabulary as well as grammar, grammar as well as discourse, and overall grammatical accuracy as well as accuracy with particular grammatical structures (e.g., Dukes, 1974; Evesham, 1977; Fey, Cleave, Long, & Hughes, 1993; Lee, Koenigsknecht, & Mulhern, 1975; Rice & Hadley, 1995).

Although some of these gains covered areas of language that were not explicit goals of the treatment plan, the nature of treatment in these studies usually involved general language stimulation as well as more intensive practice with particular forms. For example, in the investigation of Lee, Koenigsknecht, and Mulhern (1975), stories were presented to the children that not only emphasized particular grammatical forms but also served as a means of teaching new lexical items, narrative cohesion, and other grammatical forms. In some of these studies, treatment on specific language targets was supplemented by group activities such as drawing and having snacks that provided the children with additional language stimulation.

Does Treatment Allow Children with SLI to Catch up to Peers or Not Lose Further Ground?

The evidence reviewed above supports the view that with treatment, children with SLI make gains in language skills exceeding those that can be expected without treatment. But how should such gains be characterized? In chapter 1, we saw that the emergence of language is late and its development protracted in children with SLI. That is, the acquisition slopes for many of these children are shallower than normal, such that these children fall further behind their age peers across time. Given this state of affairs, gains attributable to treatment might constitute a catching up to same-age peers or, more modestly, a change in the slope of the acquisition curve, so that the children begin to progress at a normal rate, no longer falling further behind with each passing month.

There is an obvious sense in which catching up is a common occurrence. Seven-year-old children with SLI who, through treatment, learn to produce articles consistently in obligatory contexts have in a sense caught up to peers. However, their acquisition rate during treatment hardly matters, because article use by normally developing children is already mastered—and is thus at ceiling levels—by four years of age.

The more important sense of catching up is seen when the child's rate of acquisition in treatment exceeds the rate seen in normally developing children without treatment. This is illustrated in figure 10.1. The alternative is shown in figure 10.2. Here, there is an adjustment in the initially slow rate of development once treatment begins, so that the rate approximates that seen in normally developing children without treatment. However, the late start prevents the child from achieving the ability levels seen in age-mates—assuming, of course, continuous development of language material by the latter.

There is evidence for both of these outcomes. Several case studies report findings of accelerated growth. For example, Crystal, Fletcher, and Garman (1976) describe a child with SLI who progressed from producing single-word utterances at age 3;5 to multi-word utterances such as *Got a Land Rover here* in a four-month period. The

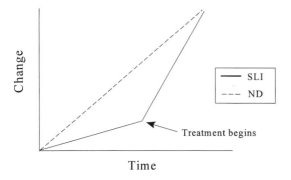

Figure 10.1
Treatment-related acceleration of language abilities in children with SLI that results in levels commensurate with those of normally developing (ND) children.

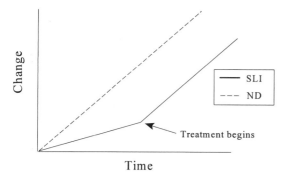

Figure 10.2
Treatment-related acceleration of language abilities in children with SLI that enables the children to proceed at a rate typical of normally developing (ND) children, though age-appropriate levels are never achieved.

strongest evidence comes from studies employing standardized tests, for these document how children with SLI compare with the norm, both before and after treatment. The results are clear; nearly all studies of this type present evidence of children with SLI closing the gap (e.g., Cole & Dale, 1986; Craig & Koenigsknecht, 1973; Dukes, 1974; Evesham, 1977; Fey, Cleave, Long, & Hughes, 1993; Friedman & Friedman, 1980; Rice, Sell, & Hadley, 1990; Warren & Kaiser, 1986). In some of these studies, raw scores were converted to language ages. For example, Lee, Koenigsknecht, and Mulhern (1975) reported a mean gain of nearly 11 months after 8 months of treatment. Cooper, Moodley, and Reynell (1979) observed gains of at least 16 months after a 12-month treatment period.

Language ages are not the ideal metric of language change, because they are extrapolations or interpolations rather than comparative statements about how a child performs relative to age peers in the standardization sample. Standard scores are preferable. And here, too, the evidence is impressive. An especially good example comes from the work of Rice and Hadley (1995). Children with SLI were seen in a classroom-based language intervention setting for either a 10-month or a 22-month period. Their standard scores on several different language tests revealed increases.

For example, the average standard score on the Peabody Picture Vocabulary Test-Revised (PPVT-R) (Dunn & Dunn, 1981) at entry was 83; following completion of the program, the average was 95. Similarly, standard scores on the expressive subtest of the Reynell Developmental Language Scales (Reynell & Gruber, 1990) averaged 72 initially and 82 at exit. The children enrolled for a longer period made greater gains. However, averages for the children enrolled for only ten months also were higher in the end for each test employed. Because these are standard scores, anchored to the child's chronological age, they would be expected to remain the same if the child merely kept pace. Gains in these scores, then, reflect an increase in the rate of development.

In a few studies, the gains are not only substantial, the posttreatment scores fall squarely within average levels. The PPVT-R results of Rice and Hadley (1995) discussed above represent one such example. Standard scores on this test have a mean of 100 and a SD of 15. The observed posttreatment average of 95, then, is well within the norm. Evesham (1977) also has reported treatment changes leading to age-appropriate standard scores—in this case, verbal IQs on the WISC falling in the normal range.

Not all investigators have found evidence of children with SLI closing the gap. Haynes and Naidoo (1991) examined the language test scores of children who were enrolled in a special school for children with SLI. The children varied in the length of time at the school. The average duration from test at entry to final test was approximately three years. Standard scores on vocabulary comprehension measures actually dropped; mean scores went from 86 at entry to 80 at departure. Scores on grammatical comprehension measures increased slightly, however, from approximately one SD below the mean for the children's ages to just under the mean.

It does not seem likely that the extensive gains reported in the literature are mere regressions toward the mean, statistical artifacts due to pretreatment scores that underestimated the children's true language ability. Gains of similar magnitude have been reported for tests on which the children were initially close to, or within, the norm (e.g., Lee, Koenigsknecht, & Mulhern, 1975), as well as for tests on which the children performed poorly at the outset. In some studies, groups of normally developing children also received treatment, and their scores, too, showed impressive gains even though their initial scores were already relatively high (e.g., Dukes, 1974).

Again, the Rice and Hadley (1995) study proves informative. Typically developing children were enrolled in their language intervention program along with children with SLI, and these children, quite naturally, showed standard scores well within the norm at entry. These typical children made gains during intervention. For example, standard scores on the PPVT-R increased from 101 to 107. Furthermore, some of the children with SLI had deficits seemingly limited to language production. For these children, scores on the comprehension tests were at age level from the start. However, these scores increased along with the scores for expressive language. For example, standard scores on the receptive subtest of the Reynell Developmental Language Scales increased from 100 to 108.

The study by Whitehurst, Fischel, Arnold, and Lonigan (1992), mentioned earlier in this chapter, provides another form of evidence that treatment effects are real. Following a period of treatment, 28-month-old children with significant limitations in expressive language were found to make larger gains in expressive language ability than a comparable group of children not receiving treatment. By 34 months of age, these children approximated age-level expectations. One of the noteworthy aspects of this study is that the control children, too, caught up to age level somewhat later, by 44 months of age. Of course, the latter finding raises the question of whether treatment was really necessary for the children receiving it. But it does seem to show that language gains in a group of children likely to catch up can be accelerated further with treatment.

The acceleration of acquisition through treatment notwithstanding, many children with SLI do not seem to reach a level of language ability that can be regarded as socially or educationally adequate. These children do not seem to catch up. This outcome obviously holds for the children studied by Haynes and Naidoo (1991) reviewed above, given that these children did not even show relative gains in one of the areas of language studied, vocabulary comprehension. But when an acceleration is observed, the gains often fall short of levels assumed for normal functioning. An investigation by Ripley (1986) provides an illustration. The children in this study had attended a school for children with SLI since the age of seven or eight years. During their period of enrollment in the program, they showed substantial gains in verbal IQ scores, with typical increases on the order of 15 to 30 points. However, almost half of these children remained at this school until the statutory age of departure, and their average verbal IQ never rose above 1 SD below the norm.

A study by Huntley, Holt, Butterfill, and Latham (1988) provides another example. These investigators tested the abilities of children with SLI five years after they had completed a two-year period of language treatment. Standard scores for these children decreased slightly in the five years following termination of treatment, though they were still considerably higher than the scores seen before treatment. However, nearly half of the children did not yet qualify for normal school placement, or had been in a regular school setting but did not succeed. Padget (1988) provides data from younger children that make a similar point. In this study, only approximately one in four children with SLI who received treatment in preschool were ready for regular kindergarten placement.

It can be seen, then, that the literature on treatment provides room for both optimism and concern. On the one hand, treatment seems to accelerate language learning in many children with SLI. On the other hand, for some children, this acceleration does not carry far enough to lead to normal language functioning. For such children, language problems, though mitigated, will remain as obstacles to social and academic success.

What the Pattern of Learning Can Tell Us About SLI

It was noted in chapter 2 that treatment designs can also be used to evaluate theoretical accounts of SLI. The studies employing these designs specifically for this purpose will be reviewed in chapters 11 and 13, where the accounts they were designed to test are reviewed. However, some of the findings covered in the present chapter have theoretical as well as clinical management implications.

For example, it does not seem plausible that the children participating in these studies could have had deficits involving the permanent absence of particular morphosyntactic features from their underlying grammar or the absence of a rule-making capacity. These children may learn slowly, but the product of their learning seems to be the creative application of rules involving a variety of morphosyntactic features. It is not clear if the degree of application of these rules is as extensive as we see in normally developing children. But, rules they are.

The patterns of generalization seen in the efficacy studies can be used to argue that the children with SLI who participated had fundamentally normal underlying grammars. Children with SLI seemed to acquire copula *is* as a result of treatment on auxiliary *is*, whereas treatment on a form such as noun plural -*s* did not lead to gains in the third-singular -*s* inflection. It is not difficult to imagine how this might have worked. Children with SLI may be slow in registering and incorporating the specific elements in the ambient language that are needed by the grammar. In treatment, the children were assisted in this process. It seems likely that the children noted the phonetic properties of the newly introduced form, then formed hypotheses based on its grammatical role. Parallels between auxiliary and copula *is* could be drawn based on the fact that these two forms agree with the subject and move within the sentence in the same way. In contrast, plural -*s* differs from third-singular -*s* because it marks number on nouns, without regard to the number (or, of course, person) of other lexical items. In the next two chapters, we'll see that these findings play a role in the evaluation of some of the competing accounts of SLI.

PART V

Theoretical Issues

Chapter 11
SLI as a Deficit in Linguistic Knowledge

Problems with morphosyntax are notorious in SLI, and it is therefore no surprise that many accounts of this disorder are centered around grammar. Of the various accounts that focus on grammar, six of them treat grammatical deficits as a knowledge problem. That is, it is assumed that the weaknesses seen in the production and comprehension of grammatical details by children with SLI are the result of incomplete knowledge of particular rules, principles, or constraints. Although the accounts differ in the specific type of knowledge the children are assumed to be lacking, each permits predictions about the pattern of relative strengths and weaknesses that should be seen in the children's sentence production or comprehension as the result of the assumed limitation in grammatical knowledge. Because most of these predictions are heavily grounded in a linguistic framework, we shall begin with an overview of a framework that can accommodate each of the accounts to be reviewed here.

An Overview of Grammatical Structure Within the Principles and Parameters Framework

The particular framework we will adopt is that of "principles and parameters" (Chomsky, 1981, 1986), a framework that has its roots in earlier versions of transformational grammar. The primary goal of the principles and parameters approach is to explain how language is learned. It departs from earlier approaches in assuming that what the child learns must be constrained in such a way that only those variations seen in natural languages are considered; hypotheses that are alien to natural languages are never entertained. The constraints that are responsible for this are the "principles" from which the framework gets part of its name. Of course, it is not sufficient to have principles that limit a child's hypotheses to those details that appear in natural languages. Languages differ a great deal from one another. The framework must also show how children can acquire very different languages so quickly. This is possible because languages do not differ haphazardly, but rather in terms of sets of characteristics that vary systematically from language to language. That is, there are "parameters" along which languages vary.

In this framework, as in its predecessors, a distinction is made between the underlying structure of a sentence and its surface form. The underlying structure provides the abstract representation of a sentence, with grammatical relationships specified. This structure is related to the surface form of the sentence through highly constrained movement rules. The underlying structure assumed for English is shown in (1).

(1)

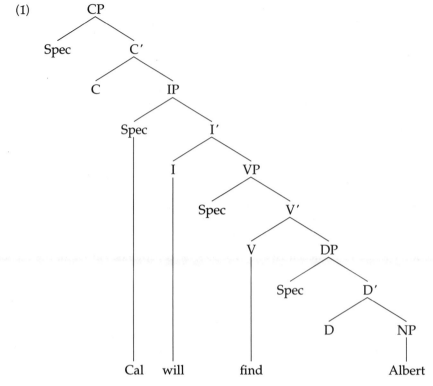

The categories shown in (1) can be divided into lexical categories and functional categories. The familiar categories noun (N) and verb (V) and their maximal projections noun phrase (NP) and verb phrase (VP) are examples of lexical categories. Other familiar examples (not shown above) are preposition (P) and adjective (A) and their phrasal projections PP and AP. The functional categories are complementizer (C), inflection (I), and determiner (D) and their phrasal projections CP, IP, and DP. Because children with SLI have significant difficulty with aspects of grammar associated (coincidentally or not) with functional categories, these categories will receive the greatest attention.

From (1) it can be seen that all phrasal projections have the same structure, illustrated in (2).

(2)

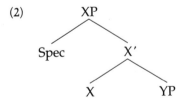

For example, the phrasal projection VP shown in (1) branches into a specifier (Spec) and an intermediate category V'. This in turns branches into V, considered to be the head, and another phrasal projection, DP.

The I-System

The properties of grammar that are associated with I and IP are sometimes referred to collectively as the "I-system." Many of the grammatical elements considered to be parts of the I-system are shown in (3).

(3)

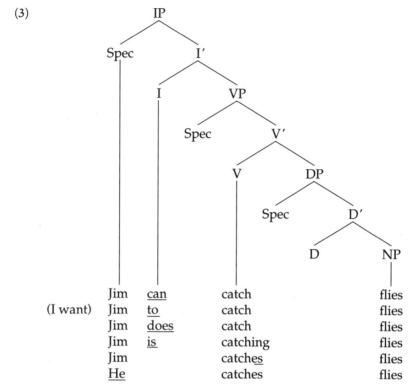

Modal auxiliaries (e.g., *can*, *will*) and infinitival *to* are assumed to be base-generated (generated in the underlying structure) in the head I position of IP. If I is finite (carrying features of tense and agreement), it is filled with a modal ("Jim *can* catch flies"); if it is nonfinite, the I position is occupied by *to* ("I want Jim *to* catch flies") If I is underlyingly empty, its tense and agreement features can be discharged by filling I with the dummy auxiliary *do* ("Jim *does* catch flies"). Another way to fill an empty I is through movement of auxiliary *be* or *have* or copula *be* from the head V position in VP ("Jim *is* catching flies"). In structures lacking an auxiliary in VP and in which I is underlyingly empty, the tense and agreement features of I can be moved to the head V and take the form of an inflection ("Jim catch*es* flies").

The role of the I-system is not limited to details traditionally associated with verbs. It can be seen in (3) that the grammatical subject appears in the specifier position of IP. The subject occupies this position through movement from the specifier position of VP. In its new position, the subject can receive nominative case assigned by I ("*He* catches flies").

In passive sentences, the specifier position of IP serves as the landing site for the theme that is located in postverbal position in the underlying structure. When the theme vacates its original position, it leaves a "trace" (*t*) that is coindexed with itself; this coindexing is designated with the subscript $_i$. An example is shown in (4). The auxiliary required as part of the passive morphology is located in I.

(4)

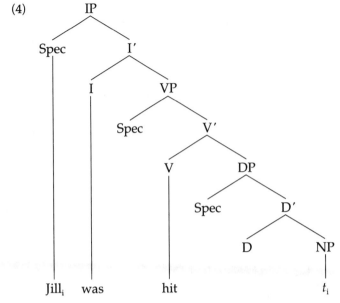

Jill$_i$ was   hit   t$_i$

For most of the accounts of linguistic knowledge deficits discussed in this chapter, the structure shown in (3) is sufficiently detailed to capture all predictions concerning verb morphology and case. However, some accounts follow proposals that assume a structure in which tense and agreement are given category status (designated as T and AGR, respectively) and replace I. Furthermore, a distinction between AGR$_s$ and AGR$_o$ is sometimes made; the former licenses the nominative case seen in subjects, the latter, the accusative case seen in objects. This structure is illustrated in (5).

(5)

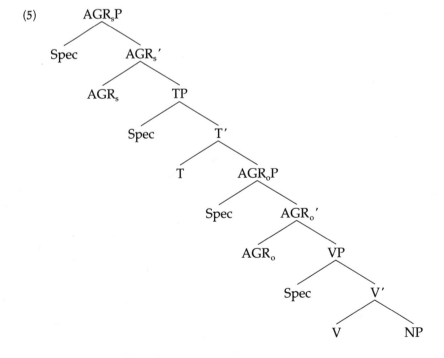

Following Chomsky's minimalist program (1993), it is possible to view the precise role of the functional categories somewhat differently. Chomsky has proposed that nouns, adjectives, and verbs are fully inflected in the lexicon. The functional categories T, AGR_s, and AGR_o are loci for tense, agreement, and case features that check off the corresponding features of the lexical items that move up to these categories. Specifically, verbs will raise to AGR_o, to T, and to AGR_s to check their tense and agreement features. Nouns will move up to the specifier position of AGR_oP to check accusative case and to the specifier position of AGR_sP to check nominative case. The difference between this formulation and the preceding one, then, is that instead of functional categories being the source of features that get inserted on lexical items, the features carried by these categories merely check off the same features on nouns and verbs.

An important implication of these latest developments in the principles and parameters framework for the accounts of linguistic knowledge deficits discussed in this chapter is that accusative case is now associated with a functional category, AGR_o. That is, the direct object is assumed to move to the specifier position of AGR_oP, where it receives accusative case (or checks accusative case). We will have recourse to these latest formulations of principles and parameters when they are really needed. Otherwise we will retain the more general functional category I and our original description of its operations.

The C-System

Another functional category system reflected in (1) is the C-system. In English, the C-system is most relevant for questions and certain types of complex sentences. From (6) it can be seen that the head C position serves as the landing site for auxiliaries that are moved from I in questions ("*Is* Ike watching TV?"). The specifier position of CP hosts the *wh-* word moved from its base position ("*What* is Ike watching?"). When the *wh-* word vacates its original position, it leaves a trace. It is also assumed that overt complementizers such as *that* and *if* are base-generated in C ("I know *that* Ike is watching TV").

(6)

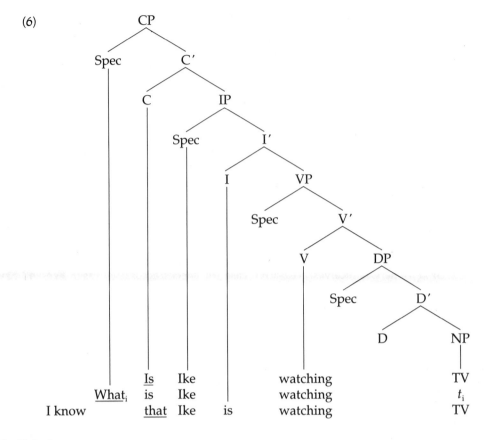

	Is	Ike		watching		TV
	What$_i$	is	Ike		watching		t_i
I know		that	Ike	is	watching		TV

The D-System
The remaining functional category system is the D-system. The elements associated with this system are illustrated in (7). The articles *a* and *the* and prenominal determiners such as *this* and *that* are assumed to be generated in D ("*the* potatoes," "*these* potatoes"). The inflection *'s* is also assumed to be in the head D position, assigning genitive case to the lexical item located in the specifier position of DP ("Ma*'s* potatoes"). Pronominal possessives ("*her* potatoes") are assumed to be in the specifier position of DP, receiving genitive case from an empty D.

(7)

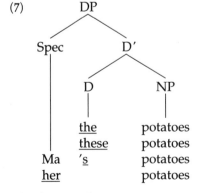

Clearly, functional categories constitute a vital part of morphosyntax. The grammatical elements associated with the I-, C-, and D-systems cover a broad range of grammatical morphemes. These include modals, auxiliary *do*, copula and auxiliary *be*, verb inflections marking tense and agreement, infinitival *to*, complementizers such as *that*, inverted auxiliaries, transposed *wh-* words, articles and other prenominal determiners, genitive *'s*, and pronominal possessives. In addition, functional category systems are required for movement of grammatical subjects and objects to positions in which they are assigned case.

Functional Category Deficits

Several investigators have proposed that individuals with SLI have special problems with the acquisition of functional categories (Eyer & Leonard, 1995; Guilfoyle, Allen, & Moss, 1991; Leonard, 1995), though some of these proposals have focused on the I-system in particular (Loeb & Leonard, 1991). One rationale for this type of proposal comes from work with young normally developing children. Some researchers have argued that in the earliest stages of grammatical development, the grammars of children acquiring languages such as English and Swedish contain lexical categories only (e.g., Guilfoyle & Noonan, 1992; Platzack, 1990; Radford, 1988, 1990). For example, according to Radford (1990), an utterance such as *Dogs like bones* would have the structure shown in (8) in early child grammars.

(8)

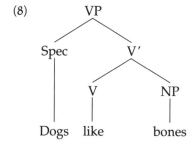

Specifically, it is assumed that there is no IP, and therefore no movement of the subject. Furthermore, the verb receives no tense and agreement features, given the unavailability of I. According to proposals of this type, children's grammars do not expand to accommodate functional categories until children reach approximately two years of age. Before that time, children's utterances occasionally contain forms that might be taken to be elements of functional categories. However, it is proposed, these are unanalyzed or misanalyzed forms that are incorporated into a grammar consisting of lexical categories only. For example, in a young child's utterance *What's his name?*, the *wh-* word might represent a base-generated subject in the specifier position of VP, *'s* might serve as an optional suffix of the *wh-* word with no agreement features, and *his* might have been misanalyzed as an adjective. Given this view, it is important to consider the contexts in which grammatical elements are used, as well as their frequency and diversity, before concluding that attested forms in the children's speech actually constitute the presence of a functional category.

Given the possibility that functional categories are later in appearance than lexical categories in normally developing children's grammars, it seemed reasonable to suspect that unusually slow development of functional categories was responsible for much of the difficulty seen in SLI. Consider the following. The grammatical morphemes most likely to be problematic for English-speaking children with SLI relative to MLU controls (see chapter 3) are all associated with functional categories. In addition, several grammatical morphemes, such as infinitival *to* and the complementizer *that*, have been examined only as part of a composite measure; but differences favoring MLU controls were the consistent outcome in these comparisons as well. One grammatical morpheme whose status in the speech of children with SLI is the subject of debate is noun plural *-s*; not all studies have found lower degrees of use of this form by children with SLI relative to MLU controls (e.g., compare Oetting & Rice, 1993 and Leonard, Bortolini, Caselli, McGregor, & Sabbadini, 1992). This morpheme is associated with a lexical category, not a functional category.

The use of functional categories to guide the study of the grammatical abilities of children with SLI has resulted in the discovery of possible weaknesses that might not have been observed otherwise. For example, the nonthematic particle *of* in *a slice of bread* is considered part of the D-system because it has only the purely grammatical function of assigning case to its complement. This form appears to be used with lower percentages in obligatory contexts by children with SLI than by MLU controls. To cite another example, even when children with SLI use auxiliary *be*, they are less likely than MLU controls to show movement of this element from I to C in *wh*-questions (e.g., *What Chris is buying?* instead of *What is Chris buying?*). Reports of a relationship between these children's use of finite verb forms (which require I) and the use of nominative case (which is licensed by I) also speak to the utility of analyses informed by functional category considerations.

The contribution of functional category analyses extends to other languages. Children with SLI acquiring Italian are more likely than MLU controls to omit direct object clitics. In the principles and parameters framework, these clitics are assumed to move from the NP in D' to the I position, as in (9). (The finite verb form is also located in I; in languages such as Italian, it is assumed that verbs move from V to I to assume tense and agreement features.) Thus, what might seem to be a surprising tendency to omit direct objects could actually turn out to be a problem with the I-system.

(9)

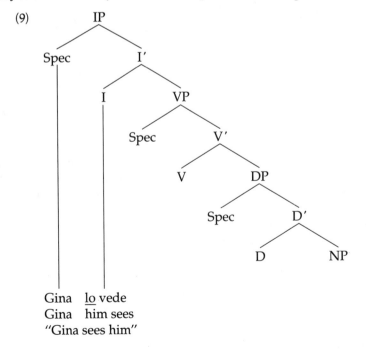

```
                        IP
                   /         \
              Spec           I'
               |          /      \
               |         I        VP
               |         |      /     \
               |         |   Spec     V'
               |         |          /    \
               |         |         V      DP
               |         |             /      \
               |         |          Spec      D'
               |         |                   /   \
               |         |                  D     NP
             Gina      lo vede
             Gina      him sees
             "Gina sees him"
```

Even though many differences are observed between children with SLI and MLU controls in the use of functional category elements, the evidence does not justify the conclusion that functional categories are absent from the grammars of children with SLI. All of the differences that have been reported are differences in degree of use; children with SLI have been found to use elements associated with functional categories with lower percentages in obligatory contexts, but they do use them.

As noted above, the occasional appearance of elements that resemble functional category elements does not ensure that these elements actually appear in functional categories in the children's grammars. However, even when conservative criteria are applied, the evidence for the presence of functional categories is plentiful. Employing such criteria in a study of ten English-speaking preschoolers with SLI, Leonard (1995) found that all of the children showed evidence of at least (1) articles; (2) the prenominal determiners *this* and *that*; (3) three different pronominal possessive forms; (4) the third-singular -*s* and/or regular past -*ed*; (5) some form of copula *be*; (6) some form of auxiliary *be*; (7) the modal forms *can* and *can't*; (8) the auxiliary form *don't*; (9) three different pronouns reflecting nominative case; (10) auxiliary inversion; and (11) an utterance-initial *wh-* phrase that could not be construed as the subject of the sentence (e.g., *What we pretend cook in here?*). The forms reflected in (1)–(3) serve as evidence for a D-system; those in (4)–(9) suggest the presence of an I-system; and the forms in (10)–(11) make a good argument for a C-system. Thus, the strongest conclusion that seems justified is that English-speaking children with SLI have extraordinary difficulty with a wide range of elements associated with functional categories. The functional categories themselves are present.

Evidently, not all functional category elements give these children special problems. Prenominal demonstratives are part of the D-system, yet they seem to be used as frequently and accurately by children with SLI as by MLU-matched control children. Likewise, pronominal possessives do not appear to cause special difficulties. The functional category system with the least frequently used elements is the I-system. However, the modals *can* and *can't* appear to be used as often by children with SLI as by MLU controls.

In other languages, it is not even proper to say that most grammatical elements associated with functional categories are especially problematic for children with SLI. Verb inflections, which make up an important part of the I-system, are generally used to similar degrees by children with SLI and MLU controls in Hebrew and Italian. Likewise, in Dutch, auxiliary use reveals no differences between children with SLI and MLU controls. Group differences certainly extend well beyond English, of course. Differences between children with SLI and MLU controls in German, for example, are quite striking, especially when omissions (rather than substitutions) of finite inflections and determiners are considered. However, the high degree of interlanguage variability in the findings for functional category elements raises serious questions about the accuracy of functional category deficit accounts.

It is true that languages are assumed to differ in certain operations within functional category systems. For example, we saw earlier that in English, tense and agreement features are assumed to move to V when there is no auxiliary, whereas in Italian and Spanish, the verb in V moves up to I to assume these features. However, if considerations of this sort (direction of movement) are responsible for the cross-linguistic differences observed, an account other than a proposal of a weak or missing functional category is in order.

The Extended Optional Infinitive Account

Rice and her colleagues have offered an account of grammatical difficulties in SLI that is called the "extended optional infinitive" account (Rice & Wexler, 1995a, 1995b; Rice, Wexler, & Cleave, 1995). This proposal has its basis in the work of Wexler (1994), who argued that young normally developing children go through a stage during which they fail to obligatorily mark tense in main clauses. According to Wexler, in spite of their inconsistency in marking tense, children at this stage know the grammatical properties of finiteness. For example, when young French-speaking children mark tense (and hence finiteness) on a verb containing a negative particle (*pas*), this marker will be correctly positioned after the verb. However, when the verb is produced in nonfinite form, the negative particle will appear in front of the verb, its proper position in infinitive constructions. Two examples from Pierce's (1992) review of the French language acquisition literature illustrate this distinction:

(10) a. elle roule pas
          it rolls not
      b. pas rouler en vélo
          not roll on bike

In a language such as French, the failure to mark tense is readily identifiable because the infinitive form used in place of the finite form has an overt suffix (e.g., *-er*, as seen in *rouler*, above). In English, the infinitive is a bare stem. Consequently, it is not obvious if the child is using a null form in a finite manner (analogous to *they run*, for example), or using an infinitive. It should be noted in this regard that *to* need not precede the infinitive; there are many instances in which infinitives are used without this form (e.g., *leave* in *she made him leave, I must leave*).

When young English-speaking children in the optional infinitive stage do produce finite forms, these forms are usually used correctly. *Nona sees me* might alternate with *Nona see me*, for example, but utterances such as *They sees me* will be rare. Similarly, when copula or auxiliary forms are produced, they will typically reflect appropriate person and number. If *are* is used, for instance, it is likely to be used with a plural subject (e.g., *They are big*).

There is one fact about young children's treatment of copula and auxiliary forms that on first appearance seems problematic for this account, yet it is handled in a plausible manner. Nonfinite forms such as *be* almost never appear in finite contexts. Either a finite form (usually the correct one) is used, or the copula or auxiliary is omitted. Thus, we see *Ginger is pretty* or *Ginger pretty* but not *Ginger be pretty*. Ordinarily, this fact would seem to pose a problem for the notion of an optional infinitive stage. Presumably, in a production such as *Rob run*, the child chose the nonfinite form *run*. It does not seem reasonable to assume that the option of an infinitive was derived from first selecting *runs* and then deleting the inflection. Assuming the direct selection of a nonfinite form, then, why shouldn't the child select the nonfinite *be*? The reason, according to Wexler, is that there is no syntactic motivation for selecting a copula (or auxiliary) form without tense. On the other hand, an overt main verb form—even a nonfinite one—must occupy (or originate in) V for thematic role purposes.

By the time normally developing children reach five years of age, they have proceeded to the more advanced stage of using finite forms consistently where these are required in the adult grammar. Children with SLI, in contrast, are assumed to remain in this earlier stage for an extended stay. "We do not know, in fact, if individuals with SLI will ever fully leave this stage" (Rice, Wexler, & Cleave, 1995, pp. 852–853).

There is a great deal of evidence that is consistent with the proposal that children with SLI have extraordinary difficulties with finite forms. As we saw in chapter 3, verb morphology is especially weak in English-speaking children with SLI. Problems of this type have been documented since the first systematic studies of grammar in these children (e.g., Leonard, 1972; Menyuk, 1964), and more recently, investigators operating from a variety of theoretical viewpoints have observed significant difficulties in this area (e.g., Gopnik & Crago, 1991; Leonard, Bortolini, Caselli, McGregor, & Sabbadini, 1992; Marchman & Ellis Weismer, 1994). Rice and her colleagues have reported some of the most detailed evidence of this sort. Data come from longitudinal studies (Hadley & Rice, 1996; Rice & Wexler, 1995b) and studies in which children with SLI are matched to normally developing children according to MLU and age (Rice & Oetting, 1993; Rice, Wexler, & Cleave, 1995). In two different samples of children, percentages of use of a range of finite forms were lower for children with SLI than for MLU controls and age controls. The forms in question were the third-person singular verb inflection -s, regular past -ed, copula and auxiliary be, and auxiliary do. Although these forms emerged late in children with SLI and their optional use persisted, when they were used, they were usually used correctly. This was true even for inverted copula and auxiliary be forms in questions. The children produced utterances such as *Are you happy?* and *Is he sleeping?* rather than, say, *Is you happy?* and *Are he sleeping?* Some errors of agreement were seen, but they were too few to contradict the conclusion that these children knew what they were doing. Corroborating data for verb inflections and copula be were reported by Leonard, Bortolini, Caselli, McGregor, and Sabbadini (1992).

The extended optional infinitive account does not take a stand on how long children with SLI remain in the optional infinitive stage. However, serious problems with tense and agreement can persist through the school years and even into adulthood (e.g., Marchman & Ellis Weismer, 1994; Ullman & Gopnik, 1994). There is little doubt, then, that this account is centered on a core problem of SLI.

It seems that the extended optional infinitive account might also provide an explanation for the frequent co-occurrence of preverbal pronominal case errors and problems with verb morphology. Loeb and Leonard (1991) and Leonard (1995) found that children with SLI were more likely to use nominative case when a finite verb inflection or auxiliary was produced than when the finite form was missing. Hence, utterances such as *She is going* and *Her going* appeared; those such as *Her is going* were less frequent. This relationship was attributed to problems with the functional category I. However, Schütze and Wexler (1995) proposed that this pattern can be attributed to whether or not tense is expressed. When the verb has tense, the subject must have nominative case; otherwise, the default accusative case may appear.

The fact that the extended optionality account provides a rationale for the co-occurrence of verb morphology and case errors is no small matter, for this type of pattern runs counter to common notions of performance limitations. For example, trade-off effects are sometimes seen when, in attempting to produce an utterance with increased grammatical complexity, the child shows some simplification elsewhere (see chapter 12). In the present example, one might expect that if children added a finite verb morpheme to an utterance, they might be prone to use an earlier form of preverbal pronoun. Instead, the use of finite verb morphemes seemed to promote the use of a more mature (nominative case) pronoun.

According to the extended optional infinitive account, the problem is one of knowledge; these children do not know that tense is obligatory in main clauses. Sometimes converging evidence of knowledge limitations can be obtained through comprehension tasks such as picture pointing tasks or tasks requiring the children to act out a sentence by using toys. However, the nature of the knowledge deficit assumed in this account cannot be tapped with such tasks. Assume, for example, a task in which the child is shown a picture of one deer looking toward a man and another picture of several deer looking toward a man. The child is then asked to point to the picture that corresponds to *The deer sees the man*, and then the picture that represents *The deer see the man*. Responses to the first sentence are expected to be correct because it is assumed that children with SLI understand tense and agreement. However, in response to the second sentence, the same picture might be selected. This does not mean that tense and agreement are not understood, only that the children treat them as optional. Given this state of affairs, a picture with a singular subject is as plausible as one with a plural subject.

To assess understanding in a manner that is sensitive to the assumptions of this account, a test of grammaticality is more appropriate. Children still in the optional infinitive stage should accept as grammatical sentences such as *The girl run every day* as well as those such as *The boy runs every morning*, but should reject sentences such as *The girls walks to school every day*. Judgment task data of this sort will strengthen the empirical basis of this account.

The extended optional infinitive account is among the most promising accounts of grammatical difficulties in children with SLI. It has a means of handling low degrees of use of several different grammatical forms while at the same time it offers a reason for the high accuracy of these forms when they do appear. It does not deal with problems that fall outside the sphere of influence of tense, but this is a limitation only to the extent that other accounts can handle a wider range of problems. This remains to be seen. And if an approach were to focus on a single type of problem, the one serving as the target of concern in this account is probably the most appropriate.

There is a logical problem in the way an extended optional infinitive stage is documented in the speech of children with SLI. The principal evidence takes the form of reliable differences between children with SLI and MLU controls in the use of finite verb forms. For example, Rice, Wexler, and Cleave (1995) found that MLU controls used third-singular -s in 45% of obligatory contexts, whereas children with SLI used it in only 30% of such contexts. Likewise, regular past -ed showed a difference (in the same direction) of 53% to 23%. For copula and auxiliary *be* forms, the corresponding figures were 65% and 45%. It is clear from these figures that the control children, too, were in the optional infinitive stage. This means that the notion of an extended optional infinitive stage must rely on the assumption that higher degrees of use that fall well short of mastery reflect something different than lower degrees of use. This is no doubt true, but the optional stage–mastery level dichotomy does not capture it.

Not all of the empirical evidence is consistent with the extended optional infinitive account. Leonard and his colleagues found that both Hebrew- and Italian-speaking children are comparable with MLU controls in their use of most finite forms, and when errors are made, finite rather than nonfinite forms are usually produced (Bortolini & Leonard, 1996; Dromi, Leonard, & Shteiman, 1993; Leonard, Bortolini, Caselli,

McGregor, & Sabbadini, 1992; Leonard & Dromi, 1994). Bottari, Cipriani, and Chilosi (1995) also reported only infrequent use of nonfinite forms where finite forms were required in the speech of Italian-speaking children with SLI.

The data for Hebrew and Italian might seem less problematic for this account if we assume that certain forms giving the appearance of being finite are in reality nonfinite. The most likely candidate in Italian is the third-person singular form; for Hebrew it is the masculine singular for present tense and third-person masculine singular for past tense. In fact, these forms are the most frequent substitutes when errors occur. However, children with SLI seem to be as proficient as their MLU-matched compatriots on most finite verb inflections, not just the ones that could plausibly serve as nonfinite forms.

In this regard, the data from Hebrew and Italian are somewhat different from data gathered from children with SLI who speak German. In this language, overt infinitives or participles are sometimes used in utterances requiring finite forms (e.g., Clahsen, 1989). However, close inspection of the data reveals that bare stems can also appear. When they do, it is not clear how they should be interpreted.

For example, S. Roberts (1995) found that when finite verb forms were not marked in the speech of German-speaking children with SLI, the error was as likely to be a bare stem as an infinitive or participle. When the infinitive or participle was produced, it appeared in final position, consistent with the predictions of the extended optional infinitive account. However, when a bare stem was used, it usually occupied the sentence position expected for the finite form. This suggests that some of the problems with verb morphology were not instances of choosing the nonfinite option. Rather, the children seemed to be simply omitting the finite inflection. If this proves to be the case, it opens up the possibility that some proportion of the bare stems produced by English-speaking children with SLI could likewise be examples of omissions of the inflection from finite verbs.

Implicit Grammatical Rule Deficit

Gopnik and her colleagues (e.g., Gopnik, 1990a, 1990b, 1994a, 1994b; Gopnik & Crago, 1991; Ullman & Gopnik, 1994) have described the grammatical problems seen in SLI as a serious and possibly permanent inability to acquire implicit rules to mark tense, number, and person. In its original formulation, the problem was characterized as a feature blindness (Gopnik 1990a, 1990b). That is, it was assumed that features of number, person, and the like were missing from the underlying grammars of individuals with SLI. As a result, morphophonological rules and rules that match features in the syntax were absent. Although productions having the appearance of inflected words were occasionally noted in the speech of individuals with SLI, it was assumed that these forms were simply phonological variants with no grammatical significance. This view was based on the observation that the forms sometimes appeared in inappropriate contexts. For example, the utterance *You make one points* was noted in the speech of the same child with SLI who produced *He only got two arena* (Gopnik, 1990b).

The feature blindness hypothesis was based on a school-age child with SLI (Gopnik, 1990b) and the British family described in chapter 3 (Gopnik, 1990a). Although the severity of the grammatical deficits in these individuals was noteworthy, the frequent omissions of grammatical morphemes from their speech represented a phenomenon quite familiar to investigators of SLI. Less typical was the characterization of seemingly random additions of inflection-like forms. Earlier studies (Leonard, 1972; Menyuk, 1964) had suggested that errors of the type *a cars* and *they runs* were infrequent in SLI. Indeed, more recent work suggests that these errors are no more likely in the speech of children with SLI than in MLU controls (Leonard, Bortolini, Caselli, McGregor, & Sabbadini, 1992; Rice, Wexler, & Cleave, 1995).

Based on additional data from the British family and other children with SLI, Gopnik and her colleagues now view the problem as one of an inability to formulate implicit grammatical rules. Without access to implicit rules, individuals compensate in one of two ways. One option available to them is to memorize inflected forms as unanalyzed lexical items, comparable with the learning of *went* as the past form of *go*. The second option is to employ explicit rules that have been taught to them, such as "add *-s* for more than one" or "add *-ed* to describe past events" (Gopnik, 1994a).

The lower percentages of use of grammatical morphemes typically reported for children with SLI relative to MLU controls could certainly be interpreted as reflecting the absence of implicit rules. That is, the lower percentages for children with SLI could in principle reflect a protracted period of accumulating stem-plus-affix forms because these forms must be learned by rote as separate lexical items. This is an inefficient way of acquiring new forms, for even in a language with relatively few inflections, such as English, the number of lexical items to be learned could double (e.g., *cats* as well as *cat*) or even quadruple (e.g., *playing*, *plays*, and *played* as well as *play*). So even though children with SLI are older than MLU controls and therefore have had more time to learn lexical items, it would not be surprising that their method of accumulating these forms could cause them to trail behind.

Yet lexical learning proceeds throughout life, even though the rate decelerates across time. Accordingly, the stem-plus-affix versions of most of the frequently occurring nouns and verbs should be acquired eventually even with this inefficient means of learning. However, the older adults in the family studied by Gopnik and her colleagues seemed no better off than the younger adults or adolescents (e.g., Gopnik & Crago, 1991).

But why doesn't the inefficiency of learning stem-plus-affix forms by rote slow down children with SLI acquiring Hebrew or Italian as much as it does children with SLI acquiring English? These children use most noun, verb, and adjective inflections as frequently and appropriately as MLU controls. If a crosslinguistic difference were to be expected, it would be in the opposite direction. The nonfinite and present tense forms of the Italian verb "play," for example, are *giocare*, *giocando*, *gioco*, *giochi*, *gioca*, *giochiamo*, *giocate*, and *giocano*; the number is even larger when other tenses are considered. Therefore, relative to their normally developing compatriots, children with SLI acquiring Italian (and Hebrew) should have an even larger proportion of forms to learn by rote, and differences between these children and their MLU controls should be even greater than is seen for English.

Another troublesome aspect of the rote learning assumption is that the low percentages of use of inflections by English-speaking children with SLI is not due to the consistent appearance of the appropriate inflection on certain lexical items and the consistent absence of the inflection from other lexical items. Rather, children with SLI are inconsistent in using inflections with the same lexical item (e.g., *She likes me; She like him*) (Bishop, 1994; Leonard, Eyer, Bedore, & Grela, 1997). A phonological explanation might handle some of these inconsistencies; for example, a child might have difficulty using an inflection when the following word begins with a particular consonant. For the inconsistency that remains after phonetic context is controlled, errors of retrieval seem to represent the only plausible explanation.

Grammatical errors attributed to word retrieval appear in the literature on normal language development. Marcus, Pinker, Ullman, Hollander, Rosen, and Xu (1992) explained young normally developing children's overregularizations of past (e.g., *throwed*) as momentary retrieval failures. Specifically, because the correct lexical item (*threw*) is not available in the instant it is needed, the default (and productive) regular past rule is applied to the present stem.

If rotely learned inflected forms are not retrieved successfully, children with SLI cannot apply a default rule, given the unavailability of implicit rules assumed in this account. And, given the high frequency of bare stems in English, it is likely that these forms would be used as the substitute. This frequency effect must be highlighted because, in this account, inflected forms have the same unanalyzed status as bare stems for these children. Thus, in principle, productions such as *plays* for *played* or vice versa can also occur.

The retrieval error idea seems less plausible when one considers the frequency with which such errors would have to occur to account for the data. Marcus, Pinker, Ullman, Hollander, Rosen, and Xu (1992) found that overregularizations of past—and hence failures to retrieve irregular forms—occurred much less frequently in young normally developing children than previously assumed. This figure averaged less than 5% across 25 children. Children with SLI are known to have word-finding difficulties (as reviewed in chapter 3), and therefore higher percentages might be expected in this population. However, given the fact that inflections are omitted from approximately 35% to 65% of obligatory contexts (depending on the inflection), one would have to assume a severity of word-finding problem not yet documented in the clinical literature. This couldn't have gone unnoticed because presumably the same children having problems retrieving rotely learned inflected forms would also have grave difficulties retrieving bare-stem verbs, nouns, and adjectives.

As Gopnik and her colleagues point out, individuals with SLI have another option—they can apply explicit rules that they were taught (e.g., "add -*ed* to describe past events"). Of course, such an option is available only to those reaching the age at which such instruction is appropriate. However, it could easily be included in treatment activities for children by kindergarten age.

This option allows for productions such as *throwed*. It would be assumed that the irregular form is unavailable either because the individual never learned it to begin with, or because it was not successfully retrieved at the moment it was required. Retrieving *throw* instead, the child then consciously applies the explicit -*ed* rule.

The same process operates when individuals with SLI participate in tasks requiring them to add inflections to nonsense words. For example, the child would, after conscious reflection, produce *zoops* in response to pictures and the prompt "This is a zoop. These are _____." It seems reasonable to assume that the same type of application can lead to the juxtaposition of function words with nonsense words, as in *to zoop* and *the tiv*. Perhaps because application of explicit rules is more taxing than that of implicit rules, individuals with SLI add endings less reliably than do controls (Gopnik & Crago, 1991; Ullman & Gopnik, 1994).

The interpretation that such productions are the result of explicit, metalinguistic rules rather than implicit rules is based on overt signs of working through the problem on the part of the individuals with SLI, such as hesitations or specific mention of the rule (e.g., one individual whispered "add an *s*" to herself). In addition, Goad and

Rebellati (1994) and Gopnik and Crago (1991) observed that endings were some-times added that failed to obey the phonotactic requirements of the language. For the nonsense word *sas*, for example, the stem-final [s] was lengthened. In other instances, [əz] was added to a stem requiring [s] (e.g., *zoop*) or [z] (e.g., *tob*), and [s] was applied to stems requiring [z] (e.g., *wug*).

As we saw in chapter 3, Fee (1995) also examined the productions of many of the same family members studied by Gopnik and her colleagues. Fee discovered that the family members had difficulties with word-final consonants and clusters. Of the eight individuals with SLI, seven showed syllable-final devoicing (e.g., final /d/ pronounced as [t]) and final consonant deletion, and all eight showed word-final cluster reduction. This raises the possibility that the peculiar use of inflections was the result of a serious phonological deficit rather than the deliberate and conscious selection of in-appropriate grammatical endings. It should be pointed out that Fee interpreted her findings as suggesting an inability to construct learned, language-specific phonologi-cal rules. If Fee is correct, the problem could still be one of absent implicit rules, as Gopnik and her colleagues argue. However, the responses to nonsense word items could have been reflecting the inability to apply phonological rather than inflectional rules. This seems to be an important topic for future research.

Other investigators have observed overregularizations of past in children with SLI (e.g., R. King, Schelletter, Sinka, Fletcher, & Ingham, 1995; Leonard, Bortolini, Caselli, McGregor, & Sabbadini, 1992; Rice, Wexler, & Cleave, 1995). Leonard, Eyer, Bedore, and Grela (1997) found that children with SLI and MLU controls were similar in the number of children from each group who applied grammatical morphemes to non-sense words. The degree of such use was slightly lower for the children with SLI (as was the use of these grammatical morphemes with real words). However, all of the children showed use of this type to some degree. Importantly, auditory perceptual judgments by experimentally naive listeners revealed no differences between the productions of the children with SLI and the control children. Evidence of compensa-tory activity such as comments or hesitations was also absent.

Swisher, Restrepo, Plante, and Lowell (1995) found that the children with SLI par-ticipating in their grammatical morpheme learning study made greater gains with a teaching approach that required implicit rule learning than one in which the rule was pointed out to the children. Such a finding is inconsistent with the idea that explicit rules are used in place of implicit rules. Finally, it should be acknowledged that several treatment studies reviewed in chapter 10 found evidence for the generaliza-tion of grammatical forms during posttesting. The designs of these studies seemed to eliminate factors other than grammatical rule learning as the basis for these findings.

In summary, there are some logical and empirical obstacles to the proposal that implicit rules are unavailable to children with SLI. The evidence for this account seems more plausible in the case of the individuals studied by Gopnik and her col-leagues. However, this impression could be due as much to the unusual severity levels seen in these family members as to how well the data fit the interpretation. The data are not so convincing as to rule out other possibilities.

Narrow Rule Learning

D. Ingram and his colleagues (Ingram & Carr, 1994; Morehead & Ingram, 1973) have proposed that the grammatical deficits seen in children with SLI are due principally to a restriction in the range of contexts to which rules are applied. Morehead and Ingram (1973) described the problem this way:

> In summary, linguistically deviant children do not develop bizarre linguistic systems that are qualitatively different from normal children. Rather, they develop quite similar linguistic systems with a marked delay in the onset and acquisition time. Moreover, once the linguistic systems are developed, deviant children do not use them as creatively as normal children for producing highly varied utterances.   (p. 344)

The words "as creatively as" are important in distinguishing this account from the account of Gopnik and her colleagues. It is assumed that rules can be applied productively. The problem is that these rules (when they finally appear) will apply over a more limited range of possible exemplars than they should.

There are two types of evidence that Ingram and his colleagues used to support their position. First, Morehead and Ingram (1973) found that a group of children with SLI produced grammatical constructions using a narrower range of applicable syntactic categories (e.g., NP, PP) than was seen for a group of MLU controls. Likewise, Ingram and Carr (1994) observed a restriction in the types of verb complements used in the speech of the child with SLI in their case study. These findings point to a rather narrow scope for the grammatical rules used by children with SLI.

Second, the children with SLI studied by Morehead and Ingram (1973) and Ingram and Carr (1994) were more limited in the range of their syntactic rules than in their use of grammatical morphology—contrary to many other investigations of grammatical morpheme use in English-speaking children with SLI. The children studied by Ingram and his colleagues were older than those participating in most other studies. According to Ingram and Carr (1994), this gave the children more time to expand the range of application of each grammatical morpheme in small increments, and/or to learn each word-plus-inflection as a lexical item. Presumably, expansion of syntactic rules to their proper scope of application requires even more time.

There are many findings in the literature that are consistent with this type of account. Patterns of use that can be viewed as the result of narrow rules are quite common, and can be found in every area of language. For example, it was seen in chapter 3 that phonological rules are not applied as widely as expected by individuals with SLI (e.g., Fee, 1995; Leonard, Schwartz, Swanson, & Loeb, 1987), two-word utterances reflect rather narrow relational meanings (e.g., Leonard, Steckol, & Panther, 1983), and verb alternation rules are more restricted in scope (Loeb, Pye, Redmond, & Richardson, 1994). The treatment literature reviewed in chapter 10 also contains evidence consistent with this type of account. Specifically, the generalization data showed that children with SLI extended the rules introduced in treatment in small steps, rarely departing dramatically from the examples used in the treatment sessions.

There are also areas of difficulty ordinarily attributed to some linguistic operation that might instead be due to the narrow application of rules. For example, Grela and Leonard (1997) observed higher degrees of subject omissions in utterances containing unaccusative verbs than unergative and transitive verbs. One obvious interpretation of this finding is that unaccusative verbs can be viewed as requiring movement of the theme to subject position. However, the problem might not be related to movement; the children may have developed a narrower subject rule that permitted agents but not themes.

Some accounts that assume limitations in linguistic knowledge have no way of explaining how the use of comparable grammatical inflections can differ so much across languages. These crosslinguistic differences in the speech of children with SLI are not problematic for the narrow rule account. Morphologically rich languages—the languages in which children with SLI show greater use of inflections—provide children with many more exemplars of the rules to be learned. Therefore, even though children with SLI will show narrower versions of these rules than normally developing children, they should show more extensive use of these rules than is seen in the speech of children with SLI acquiring languages with a sparse morphology.

This account has no difficulty with the fact that children with SLI sometimes produce overregularizations of past or succeed in using an inflection with a nonsense word. The fact that these children can acquire rules—and thus apply an inflection to new instances—is not in debate. Rather, the problem is in the generality of the rules.

On the other hand, considering the fact that English-speaking children with SLI show lower degrees of use of grammatical inflections in obligatory contexts than do MLU controls, it is surprising that the children with SLI show as much overregularization and use of the inflections with nonsense words as they do. That is, if lower percentages of use in obligatory contexts reflect the narrow scope of the rule, relatively few new (in this case, nonsense) words would be expected to fall within the range of the rule's application.

There are patterns of use that seem to elude this account. For example, children with SLI are inconsistent in their use of an inflection with the same lexical item (e.g., *Mommy likes it; Daddy like it*). There is nothing in principle that prevents such inconsistencies (whatever their source) from co-occurring with narrow rules, but it is not clear how narrow rules can cause them.

The restricted use of inflections seen in the speech of the family members studied by Gopnik and her colleagues (e.g., Gopnik & Crago, 1991) seems to contradict one of the assumptions of this account. Based on the findings of Morehead and Ingram (1973) and Ingram and Carr (1994), the oft-cited limitations in the use of inflections seen in children with SLI are actually restricted to younger ages. Older children with SLI have had time to acquire the rules for these inflections in small steps or, if necessary, to learn inflected words as separate lexical items. Many of the family members studied by Gopnik and her colleagues were adults; yet they showed the kind of limitations in inflectional use ordinarily observed in preschool-age children with SLI.

Given the many hints in the literature that children with SLI employ patterns of use that are more restricted than those of their normally developing peers, investigators should begin to employ methods designed to uncover whether this restricted use is principled. Specifically, it seems important to determine whether instances of production and omission cluster in ways that create an outline of the boundaries of a sensible if narrow rule.

Problems with Structural Relationships

The Missing Agreement Account
Clahsen (1989) proposed that the grammatical deficits of children with SLI are due to a selective impairment in establishing the structural relationships of agreement. It is assumed that these children lack the knowledge of asymmetrical relations between categories, where one category controls the other. This account—called the missing

agreement account—first applied to German, but Clahsen (1993) later identified characteristics in the speech of children with SLI learning other languages that conform to his proposal.

Grammatical details that are adversely affected by this agreement deficit include: verb inflections, auxiliaries, and copula forms that must agree with the subject according to person and number; gender and number agreement between determiners and nouns and between adjectives and nouns; the possessive suffix; and case markings on determiners. All of the forms assumed to be affected by this deficit are elements of functional categories. However, this proposal is not identical to the functional category deficit account. Past tense verb inflections, for example, are not assumed to be especially problematic for children with SLI because they do not involve agreement. Similarly, difficulties with determiners should not extend to definiteness, as this feature is not related to agreement in Clahsen's framework.

Because children with SLI do not have a paradigm of person and number verb inflections, they are unable to generate appropriate finite forms. When correct forms are seen, these are the result of the children's having learned them on a rote basis, as separate lexical items. Similarly, correct instances of gender marking on determiners and adjectives occur because the children memorized the forms for specific determiner–noun and adjective–noun pairs. Because these correct instances are rotely learned, percentages of correct use in obligatory contexts are lower than for MLU controls.

In a language such as German, Clahsen's missing agreement proposal has major implications for word order. This can be illustrated by using the underlying structure assumed for German in the principles and parameters framework, shown in (11). Translations appear in (12).

(11)

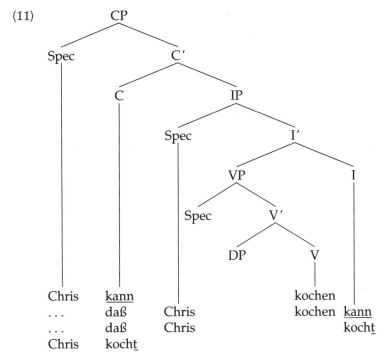

(12) a. Chris kann kochen
          Chris can cook
          "Chris can cook"
      b. ... daß Chris kochen kann
          ... that Chris cook can
          "... that Chris can cook"
      c. ... daß Chris kocht
          ... that Chris cooks
          "... that Chris cooks"
      d. Chris kocht
          Chris cooks
          "Chris cooks"

In German, as in English, nonfinite verb forms ("kochen," in the example above) occupy V in both main and subordinate clauses, and modal auxiliaries ("kann") are generated in I. It is assumed that auxiliaries move into C and the subjects move from the specifier position of VP to the specifier position of IP and then to the specifier position of CP. However, if C is already occupied by a complementizer ("daß"), the auxiliary will remain in I and the subject stays in the specifier position of IP. When auxiliaries are not in the structure, the main verb in V moves to I, where it acquires tense and agreement in the form of an inflection ("kocht"). The inflected verb then moves to C if the latter is not already occupied, and the subject moves to the specifier position of CP.

Adverbials of time and place and topicalized objects can also appear in sentence-initial position. In such instances, they appear in the specifier position of CP. The subject cannot occupy the same position, and therefore remains in the specifier position of IP. The finite verb is in C, and therefore in second position even though something other than the subject is in the specifier position of CP. This pattern illustrates the "verb-second" rule of German and related languages.

It can be seen, then, that second position is occupied by finite verb forms—auxiliaries or inflected verbs. If the verb is nonfinite, it remains in V, which, as can be seen in the structure above, is to the right of object position. Finite verb forms can appear in final position, but only in subordinate clauses. Not surprisingly, such clauses are quite rare in the speech of children with SLI.

Clahsen (1989) found that the sentence position of verbs in the speech of a group of German-speaking children with SLI patterned with whether or not a finite form was produced. When the children produced a nonfinite form—even if a finite form was required—they usually placed the verb in final position. Similarly, finite forms typically appeared in second position. According to Clahsen, this finding suggested that the word order errors seen in children with SLI are attributable primarily to problems with agreement. When children managed to produce a finite form, even if through memorization, it usually was correctly located in second position. This interpretation gained additional support from Clahsen's (1991) finding that the same children did not commit word order errors within constituents such as NP and PP. If word order constituted an independent problem, errors of this type might have been expected.

Clahsen and Hansen (1993) conducted a treatment study with the aim of testing the missing agreement account. They provided four German-speaking children with multiple examples of verb agreement inflections in verbs presented in isolation and in simple subject + verb sentences. The exposures included the use of the same verb with different inflections (e.g., *Peter tanzt*, "Peter dances"; *Ich tanze*, "I dance") to facilitate identification of the agreement function of each inflection. Clahsen and Hansen found that treatment led not only to an increase in the children's use of agreement inflections but also to the correct use of the verb-second rule, a rule the children lacked prior to treatment. Movement of the verb to second position is licensed only if it marks agreement with the subject. According to Clahsen and Hansen, the appearance of the verb-second rule after treatment focusing only on the inflections served as evidence that the underlying problem for these children was one of agreement.

In chapter 4, it was seen that some investigators (e.g., Grimm & Weinert, 1990) reported that finite verb forms can be found in sentence-final position in the speech of German-speaking children with SLI, contrary to the claim that word order errors are linked to agreement problems. However, Clahsen and Hansen (1993) argued that the frequency of such occurrences was not reported and that, on occasion, young normally developing children also produce utterances of this type.

On the other hand, Clahsen (1991) acknowledged that a minority of children with SLI do not conform to the predicted pattern. He described one child, for example, who reserved second position for a restricted set of finite forms (usually modals); all other verbs were used in final position in main clauses, sometimes in finite form, sometimes not.

Some of the findings for English are readily explained by the missing agreement account. Auxiliary and copula *be* forms, the third-person singular verb inflection *-s*, and the possessive suffix *'s* are all difficult for children with SLI and all involve agreement. Problems with nominative case can likewise be attributed to difficulties with agreement relations.

As with most accounts, there are findings that seem to defy explanation by this account. S. Roberts (1995) found that some of the children with SLI studied by Clahsen (1989) showed higher percentages of appropriate use of verb agreement inflections and copula forms than a group of English-speaking children with SLI matched according to MLU in words. If such forms must be memorized, it is not clear why the children acquiring German were better at it than the children acquiring English. The German-speaking children were a few months older on average, but their morphologically richer language would have required them to learn many more inflected forms by rote.

In English, the regular past verb inflection is usually used with lower percentages by children with SLI than by MLU controls. The missing agreement account of Clahsen (1989) provides no basis for expecting this difference. Other English forms unrelated to agreement, such as articles and infinitival *to*, are also difficult for these children.

Clitics are often omitted by Italian-speaking children with SLI, as are articles, which must agree with the noun in number and gender. However, when these forms are produced, they almost always show the correct marking. To interpret these findings as consistent with the missing agreement account, it would have to be assumed that each time a nonmemorized form is required, omission occurs. This seems unlikely.

More serious is the finding that verb agreement inflections in languages such as Hebrew and Italian do not show differences between children with SLI and MLU controls, with the exception of the third-person plural inflection of Italian. It is not clear how, without changing the nature of this account, children acquiring these languages can have access to knowledge of agreement relations that is unavailable to children with SLI learning Germanic languages.

Problems with Specifier-Head Relations

Clahsen's account is not the only one to posit problems in agreement relations. Rice and Oetting (1993) proposed that children with SLI have difficulties with specifier-head relationships, such as between subjects (in the specifier position of IP) and finite verbs (that receive tense and agreement from I). In many respects, this account resembles that of Clahsen in its predictions, though past tense verb inflections constitute an important exception. Because tense features originate in I, the Rice and Oetting account predicts problems with past tense inflections as well as agreement inflections. Clahsen, as noted above, does not, because there is no asymmetrical, controlling relationship between the subject and finite verb in this instance. Given that problems with past tense forms are seen in children with SLI acquiring languages such as English, the specifier-head deficit account therefore holds an advantage over the missing agreement account. Because the two accounts are quite similar in other respects, however, most of the strengths and weaknesses pointed out for the missing agreement account are applicable as well to the proposal that the problem centers on specifier-head relations.

Representational Deficit for Dependent Relationships

Another account that assumes difficulty with structural relationships was proposed by van der Lely (1994, 1996). This account was designed to explain the pattern of difficulties exhibited by those children with SLI who show poor grammatical comprehension as well as poor grammatical production abilities. As in the proposal of Clahsen (1989), van der Lely's account assumes problems with grammatical agreement. In addition, this account shares with Rice and Oetting (1993) the assumption that specifier-head relations not involving grammatical agreement, such as past tense, also fall within the domain of difficulty for children with SLI. According to van der Lely, these children's problems can be characterized as a representational deficit for dependent relationships. The problems expected as a result of this deficit cover a wide range. In addition to problems with tense, agreement, and case, the children will have difficulties assigning thematic roles to NPs when given only the syntactic structure. This was seen in some of the studies reviewed in chapter 3. For example, children with SLI were found to have great difficulty in the comprehension of reversible passives. More dramatically, they seemed to have difficulty when asked to act out sentences containing a nonsense verb (e.g., *The dog is slooving the cat to the bear*) whose action they had never seen before.

As noted in chapter 3, O'Hara and Johnston (1997), like van der Lely (1994) found that children with SLI had difficulty with sentences containing a novel verb. However, their analyses revealed the possibility that processing factors rather than thematic role assignment were the source of the children's difficulty. Bishop (1992b) has offered a similar interpretation for difficulty with passives. In the next chapter, the question of processing demands will be taken up in more detail.

The fact that there are differing views on the factors responsible for these comprehension findings is not problematic for van der Lely's account if the remaining data fall into place. Certainly much of the data do conform to the predictions that would be based on this account; these are essentially the same findings that lend support to the missing agreement account and the specifier-head deficit account as well. However, as was the case for these other accounts, some of the English data suggest that problems are not confined to those reflecting difficulties with structural relationships. Infinitival *to*, for example, has no relational status, yet it is often omitted by children with SLI. Then there is the problem of explaining why, when verb inflections and auxiliaries are used, they are usually correct.

More important, some of the findings for English seem to contradict predictions based on this account. Children with SLI are as proficient as MLU controls in their selection of prenominal determiners that agree with the following noun in number (e.g., *this boy, these cats*). Productions such as *those kite* and *this pictures* are quite rare (Leonard, 1995). In the linguistic framework used by van der Lely (1996), accusative case assignment is carried out by the functional category AGR_o, which constitutes a dependent relationship. Yet, direct objects are reliably marked for accusative case in the speech of children with SLI, if pronouns are any indication (and, in English, pronouns are the only way of knowing). Finally, data from non-Germanic languages don't seem to fit the pattern expected in this account. Most notably, verb and adjective agreement inflections are used as accurately by children with SLI as by MLU controls.

Presumed Deficits of Linguistic Knowledge and the Problem of Crosslinguistic Differences

It is fair to say that the proposals of deficits in linguistic knowledge have advanced the precision of research in the area of SLI. Specific predictions follow from most of these accounts, allowing investigators to zero in on particular structures or clusters of presumably related structures. The level of precision has been aided greatly by the linguistic frameworks adopted by the authors of these accounts; seemingly disparate grammatical forms (e.g., nominative case and auxiliary *be* forms; auxiliary inversion and overt complementizers) are sometimes integrally connected in these frameworks, giving researchers an opportunity to examine the comprehension or production of combinations of grammatical forms that probably would not otherwise be examined in the same investigation.

The Achilles' heel of most of these accounts is that their predictions hold for only a limited range of languages. Evidently, potentially problematic notions are easier to puzzle out in certain languages and hence are more learnable in these languages. There seems to be no universal gap in these children's linguistic knowledge. This means that the knowledge the children with SLI are lacking must center on how particular linguistic notions are instantiated in the language being learned. For example, something about verb inflections and auxiliaries in English might make them less identifiable as exemplars of an I-system, or as agreement markers, or as signs that tense is obligatory in main clauses. Until children discover these facts about such forms, it is proper to characterize the problem as a problem of knowledge.

Ironically, it does not follow that knowledge problems of this type are by necessity linguistic in nature. Let's consider two possible sources of difficulties, using English as the example. First, it might be the case that children with SLI acquiring English have difficulty identifying agreement markers because they think that the presence of bare stems eliminates the possibility that inflections are also used. It seems reasonable to characterize such a problem as being linguistic in nature.[1] An alternative possibility is that something else about English morphology makes it difficult to identify agreement markers, such as their lower frequency of occurrence relative to other languages, or that some of them are consonantal and thus briefer in duration, or that they do not share the same phonetic suffix as the constituent with which they must agree (as is the case in Hebrew, *ha-yeladim roxvim*, "the children ride," or Italian, *la macchina rossa* "the red car"). These factors of frequency, perceptual salience, and phonological redundancy are not independent of the typology of the language, but they are not linguistic either.

The quintessential evidence used to support accounts of deficits in linguistic knowledge is a change in all relevant places in the grammar once the missing piece of information is correctly identified. For example, the children might begin to mark nominative case in English (or show evidence of the verb-second rule in German) once finite verb forms are finally used. However, what such data actually demonstrate is that the children's grammars were organized in a normal manner, such that once the relevant information was ascertained, the grammar readily accommodated it. They don't tell us why the children were so slow in identifying the missing pieces of information in the first place. In the next two chapters, we will examine some attempts to answer this question from a very different perspective.

Note

1. Although this type of problem is linguistic in nature, it does not accurately characterize the problems of children with SLI. As pointed out by Hyams (1987), if children mistakenly treat the language as one with morphological uniformity (in this case, as a language with no grammatical inflections), they should mistakenly conclude that the language allows null subjects (sentences without an overt subject, as in Spanish and Italian). Loeb and Leonard (1988) explored this possibility with a group of English-speaking children with SLI and found no relationship between the use of verb morphology and the degree to which subjects were absent from the children's utterances.

Chapter 12

SLI as a Limitation in General Processing Capacity

In chapter 5, it was noted that many children with SLI do not have a clean bill of health when it comes to nonlinguistic abilities. The deficits seen in nonlinguistic areas are not usually severe; if they were, it is doubtful that these children could have met the criteria for SLI in the first place. However, findings indicating subtle nonlinguistic weaknesses in these children are by now so commonplace that no theory of SLI can be truly comprehensive without taking them into account.

Some investigators have interpreted these findings as more than a sign that children with SLI have concomitant weaknesses outside of language. To these investigators, the nonlinguistic deficits are a fundamental part of the children's problem. Many of the proposals that attempt to integrate linguistic and nonlinguistic findings to account for SLI treat the problem as one of a limitation in information-processing capacity. It seems useful, therefore, to begin with a brief overview of this notion.

Any proposal of limited processing capacity carries the assumption that within some domain, the specific nature of the material is less important than how this material is mentally manipulated. In the cognitive processing literature, the notion of limited processing capacity is discussed in three different ways: in terms of space, energy, or time (Kail & Salthouse, 1994; Roediger, 1980; Salthouse, 1985). According to interpretations based on space, it is assumed that there is a restriction on the size of the computational region of memory; there is insufficient work space. Limitations of energy refer to inadequate fuel to complete a cognitive task. Here, a mental task is begun but all of the energy available is expended before the task is completed. Finally, time restrictions refer to limitations dictated by the rate at which information can be processed. If the information is not processed quickly enough, it will be vulnerable to decay or interference from additional incoming information.

These ways of characterizing limited processing capacity are not mutually exclusive; it is often possible to discuss the same task in terms of some combination of them. For example, it might be assumed that certain types of lexical items are located deeper in memory stores than others. Retrieving these items, then, might require greater expenditure of energy and more time.

Investigators differ on whether it is proper to conceive of a single, general resource underlying cognitive processing or whether multiple resources should be assumed (Kail & Salthouse, 1994). The accounts of deficits in children with SLI discussed in this chapter and the next differ on this very point. In some cases, limitations in processing are assumed only within particular domains; in other accounts, the limitations are assumed to be broad in scope.

Finally, there is nothing in the processing capacity approach that precludes the coexistence of problems of an entirely different nature. It is highly plausible that children might lack knowledge of some aspect of language and at the same time show inconsistency in making use of those aspects they do know, due to limitations in processing capacity. Of course, for sake of parsimony, it is preferable to try to account for as many problems as possible by the same factor—limited processing capacity in this case—before other factors are incorporated into the theory.

In this chapter, we focus on attempts to account for SLI as a more general limitation in processing capacity. We begin with a discussion of the range of problems that seem amenable to this type of explanation.

Explaining the Global Deficit

The Breadth of Difficulties That Can Be Viewed from a Limited Processing Capacity Perspective

The most obvious problems that might be cast in terms of processing capacity limitations are those requiring word recall and retrieval. For example, Kirchner and Klatsky (1985) employed a task in which children were presented with lists of 12 pictures of objects coming from four different categories (e.g., vehicles, animals). The children were asked to rehearse the names of the pictures as they appeared, and then to recall them once the entire list had been presented. The main factor that differentiated the children with SLI from a group of age controls was the number of items recalled. As noted in chapter 3, Kail, Hale, Leonard, and Nippold (1984) and Kail and Leonard (1986) reported similar findings.

However, many other types of activities can also be viewed from a processing capacity perspective. Some of these fall in the realm of pragmatics, others in morphosyntax, still others in phonology. A few seem to be nonlinguistic.

Johnston, Smith, and Box (1988) asked children with SLI and mental age controls to play a communication game in which they described toys to a blindfolded puppet. For each item, the children were shown three objects that varied in properties such as size and color. For example, the three objects in one set were a large green peg, a small green peg, and a large purple peg. The examiner then pointed to two of the three objects, and the child's task was to describe the selected objects to the puppet so that the puppet could choose the correct two. Johnston et al. found that all of the children could provide communicatively adequate messages; however, the children with SLI produced far fewer descriptions termed "quantitative groupings." These responses (e.g., "two green ones") required greater processing capacity because to make them, children had to identify and store the attributes shared by the target objects but not by the remaining object, and to quantify the objects in terms of this shared attribute. The children were capable of identifying attributes without quantification (e.g., "a green big one and a green little one"), and during control items, they could quantify when attributes were not involved (e.g., "the two house").

Bishop (1992b) noted that findings reported by Bishop and Adams (1991) could be interpreted in a similar manner. In that study, children with SLI and age controls were asked to describe one of eight pictures to an adult who was not looking, so that the adult could locate the object a moment later. The eight pictures represented all combinations of three binary variables. For example, the child might be asked to describe a picture of a boy riding a red bicycle with little wheels from an array that included a girl riding a red bicycle with little wheels, a boy riding a red bicycle with big wheels, and a boy riding a yellow bicycle with little wheels. The children with SLI showed evidence of recognizing and naming the attributes; their problem seemed to lie in holding each relevant attribute in mind as they scanned the array, and retaining all three attributes as they formulated their message.

The findings of a study by Johnston and Smith (1989) illustrate that information-processing factors can sometimes be more important than language factors in determining the performance of children with SLI. In this study, children with SLI and mental age controls played a follow-the-leader game with two adults. The game began with the first adult selecting one or two objects from an array of three objects—for example, two red houses of different sizes from an array that included a large blue house. The second adult, whose array differed from the first adult's array, then selected objects that could confirm the basis of the first adult's choice—for example, two yellow houses of different sizes from an array that included a small red house. The child was then encouraged to make a selection from a third array that differed from the first two but possessed the relevant attributes to conform to the pattern illustrated in the adults' responses.

Johnston and Smith (1989) also employed a task in which the children were told which items to select from the array (e.g., "Take the two that are the same color"). The results indicated that the children with SLI had great difficulty with the non-verbal task items involving the attribute of size; their performance on the verbal equivalents (e.g., "Take the two that are the same size"), however, was relatively good. For the control children, neither the verbal task nor the nonverbal task presented particular difficulty. According to Johnston and Smith, the nonverbal tasks involving size required the greatest processing capacity because in this task the children had to infer size as the relevant attribute (in the verbal task they were told this information), and size, in contrast to an attribute such as color, is inherently ordinal.

In chapter 5, the haptic recognition abilities of children with SLI were discussed. In the usual version of this task, children wearing blindfolds feel geometric shapes and then, from an array of pictures, identify the picture that depicts the shape they felt. In such a task, children with SLI perform relatively poorly (Johnston & Ramstad, 1983; Kamhi, 1981; Kamhi, Catts, Koenig, & Lewis, 1984). Montgomery (1993) examined the possibility that the low performance levels of these children were due not to problems in the recognition of shapes but to limitations in processing capacity. That is, it seemed possible that the requirement of holding the image of the shape in mind long enough to compare this image with each of a number of depicted shapes simply overtaxed the children's processing capacity.

In his study, Montgomery employed the usual haptic recognition task as well as two others. In the simpler of these, the children remained blindfolded after the shape had been felt. A single shape was then placed in their hands and the children were asked whether or not the shape matched the one felt earlier. In the more difficult task, the children remained blindfolded and were handed a series of shapes one at a time and had to judge whether each one matched the original. Montgomery found that the children with SLI were comparable with mental age controls on the tactile task in which they were presented only a single shape and had to judge whether or not it matched the original shape. On both of the remaining tasks, the performance of the children with SLI was lower than that of the control children. Although the more difficult tactile task did not require a change in modalities, the children were required to retain the original image for a longer period, until a match appeared in the sequence. Through an examination of accuracy by serial position, Montgomery discovered that the performance of the children with SLI dropped earlier in the sequence than was true for the control children, suggesting a more limited capacity.

The available evidence on the comprehension abilities of children with SLI also seems to implicate processing capacity as an area of weakness in these children. Tallal (1975) examined the ability of children with SLI to manipulate tokens of varying sizes, shapes, and colors in response to sentence instructions. She found that these children had the greatest difficulty with the sentences that required recall of the largest combinations of attributes—two different tokens, each specified for size, color, and shape (e.g., "Point to the large white circle and the small green rectangle")—even though these were not the most complex sentences used in terms of grammatical structure. It should be noted that the grammatically complex sentences giving the children the greatest difficulty seemed to place significant processing demands on the children. For example, the most difficult item, "Before touching the yellow circle, pick up the red square," required the children to register two attributes (color, shape) and maintain this information in working memory until the response to the second clause was carried out.

In a study by Bishop (1979), children with SLI showed lower comprehension performance on grammatically complex sentences than on grammatically simple sentences of the same length. On close inspection, it was seen that the former were complex not only in terms of presumed underlying structure but also in terms of the number of words that had to be retained before a correct interpretation was possible (Bishop, 1992b). For example, in "the girl is kissed by the boy," correct interpretation requires that the child suspend judgment on the role of the first NP until the grammatical morphology associated with the passive construction appears. In this example, the critical cue is the appearance of -ed rather than -ing at the end of the main verb.[1] As we saw in chapter 3, several studies have documented difficulties in the comprehension of passive sentences by children with SLI.

O'Hara and Johnston (1997) found that children with SLI had difficulty acting out sentences that contained a novel verb (e.g., "Show me *the cow sleems the duck to the barn*"). Problems of this sort have been interpreted as difficulties with thematic role assignment (van der Lely, 1994). However, analyses performed by O'Hara and Johnston revealed the possibility that processing factors were the source of the children's difficulty. Essential elements such as causation, contact, and object movement were consistent with the syntactic structure of the sentence. Details lost were often those that couldn't be interpreted as assignment of an inappropriate thematic role (e.g., the child had a toy bear rather than a toy cow "voofed" by a toy farmer). According to researchers who employ computer simulations to model the language comprehension performance of adults (including adults with aphasia), thematic role assignment based on syntactic structure consumes a great deal of processing capacity (Haarman, Just, & Carpenter, 1997). The retrieval of appropriate thematic roles for a given syntactic structure requires an extensive search, especially under the conditions used by van der Lely and O'Hara and Johnston, where the verb has not been heard before.

Two comprehension studies of the inferential abilities of children with SLI can be interpreted in terms of information-processing limitations. Ellis Weismer (1985) presented three-item spoken and pictorial stories to three groups of children—children with SLI, mental age controls, and language comprehension controls. The children's understanding of the stories was then assessed. Test items included questions that required the children to make inferences as well as questions that could be answered on the basis of information explicitly provided in the story. The children with SLI resembled the language comprehension controls on both types of questions, in both the spoken and the pictorial story conditions. Although the children with SLI also performed as well as the mental age controls on the questions requiring literal recall in the pictorial story condition, they performed significantly worse on the inference questions in the same condition.

Bishop and Adams (1992) conducted a similar study, using longer spoken and pictorial stories. They found that children with SLI had more difficulty with questions requiring inferences than with questions requiring literal recall. However, these children's overall performance was worse than that of children serving as language comprehension controls. This was true even for the stories presented in picture form.

Limited processing might be implicated in two ways in the Ellis Weismer (1985) and Bishop and Adams (1992) studies. First, items that required recall of literal information were not especially difficult for the children with SLI, at least in the pictorial mode, when the stories were short, as in the Ellis Weismer study. Even these types of items became more difficult, however, when the stories were longer, as in the investigation of Bishop and Adams. In this case, a greater amount of literal information had to be recalled. Items requiring inferences were more difficult than items requiring literal recall because in the case of the former, operations in addition to recall are required to construct information that follows from, but is not contained in, the pictures (or sentences) provided.

Information-processing capacity has also been examined by means of a dual processing task paradigm. Riddle (1992) employed a task in which preschoolers with SLI and age controls identified pictures of objects having the same name but had to interrupt this process by pressing a button as soon as a buzzer was heard. Although the children with SLI showed high levels of accuracy on the picture task, their response times were slower than those of the controls on the auditory detection task. These differences were not seen when the children participated in the auditory detection task alone, suggesting that their activity during the picture task was requiring a greater expenditure of available resources than was true for the control children.

In two separate studies, Ellis Weismer and Hesketh (1993, 1996) examined lexical learning in children with SLI as a function of presentation rate. It was assumed that manipulations of presentation rate serve as an effective means of examining processing capacity. Here, limitations in processing capacity are conceived of as restrictions imposed by the rate at which information can be processed; if preceding material is not fully processed before subsequent material appears, the preceding and/or subsequent material will be incompletely processed, and performance will suffer. The two studies revealed that the learning of novel words appearing in sentences presented at a slow rate (averaging 2.8 syllables per second) was easier for children with SLI than learning words appearing in sentences presented at normal (4.4 syllables per second) or fast (5.9 syllables per second) rates. Relative to chronological age and mental age controls, the children with SLI appeared to experience considerable difficulty in the normal and fast rate conditions. This effect was most notable during a production posttest and a posttest in which the children had to recognize the novel word when presented with this word and phonetically similar foils. The children with SLI had no special problems on a posttest requiring the children to identify the target word from an array consisting of words that bore little phonetic similarity to the target word.

The results of the Ellis Weismer and Hesketh (1993, 1996) studies seem compatible with an information-processing capacity account of SLI. The two posttests giving the children with SLI particular difficulty seem to require the greatest degree of processing. That is, whether required to produce the novel words or recognize them in a context in which phonetically similar words also appeared, the children had to have stored the words in a more complete phonetic form. Accordingly, if processing proceeds slowly in these children, in the faster rate conditions the children may not have had time to process all of the phonetic details of these words before subsequent material appeared. Adequate time for such processing may have been provided in the slow rate condition. Evidence from earlier studies indicates that sentence comprehension also runs into difficulty when presentation rate is increased (McCroskey & Thompson, 1973).[2]

The fact that processing capacity increases with age (e.g., Kail, 1991) gives it the potential to refine existing proposals of SLI. Leonard (1987) argued that many children with SLI might be viewed as falling on the low end of a general language-learning ability continuum rather than as having problems resulting from a disruption in some circumscribed portion of the language-learning mechanism. Leonard did not attribute the presumed general language-learning limitation to a general processing capacity limitation, but this is certainly one possibility.

Johnston (1991, 1994) and Bishop (1992b) have pointed out the paradox in assuming a limitation in general information-processing capacity. On the one hand, the generality of this notion makes it applicable to a wide range of tasks and operations. On the other hand, because it is so general, its explanatory power can be questioned. Until the operations involved in each task are identified and their processing requirements are specified, proposals of limited processing capacity cannot be treated as definitive.

The Generalized Slowing Hypothesis
Although we are far from the point of being able to catalog all of the operations involved in processing tasks, a very useful method of evaluating the feasibility of a general limited processing capacity account has been offered by Kail (1994). Kail's method employs speed of processing as a metric of processing capacity, under the usual assumption that speed determines the amount of work that can be accomplished in a given unit of time.

This approach seems to have immediate applicability to the study of SLI, given that children with SLI have displayed slower response times (RTs) across a wide range of tasks, including mental rotation (Johnston & Ellis Weismer, 1983), picture naming (e.g., Anderson, 1965), word monitoring (e.g., Stark & Montgomery, 1995), and judgments of grammaticality (Wulfeck & Bates, 1995). In fact, even on tasks requiring minimal levels of cognitive processing, such as pressing a panel upon hearing a tone or seeing a light, children with SLI might be slower than age controls. Hughes and Sussman (1983) and Nichols, Townsend, and Wulfeck (1995) found differences favoring controls on such tasks, whereas Montgomery, Scudder, and Moore (1990), Riddle (1992), and J. Edwards and Lahey (1996) did not. Tasks requiring only a minimal increase in processing, such as moving pegs, also have produced results indicating slower RTs by children with SLI (Bishop, 1990; Bishop & Edmundson, 1987b).

The method developed by Kail (1994) involves examining the RTs of children with SLI and age-matched normally developing children across a range of tasks. According to Kail, responding to any given task is likely to involve several processes. Picture naming, for example, involves (minimally) the recognition of the picture, retrieval of the name of the picture, formulation of the word, and the actual production of the word. Thus, the RTs for normally developing children will be influenced by the time required for each of these processes, as shown in (1).

(1) $RT_{ND} = a + b + c + \cdots$

where a is the time needed to execute the first process, b is the time needed to execute the second process, and so on.

According to Kail (1994), if children with SLI are slower to execute each process by a common factor, their RTs can be represented as in (2).

(2) $RT_{SLI} = m(a + b + c \ldots)$

where m is the factor by which children with SLI respond more slowly. The equation in (3) expresses the RTs for children with SLI as a function of the RTs for normally developing children.

(3) $RT_{SLI} = mRT_{ND}$

Figure 12.1 provides an illustration of data that would support the hypothesis that children with SLI experience a generalized slowing. Plotted are hypothetical results from different experimental conditions for different tasks. If the RTs of children with SLI matched those of the control children, a slope of 45° would be seen. In the example in figure 12.1, a steeper slope is seen and the function is linear, suggesting that as RTs increase for the age controls, they also increase for the children with SLI by some extra proportion.

Kail (1994) analyzed the RTs of children with SLI and age-matched normally developing children from five different experiments involving a total of 22 different conditions reported in Kail and Leonard (1986), Leonard, Nippold, Kail, and Hale (1983), and Sininger, Klatsky, and Kirchner (1989).

The results indicated that the RTs for the children with SLI increased linearly as a function of the RTs for the control children. Consistent with the hypothesis that the slower RTs of the children with SLI represented a generalized slowing, the slope of the linear function showed that regardless of the task and condition, the RTs of the children with SLI were about 33% slower than those of the age controls.

Kail (1994) stressed that his findings provide only preliminary support for a general processing limitation in children with SLI. The data employed were group means rather than RTs for individual children, and only five different tasks were represented. That said, Kail's method offers the advantage of assessing the plausibility of an account of general processing limitations in children with SLI even if the precise mental operations involved in each task and condition are unknown. For example, in the studies considered by Kail, the differences in the normally developing children's RTs across tasks and conditions within a task presumably reflect differences in the number of processes required and/or the time required to execute some of those processes. The corresponding RTs for the children with SLI suggest that whatever these processes are, they consistently took these children about 33% more time to execute.

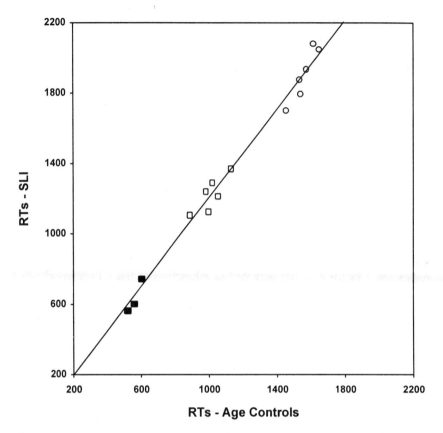

Figure 12.1
Response times (RTs) in ms of children with SLI as a function of response times of age controls. Each cluster of data points comes from a different task; the data points within each cluster come from different conditions within the same task.

If slower RTs for children with SLI were due to limitations in only one component of the response—say, in perceiving the stimulus or in the motor act of responding—their RTs would differ from those of age-mates by an absolute constant (e.g., 100 ms), not a proportional constant (e.g., 33%), and the slope of their RTs would resemble those of the normally developing children. If, instead, slower RTs were due to several independent limitations, it would be highly unlikely that these RTs would increase linearly as a function of the normally developing children's RTs.

Studies by J. Edwards and Lahey (1996), Lahey and Edwards (1996), and Taylor (1995) suggest that the generally slower RTs of children with SLI seem to be true for children with comprehension as well as production problems, but not for children with difficulties limited to production. This outcome could also be shown clearly by using Kail's (1994) method. By plotting the RT data from these two subgroups separately, the slope for the children with production problems only should approximate 45°, whereas those of the remaining children with SLI should resemble the data reported by Kail.

When Less Is More

It is natural to view a limited processing capacity as a disadvantage. However, in one important respect, it can be quite the opposite. Newport (1990) has noted that in the early stages of language acquisition, children's ability to discover the grammatical rules of the ambient language may actually be helped by a limited processing capacity. This has been termed the "less is more" hypothesis. The point has been nicely illustrated in work by Elman (1993) and Goldowsky and Newport (1993). In these studies, computer simulations of grammatical learning were run. For example, in the Goldowsky and Newport study, the system to be learned simulated a simple morphology such that for each word there were three positions in which morphological distinctions could be made, with one of three possible morphemes in each position. The meanings that corresponded to the morphemes were also provided to the computer. The computational task, then, was one of correlating the forms and meanings.

Let's consider an example that seems to correspond to the system to be learned. In agglutinating languages (e.g., Korean, Tamil, Turkish), a sequence of bound morphemes appears at the end of a word. For example, for word X, we might have X-*ge-mo-tin*, where the slot occupied by *-ge* marks tense, the slot occupied by *-mo* indicates number, and the one occupied by *-tin* reflects gender. Each position permits one of three morphemes; thus X-*ge-mo-tin* (the example above), X-*ge-mo-zet*, and X-*ge-mo-nok* might represent present singular masculine, present singular feminine, and present singular neuter, respectively.

Goldowsky and Newport (1993) presented exemplars of their morphology in two different ways. In the first, the exemplars were introduced into the computer without modification. For example, the analog to *-ge-mo-tin* was presented along with the corresponding analog to its meaning, here, -present-singular-masculine. In the second method of presentation, each exemplar was first run through a filter such that the computer received a reduced rendition of the morphology. For example, although the original input may have been *-ge-mo-tin*, the computer had access to, say, only *-ge-mo-* or *-mo-tin*. This filtering served as the functional equivalent of a limitation in processing capacity.

The results indicated that the filtered version led to more efficient learning of the morphology. Because the unfiltered version allowed for access to a much greater number of combinations of morphemes and meanings, it also led to a greater number of false leads due to the computer's sensitivity to co-occurrences that in fact were only accidental. Goldowsky and Newport (1993) pointed out that in the early going, a limited processing capacity serves the child well, for it forces the child to deal with a smaller portion of the material in a sentence. Of course, to learn more complex details of the grammar, greater capacity is required. But as children grow, so does their processing capacity (e.g., Kail, 1991).

Findings such as those of Goldowsky and Newport (1993) and Elman (1993) suggest that if children with SLI operate with a limited processing capacity, they probably do not form peculiar hypotheses about grammatical structure but, rather, arrive at the same type of grammar seen in young normally developing children. The evidence in chapter 3 is clearly consistent with this position. Instead, these children's problems have to do with the rate at which grammatical details can be learned and whether capacity ever reaches the point of enabling them to acquire the most complex aspects of the grammar. Given the long-standing nature of grammatical problems in some individuals with SLI, as seen in chapter 1, it is plausible that limitations in processing capacity persist.

Explaining Specific Profiles

Thus far, we have seen how an assumption of a general processing capacity limitation might deal with a range of problem areas in children with SLI. However, another important issue must be addressed. Children with SLI are not uniformly weak across areas. Linguistic abilities are usually weaker than nonlinguistic abilities, and within the domain of language, some areas are weaker than others. Although it seems counterintuitive, we might be able to account for such uneven profiles by assuming a broad-based processing capacity limitation. In this section, we will see how this might be done. Special attention will be placed on attempts to explain the extraordinary weakness in grammatical morphology seen in English-speaking children with SLI.

This endeavor is aided greatly by research in two other areas. First, crosslinguistic studies of normal language development reveal different profiles of language acquisition as a function of the language being learned. Factors responsible for such differences are assumed to include relative frequency, redundancy, regularity, perceptual salience, and pronounceability (see Bates & MacWhinney, 1987; Peters, 1985; Slobin, 1973). Features of language that are low in these characteristics are most vulnerable and hence are acquired later by children. For example, grammatical morphemes make a later appearance in the speech of normally developing children acquiring English than in the speech of children learning most other languages. In English-speaking children with SLI, the relative weakness in grammatical morphology seems to be an exaggerated version of the profile already associated with English child language.

Second, the grammatical profile observed in English-speaking children with SLI can be replicated in normally functioning adult speakers of English during tasks in which linguistic material must be produced or comprehended under difficult listening conditions (e.g., Kilborn, 1991) or when cognitive resources must be shared with another task (e.g., Blackwell & Bates, 1995). Computer simulation studies produce the same profile when either degrading the input (Hoeffner & McClelland, 1993) or severing a proportion of the connections (Marchman, 1993). What the literature seems to reveal is that aspects of a language that are already the most fragile are the first to show a significant drop under a variety of difficult conditions. The fact that children with SLI show grammatical profiles consistent with the input language with reductions in these areas of frailty rather than some universal profile suggests that it is not the underlying grammar itself that is directly responsible for this disorder.

The accounts that follow share the assumption that the profiles of children with SLI result from an interaction between a general processing capacity limitation and the characteristics of the language being acquired.

The Surface Hypothesis

Leonard and his colleagues (Leonard, 1989, 1992b; Leonard, McGregor, & Allen, 1992; Leonard, Eyer, Bedore, & Grela, 1997) proposed an account of the grammatical morpheme limitations of English-speaking children with SLI that has been termed the "surface" hypothesis because of its emphasis on the physical properties of English grammatical morphology. This account assumes a general processing capacity limitation in children with SLI but assumes also that, in the case of English, this limitation will have an especially profound effect on the joint operations of perceiving grammatical morphemes and hypothesizing their grammatical function.

The suspicion that perceptual properties of English grammatical morphemes play a role in the difficulties of children with SLI has been raised before. Fletcher (1983) made the point quite explicitly, noting, for example that "an explanation for the deficiencies in auxiliaries in the impaired child may lie in the phonetic character of the auxiliaries—in their brevity and relative lack of salience." (p. 22) If this is correct, Fletcher reasoned, "we should expect for the language-impaired child deficiencies in inflections just as we find problems with auxiliaries." (p. 22)[3]

Suggestions such as those of Fletcher (1983) were buttressed by a separate body of research—reviewed in chapter 6—indicating that children with SLI have serious problems processing verbal and nonverbal stimuli of brief duration.

The surface hypothesis represented an attempt to formalize and expand this basic idea. We turn first to a consideration of the important role that the physical properties of speech are assumed to play.

For some time, researchers have noted that the acoustic characteristics of many grammatical morphemes of English make the task of learning grammatical morphology relatively difficult for English-speaking children (e.g., Gleitman, Gleitman, Landau, & Wanner, 1988; Slobin, 1985). Several acoustic properties probably contribute to this difficulty; however, the one that seems characteristic of most of these morphemes is short duration, at least relative to adjacent material. This holds for grammatical inflections in the form of single consonants and unstressed syllabic morphemes (freestanding and bound) that rarely if ever appear in sentence positions in which significant lengthening occurs. This latter point makes it clear that lack of stress is not by itself a sufficient factor; utterance-final or phonological phrase-final syllables are significantly lengthened in English, even when they are not stressed (e.g., Beckman & Edwards, 1990; Klatt, 1975).

Upon reviewing the literature, Leonard (1989) observed that most of the closed-class morphemes that distinguished English-speaking children with SLI from their MLU controls were morphemes of short relative duration. These included the third-person singular -s and past tense -ed inflections, possessive 's, articles, copula and auxiliary be forms, infinitival to, and the complementizer that. In contrast, the progressive -ing seemed least likely to produce differences between children with SLI and MLU controls. Unlike the morphemes noted above, this inflection possesses considerable duration when it occurs (as it frequently does) in phonological phrase-final position.

Findings from Italian and Hebrew appeared to support the impression that relative duration is important. In these languages, inflections are syllabic and (whether stressed or not) occur frequently in utterance-final and phonological phrase-final position. And, in each of these languages, syllables in these positions are significantly lengthened (e.g., Berkovits, 1993; Farnetani & Fori, 1982; Leonard & Eyer, 1996). The available evidence from children with SLI acquiring these languages suggests that these children closely resemble MLU controls in their use of most inflections. Differences are seen, however, in the use of several freestanding closed-class morphemes. For example, Italian-speaking children with SLI show lower percentages of use of articles and direct object clitics than their MLU-matched compatriots. These monosyllabic forms are unstressed and usually occur in phrase-initial or phrase-medial position, where they are short in duration (Leonard & Eyer, 1996). It does not appear to be the case that these differences are due merely to the fact that these morphemes are freestanding forms. Le Normand, Leonard, and McGregor (1993) reported preliminary evidence suggesting that French-speaking children with SLI show percentages of use of articles that match those of MLU controls (even when *l'* is excluded). In French, the duration of articles and other unstressed syllables differs relatively little from the duration of adjacent, nonfinal stressed syllables (Fant, Kruckenberg, & Nord, 1991).

According to the surface hypothesis, children with SLI are capable of perceiving word-final consonants and weak, nonlengthened syllables, but they have a limited processing capacity—best thought of in terms of reduced speed of processing—that is severely taxed when such challenging forms play a morphological role. That is, when these forms are separate morphemes, the child must perform additional operations, such as discovering the grammatical functions of the forms and placing the forms in the proper cell of a morphological paradigm. This, of course, must be done in the press of dealing with the rest of the sentence that is being heard. It is assumed that the additional operations combined with the brevity of the morphemes will result in the morphemes' sometimes being processed incompletely and hence requiring a greater number of exposures before these brief grammatical forms are established in the grammar.

The additional operations assumed when a form plays a morphological role are essentially those discussed by Pinker (1984) in his proposal of how children build paradigms. A paradigm is a matrix representation of a set of related morphemes. Paradigms contain cells, each representing a conjunction of levels of different dimensions. Thus, a paradigm may contain the dimensions of NUMBER and PERSON, with the levels of singular and plural, and first, second, and third person, respectively. Paradigms are not limited to inflections; freestanding closed-class morphemes such as auxiliaries and articles are also assumed to enter into paradigms.

According to Pinker (1984), children initially create word-specific paradigms. In the case of inflections, these are paradigms in which each cell contains the stem as well as the inflection, as in the (present tense) paradigm for the Italian verb "drink," shown in (4).

(4)   NUMBER

		singular	plural
	first	bevo	beviamo
PERSON	second	bevi	bevete
	third	beve	bevono

Word-specific paradigms remain in the child's grammar, and for each word the child learns in the future, a new word-specific paradigm is created. Eventually, of course, children's inflectional systems become productive. Hearing a word such as *corre* (he or she runs), for example, the child is able to produce *corro* (I run) without having heard it before. This is possible because the word-specific paradigms lead to the development of general paradigms; that is, paradigms containing inflections free of stems, as in (5).

(5)   NUMBER

		singular	plural
	first	-o	-iamo
PERSON	second	-i	-ete
	third	-e	-ono

The learning of freestanding closed-class morphemes such as articles also begins with word-specific paradigms. Pinker (1984) treats articles as comparable with prefixes (e.g., *lamacchina*, "thecar"); they become freestanding morphemes only when the child registers the presence of intervening modifiers in the input (e.g., *la piccola macchina*, "the little car"). At this point, articles can be placed in general paradigms. In contrast with inflections and articles, auxiliaries of all types (*be* forms, modals, *do* forms) remain as entries in word-specific paradigms only.

It can be seen, then, that in this framework, inflected forms, even when general paradigms have not yet been developed, require computations in addition to those required for uninflected words. Thus, *laughed* requires relating *laughed* to *laugh*, hypothesizing that they belong to the same set, splitting the paradigm for *laugh* so that it includes a cell for past, and placing *laughed* in it.

According to the surface hypothesis, these additional operations can cause problems, given the reduced speed of processing in children with SLI. For example, under typical conditions, these children can detect the final consonant in both *raft* and *laughed*. However, the extra operations required in analyzing the latter increase the likelihood that it will be reduced to its bare-stem counterpart. The same fate is less likely in the case of inflected forms of longer relative duration, such as the past forms for "laugh" in Hebrew (e.g., [tsaxká], "she laughed"; [tsaxkú], "they laughed"). Consequently, English-speaking children with SLI will require a greater number of encounters with each inflected form before a sufficient number of these exposures can be fully computed and properly placed in paradigms.

According to Leonard (1989), this slow development of grammatical morphology in English-speaking children with SLI has negative consequences elsewhere in the grammar. For example, passives should be especially late in making their appearance in these children's speech, for the morphemes needed to identify that these sentences are not in canonical subject-verb-object order are brief in duration. In addition, to the extent that children make use of surrounding closed-class morphemes to determine the grammatical category of new words (e.g., *-ed* suggests a verb, *the* suggests a noun), the lexical development of children with SLI will proceed slowly.

It is assumed that there is no fundamental problem with the underlying grammars of children with SLI independent of the problem of a slow intake of relevant data due to the reduced speed of processing. The organization of these children's grammars and the order and types of hypotheses they form are no different from those seen in normally developing children. Furthermore, because the speed of processing limitation is assumed to be general rather than specific, its hazardous effects on grammatical morphology are due more to the fact that features such as morphology are quite fragile in languages such as English. In a language whose typology differs markedly from that of English, this same processing limitation can lead to another type of linguistic profile.

An example of the joint effects of brief duration and speed of processing limitations in an area outside of grammatical morphology can be seen in a study by

Leonard, McGregor, and Allen (1992). Based on findings by Tallal and Piercy (1975) and Tallal and Stark (1981), Leonard et al. reasoned that children with SLI should do poorly on a perceptual task dealing with contrasts of short relative duration when the task required more than perceiving a difference between the stimuli. The stimuli they created were contrasts such as [dab]–[i]–[ba] versus [dab]–[u]–[ba], and [dab] versus [dæb]. In the first contrast, the two members of the pair differed in a portion of the signal that consisted of 100 ms of the total duration of 775 ms. In the second contrast, the members differed in a portion that consisted of 290 ms of the total duration of 425 ms. Thus, the first type of contrast was expected to be more difficult.

Leonard, McGregor, and Allen (1992) employed the target identification task employed by Tallal and Stark (1981). In this task, the child is taught that one of the stimuli is the "special sound" (the target sound), and a panel should be pressed only when this sound is heard. Once it is established that the child understands the task, the child hears multiple presentations of the target sound and the comparison sound in random order. Leonard et al. assumed that this task would tax the children's processing abilities, given that the children not only had to perceive the difference between the two stimuli but also to remember one of them as the target and press the panel only when this one was heard—all before the next stimulus appeared. The results were largely consistent with expectations. A group of children with SLI did not differ significantly from age controls on contrasts such as [dab] versus [dæb], but differences favoring the control children were seen for [dab]–[i]–[ba] versus [dab]–[u]–[ba]. In fact, none of the children with SLI performed above the level of chance on this contrast.

Probing the Logic Behind the Surface Hypothesis   Why should stimuli of relatively short duration be especially difficult if children have speed of processing limitations? Indeed, as noted by Goad and Gopnik (1994), it is not obvious why perceptual properties should be related to processing at all. Perception has to do with whether the material is registered adequately. Processing seems to relate to mental operations performed on material that is already registered.

To examine this question, two fundamental assumptions of the surface account should be considered. The first is that already perceived material can be lost when additional processing operations are required. The second is that processing limitations place material of relatively short duration in the greatest jeopardy. According to this second assumption, the processing limitations may actually prevent brief material from being perceived, or the similarities between bare stems and the same stems with brief closed-class morphemes may promote the replacement of the latter by the former when memory decays.

Let's consider the two assumptions in turn. To see how already perceived material might get lost when processing demands increase, we can turn to models of working memory. Although models differ somewhat in the components that are assumed, all of them assume that perceived material must be held in working memory for a sufficient period to permit processing. Central to the surface account is the proposal (after Pinker, 1984) that stems with inflections or other associated closed-class morphemes (e.g., articles, auxiliaries) require operations in addition to those required for bare stems. These include relating the inflected (or otherwise modulated) form to its bare-stem counterpart, hypothesizing the grammatical function of the inflected version, and placing the latter (in its proper place) in a word-specific paradigm. All of this takes time. Thus, inflected words are assumed to require more time in working memory than the same words in bare-stem form. And, of course, there is a time limit to which material can remain in working memory before it is lost (Hulme & Tordoff, 1989).

If inflected words were typically heard in one-word sentences separated by pauses, there would be no problem. However, fast on the heels of the inflected word is the next word in the utterance that must be held in working memory and processed, and so on. Thus, processing is pressed from two directions; processing of a first item must be completed before the item fades from memory, and it must be processed in time for the next item. Given the reduced speed of processing assumed for children with SLI, sufficient processing of one item can't be completed before the next item appears. Consequently, some material is processed incompletely or not at all. In a language such as English, it is reasonable to expect that if an inflected word is incompletely processed, only the bare stem will be retained.

Why should morphemes of relatively short duration be the most vulnerable to loss? Indeed, there is reason to expect just the opposite: that material of short duration should place less of a burden on processing. For some time, it has been assumed that material of short duration can be rehearsed more quickly in short-term memory, leaving more time for rehearsal of other items that have been presented (Hulme & Tordoff, 1989). This is seen, for example, in studies requiring subjects to learn two lists of words, one list consisting of words of a single syllable, the other, of words of two or three syllables. More words are recalled in the short word condition.

In the surface account, it is assumed that words whose inflections (or other associated closed-class morphemes) constitute only a small portion of the entire word are more likely to be left as bare stems during incomplete processing than are words whose inflections constitute the bulk of the word. There are three plausible reasons for this. First, because of slow processing speed, there will be instances in which material will not be fully registered because prior material is still being processed when it appears in the input. Brief closed-class morphemes, especially prefixal, prenominal (e.g., articles), and preverbal (e.g., auxiliaries) morphemes, would seem to be the most vulnerable in such cases.

A second possible reason is that words whose inflections are relatively brief share a greater portion of their phonetic content with the bare-stem forms than do words whose inflections represent the bulk of the word. Thus, if a child abandons processing of one word as another occurs, the words with relatively brief inflections are more likely to be left in bare-stem form.

Third, words with relatively brief inflections bear a closer resemblance to their bare stems than do words with longer inflections. This could be important in cases where an already perceived inflected word starts to decay. Consider, for example, studies in which two kinds of lists are presented for recall. The first consists of words that are phonologically similar (e.g., *hat, map, tab*); the second, of words that are unlike each other. Recall will be worse in the case of the former (e.g., Baddeley, 1966; Gathercole & Baddeley, 1990). Presumably this is because, as words begin to decay in memory, phonological details will be lost, leaving open the possibility of competition from other, similar forms that have been activated. The parallel between lists of phonologically similar words and the case of inflected forms being incompletely processed as bare stems is that when inflected words are heard, the bare stems are in fact activated. It can be recalled from above that processing is assumed to entail the inflected form's being related to the bare stem and placed in a word-specific paradigm that, in English, also contains the bare stem. Thus, even if the inflected form begins to fade from memory early in the processing operations, the bare stem could be available as a replacement.

An important question that remains is how cases of incomplete processing are registered in the child's paradigm. It is generally acknowledged that during the development of paradigms, more than one entry can compete for the same cell in a paradigm (e.g., Pinker, 1984). For example, if children hear the form *He likes ice cream*, they might hypothesize that -*s* marks singular, and then, third person. These possibilities will be rejected in turn because the children will have already registered examples such as *I like cake* and *They like beer*. It is only when the children consider the conjunction of singular and third person will they arrive at the correct hypothesis. However, the well-entrenched bare-stem forms associated with first- and second-person singular, and first-, second-, and third-person plural will ensure that the paradigm is not easily split and filled by a different form for the third-person singular cell. Additional encounters with this form will be needed before bare stems are properly confined.

From this description, it can be seen that underlying representations can vary in strength. Applied to the surface account, this means that brief closed-class morphemes that are sometimes processed incompletely and lost may have weaker representations than morphemes of greater relative duration.

Weaker representations due to occasional incomplete processing of encountered morphemes are usually thought to be the functional equivalent of infrequent exposure to these morphemes (Leonard, 1994). That is, the morphemes may have been sufficiently abundant in the input but, because they were not always fully processed, fewer instances of these morphemes actually had an impact on the underlying paradigm. According to this interpretation, then, only fully processed morphemes have an effect on the paradigm. This is certainly likely to be the case when morphological analysis of a freestanding closed-class morpheme never occurred in the first place because processing was still focused on preceding material when the morpheme appeared in the speech stream, or when processing of an inflected word was abandoned prior to the inflection in favor of the next word appearing in the utterance.

However, as we saw earlier, another form of incomplete processing results in the maintenance in working memory of the bare stem only. It is possible that in this case the paradigm has no access to whether retained material is fully or incompletely processed. Consequently, each bare stem resulting from this type of incomplete processing might be treated as a bona fide entry in the paradigm. Thus, the weaker representation of the correct form in the paradigm will be due to competition between that form and the bare stem.

Each of the incomplete processing scenarios considered thus far results in omission. However, there is one more scenario whose effects might lead to substitution. In some instances, instead of completing morphological analysis before placing a morpheme in a paradigm cell, placement might occur before analysis is completed. Because this cell will probably already be occupied by an appropriate form, the strength of the inappropriate form will be relatively weak. However, it might have sufficient strength, before it is expunged from the grammar, to occasionally appear in place of the correct form. Thus, although omission is the usual result of incomplete processing in a language such as English, substitutions are possible, at least in principle.

Some Shortcomings of the Surface Hypothesis   Leonard and his colleagues have identified several limitations in their account (see Leonard, 1989, 1994; Leonard & Eyer, 1996; Leonard, McGregor, & Allen, 1992). First, although relative duration is emphasized, once absolute duration reaches a certain value, relative duration may prove less important. Second, fundamental frequency and intensity interact with duration in important ways, yet these acoustic factors have not been formally incorporated into the surface account. Third, thus far, only a binary distinction has been made between morphemes; either the morpheme has shorter duration than adjacent material, or it does not. Yet, a continuum is more likely to reflect the real state of affairs. For example, main verbs in English have greater duration than modal auxiliaries, which in turn have greater duration than copula and auxiliary *be* forms (Altenberg, 1987; Swanson & Leonard, 1994; Swanson, Leonard, & Gandour, 1992). Fourth, to date, the importance of utterance position has been considered only in terms of weak syllables. However, fricatives, too, are significantly lengthened in phonological-phrase and utterance-final position (Coker, Umeda, & Browman, 1973). This might have special relevance for noun plurals, given the tendency for nouns to appear in this position in English (Gentner, 1982).

Several investigators have criticized the surface hypothesis for its apparent failure to account for differences in degree of use of grammatical morphemes that have identical phonetic values, such as the plural -*s* and the third-person singular verb inflection -*s* (Gopnik & Crago, 1991; Lahey, Liebergott, Chesnick, Menyuk, & Adams, 1992; Rice & Oetting, 1993). It is true that this account does not offer a unique way of handling these differences, but such differences are perfectly compatible with the account. Leonard (1989) placed the surface hypothesis within the framework of Pinker's (1984) learnability theory and its set of assumptions concerning acquisition. One of these assumptions (after Slobin, 1985) is that there is a hierarchy of accessible notions relevant to grammatical morphology. Notions that are high on this grammaticizability hierarchy are those that appear in many of the world's languages and have clear semantic correlates. Those that are low appear in fewer languages and have less salient correlates.

Other factors influencing the hierarchy are phonetic substance and nonhomonymy. Thus, if grammatical morphemes in two different languages express an identical notion but one has greater phonetic substance and is not homonymous with any other morpheme in the language, it will be hypothesized and acquired earlier. Likewise, if in the same language two morphemes are identical in phonetic substance and are homonymous, the one that expresses the more accessible notion will be acquired earlier. Thus, in the surface account two phonetically identical morphemes can be acquired at different rates, because the child will hypothesize the grammatical function of one before the other. However, if these morphemes have short relative duration, the use of each should lag behind that of MLU controls. Figure 12.2 illustrates this type of outcome, using as an example the plural -*s* and third-person singular -*s*.

Of course, it is not yet clear that the order in which grammatical morphemes are acquired in normally developing children is accurately handled by a grammaticizability hierarchy. Thus, in embracing this concept the surface hypothesis may be flawed. For example, as noted in chapter 11, in the principles and parameters framework, the differences between plural -*s* and third-person singular -*s* include the fact that the former is associated with a lexical category, N, whereas the latter is associated with a functional category, I. So, even though the surface account allows for

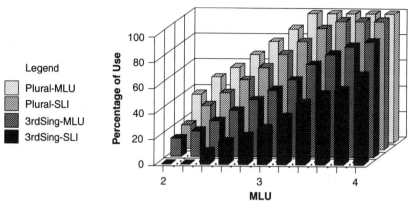

Figure 12.2
Profiles predicted by the surface hypothesis when two phonetically identical grammatical morphemes of short relative duration are acquired at different points in time by normally developing children. The relationship between the two grammatical morphemes should be the same in the speech of children with SLI as in the speech of normally developing children matched according to mean length of utterance (MLU), with lower percentages of use of both morphemes by the children with SLI.

differences in degree of use of phonetically identical forms, its level of precision in predicting such differences seems to be at the mercy of current thinking in linguistic theory and development.

There are several empirical findings that pose problems for the surface hypothesis. We shall begin with a few for which the account has no ready explanation—suggesting that the account is incomplete—and conclude with one that raises the possibility that details of the account are simply wrong. One of the findings in the first category is the observation that children with SLI sometimes produce preverbal case errors, as witnessed in utterances such as *Me take that* and *Them see the tuba*. Some investigators have found this type of error to be higher in frequency in the speech of children with SLI than in the speech of MLU controls (Loeb & Leonard, 1991); other investigators, however, have found no such difference (Moore, 1995). The surface hypothesis has the means of accounting for problems with nominative case pronouns, for such pronouns are very brief in duration (see McGregor & Leonard, 1994; Read & Schreiber, 1982). However, there seems to be no theory-internal reason to expect forms such as *me, her, him*, and *them* to replace their nominative case counterparts, even granting that they have longer durations (thanks in large part to their frequent appearance in final position and their ability to appear in isolation).

Another empirical detail at odds with the surface hypothesis is a finding by Leonard (1995) that children with SLI showed lower percentages of auxiliary inversion in wh-questions than a group of MLU controls. That is, these children showed a higher percentage of utterances such as *What mommy is making*? Because the surface account assumes a normal underlying grammar on the part of children with SLI, there is no basis for expecting lower rates of inversion when an auxiliary is available.

The surface hypothesis provides no rationale for expecting special problems with irregular past forms in children with SLI. The distinctions between the present and past counterparts in such cases are seen in the vowels of strong syllables (e.g., *throw–threw*); some pairs differ in consonantal information as well (e.g., *catch–caught*). The literature on the use of these forms by children with SLI is somewhat unclear. Some investigations have found higher percentages of use of irregular than of regular past in these children (e.g., Leonard, Bortolini, Caselli, McGregor, & Sabbadini, 1992; Oetting & Horohov, 1997). However, recent studies have reported problems with the irregular as well (e.g., Moore & Johnston, 1993).

Cleave and Rice (1995) found evidence that children with SLI produce contracted copula and auxiliary *be* forms with higher percentages than their uncontractible counterparts (but see D. Ingram, 1972b). If these findings reflect the true state of affairs, the surface account would have no way of explaining it.

There are also details from languages other than English whose explanation can't be found in the surface account. For example, Clahsen and Rothweiler (1992) found that German-speaking children with SLI produced participles as accurately as did MLU controls. Some of the components of participles (e.g., *ge-*, *-t*) are of short duration and should have produced differences between the groups. Findings from Inuktitut reported by Crago and Allen (1994) are also problematic. The child with SLI in their study showed a profile that differed from the MLU control child in ways that can't be handled by the surface hypothesis.

In addition to findings suggesting that the surface hypothesis is incomplete, there are now findings that suggest inaccuracies in the account. Some of these data come from the work of Oetting and Rice (1993) and Rice and Oetting (1993). These investigators compared children with SLI and MLU controls in the use of plural *-s* (Oetting & Rice, 1993; Rice & Oetting, 1993) and third-person singular *-s* (Rice & Oetting, 1993). For the plural inflection, differences between the groups were small and, for some measures, nonexistent. Differences were much larger for third-person singular. Because the plural might be expected to be hypothesized and learned earlier than the third-person singular inflection (see above), the finding of smaller differences for the plural would not be problematic. However, Rice and Oetting (1993) found no differences in MLU between subgroups of children with SLI and MLU controls whose percentages of use of plural *-s* were relatively low. According to the surface account, the MLUs of the subgroup with SLI should have been higher.

The importance of this finding is that it calls into question a basic assumption of the surface hypothesis. According to this hypothesis, if group differences in the use of grammatical morphemes of relatively short duration are seen for later-hypothesized notions, then at an earlier point in development, differences of similar magnitude should be seen for phonetically identical morphemes involving more transparent notions, as seen in figure 12.2. If this proves not to be the case, the surface hypothesis would be weakened, for it would suggest that duration effects operate only at particular MLU levels, or only when the notion is less transparent, taking a backseat to other factors in cases to the contrary.

Hypotheses Based on Morphological Richness
Some of the crosslinguistic findings that have lent support to the surface hypothesis have simultaneously introduced data that reveal one of its limitations. In certain languages, brief morphemes that are used with lower percentages by children with SLI than by MLU controls are nonetheless used with higher percentages than the corresponding (brief) form in English. For example, even though the third-person singular

inflection and copula forms are more difficult for German-speaking children with SLI than for MLU controls, these forms are used with higher percentages than the corresponding English forms are used by English-speaking children with SLI at the same level of MLU (S. Roberts, 1995).

Findings such as these indicate that characteristics of the language in addition to the duration of its morphemes are crucial factors. Accordingly, several investigators have proposed accounts that give these characteristics considerable importance. However, these accounts also rely on an assumption of a general processing capacity limitation. We shall take up these two points in turn. First we discuss the characteristics on which these languages differ, then we consider how processing limitations are implicated.

The specific characteristics of language that are deemed important have varied somewhat from study to study. Advantages seen for Italian- and Hebrew-speaking children with SLI over their English-speaking counterparts have been attributed to the fact that in the first two languages, nouns, verbs, and adjectives must always be inflected (e.g., Dromi, Leonard, & Shteiman, 1983; Leonard, Sabbadini, Leonard, & Volterra, 1987). In predicting greater use of grammatical morphology by German-speaking children with SLI than by English-speaking children with SLI, Lindner and Johnston (1992) emphasized the fact that German relies more on morphological means to signal basic grammatical functions such as subject and object. It is not yet known which characteristics are the crucial ones (we noted other possibilities earlier, such as redundancy and regularity), or, for that matter, whether such characteristics can vary in importance from language to language. Likewise, it is not clear whether these characteristics alter the relative ease of identifying instances of linguistic categories already in the children's grammatical systems, or whether the characteristics are in part responsible for the creation of these categories. However, it can be assumed with some confidence that whatever the characteristics are, they are associated with the morphological richness of the language.

There is some irony in this, of course. Children with SLI seem to have grammatical impairments regardless of the language they are learning, but if there are many grammatical morphemes to learn, they do better. Better, that is, than their counterparts with SLI in other languages. At best, they will do as well as their MLU matches in the same language, and sometimes they will do more poorly.

These morphological richness accounts assume that children with SLI have a general processing capacity limitation that reduces their available cognitive resources. If the language is one such as English, for example, these children will focus on word order at the expense of grammatical morphology. In languages such as Italian and Hebrew, by contrast, grammatical morphology will receive the greatest share of resources. It can be seen, then, that the relative strength of, say, Italian children with SLI in the area of grammatical morphology does not mean that they are better learners, only that they are directing more of their resources to this area.

A price is paid for the greater expenditure of resources in the area of grammatical morphology. Because resources are limited, other areas of language will be short-changed. Unfortunately, research aimed at discovering areas of weakness in children with SLI acquiring morphologically rich languages is only just beginning, and therefore this assumption has not been adequately tested. Word order is variable in Italian and Hebrew, and therefore problems with this area of grammar are difficult to identify (Leonard, Sabbadini, Volterra, & Leonard, 1988). A better test case is German (and related languages), given that it has a richer morphology than English and word order rules that permit easier identification of errors. In chapter 4, it was seen that German-speaking children with SLI have problems with the verb-second rule of the language, and their use of grammatical morphology is greater than in English-speaking children with SLI. This would seem to support the general idea that children with SLI devote their limited processing resources to a dominant property of the language and, by necessity, allow other aspects of the language to falter.[4]

Given the compelling evidence for processing limitations in children with SLI and the documented differences in these children's profiles as a function of the language being learned, future research is likely to be devoted to a more detailed assessment of morphological richness accounts and a more careful exploration of the characteristics that might be most important. High on the research agenda should be studies of languages that permit bare stems yet have a rich morphology, as well as languages that prohibit bare stems but have rather simple inflectional systems (e.g., one inflection that is used for present tense regardless of person, number, or gender; one inflection used for past tense; and so on).

Another topic for future research is a test of the limits of the advantages of a rich morphology. Grammatical morphology should be a relative strength only if it can be more consistently and more fully processed when children with SLI direct their limited resources toward this area of language. However, if the morphology itself is extremely complex in terms of its processing demands, this area of language might not look any stronger than it does in a language such as English, where grammatical morphology receives fewer resources in the first place. For example, in Hebrew, past tense verb inflections in third-person singular and second-person singular and plural are distinguished according to gender. Thus, children must discover the function of these inflections by hypothesizing the combination of tense, number, person, and gender. Such an operation might well exceed the processing speed limitations of these children. Although Hebrew verb inflections have thus far revealed no differences between children with SLI and MLU controls (Dromi, Leonard, & Shteiman, 1993; Leonard & Dromi, 1994), the third-person singular masculine and feminine past tense inflections are the only ones requiring the conjunction of four features that have been examined to date.

Reduced Speed of Processing Applied to Language Production
In the discussion of the surface and morphological richness accounts, the emphasis was on acquisition. However, if the reduced speed of processing in these children is general—the assumption made here—it should have an effect on the process of language production as well.

It has been noted that children with SLI usually differ from normally developing children in the degree to which they express particular linguistic information, not in whether they express such information. Thus, particular words that are called for in the situation seem less likely to be produced, particular argument structures are less likely to be used when needed, and, especially, closed-class morphemes are used less frequently in obligatory contexts.

In some cases, such inconsistency might reflect limitations in linguistic knowledge. For example, the child might not have sufficient command of a semantic category to recognize certain exemplars of the category, even though other exemplars are named correctly. Or the child might not know the argument structures of certain verbs even though the same argument structures are known and produced for other verbs. However, it does not seem likely that all of this inconsistency can be attributed to limitations in knowledge. This is especially clear in the case of grammatical inflections. Bishop (1994) studied a group of children with SLI who displayed considerable inconsistency in their use of grammatical inflections. On close inspection, she found instances of inconsistency involving the same lexical item. More recently, Leonard, Eyer, Bedore, and Grela (1997) found evidence of the same type for each of the children with SLI participating in their study. Bishop also found some evidence suggesting that errors in grammatical morphology increased with the amount of material already generated in the utterance.

Bishop (1994) proposed that grammatical errors such as those above reflect performance limitations on the part of children with SLI. This proposal certainly seems sensible, and had been made by others before her (e.g., Fletcher, 1992). As noted in earlier chapters, many children with SLI show higher scores on language comprehension tests than on tests of language production. However, because items on comprehension and production tests often do not show a close correspondence, such test results are only suggestive of a performance basis for the production limitations observed. Findings such as those of Bishop are especially valuable for this reason.

Bishop (1994) suggested that the types of grammatical errors she observed might be profitably considered as the result of slow processing within a speech production model such as that of Levelt (1989). Other investigators have made similar suggestions. Fletcher (1992), also citing Levelt (1989), noted that a major problem in children with SLI might be the deployment of available syntactic structure in real time. Chiat and Hirson (1987) stressed the value of examining the interaction of semantic, syntactic, and phonological operations within a production model as a means of properly interpreting the nature of errors seen in children with SLI. Proposals of this type are worth considering in some detail. In this section, we shall review some of the components that are assumed in a speech production model, and then consider how, within this model, slow processing might play out.

Figure 12.3 is adapted from the model presented by Bock and Levelt (1994). The figure highlights those details of production most relevant to grammatical encoding—lexical selection, function assignment, constituent assembly, and inflection.

Lexical Selection and Function Assignment   As can be seen in the figure, the message, reflecting the speaker's intended meaning, leads to lexical selection and function assignment. Lexical selection is the retrieval of the lexical concepts appropriate for the message along with the "lemmas" linked to these concepts. Lemmas are lexical representations carrying semantic, form class, and other lexically based grammatical information. For example, for both English and Italian speakers, the lemma associated with the concept CAR specifies the form class Noun. For Italian speakers, the lemma also specifies the grammatical gender Feminine. Although lemmas carry information, they have yet to receive morphological and phonological form.

Lexical concepts are organized in networks, such that one lexical concept will have links with other, associated concepts. An abbreviated network appears in figure 12.4. CAR, for example, will have links with VEHICLE, DRIVE, TIRES, and PASSENGER. It will also have links with AIRPLANE—if not directly, then indirectly through the connection with VEHICLE. It is assumed that when CAR is part of the speaker's intended meaning, these related concepts will also be activated. Lemmas also have

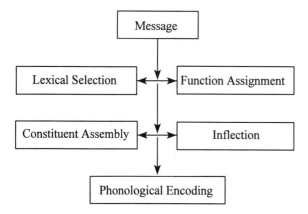

Figure 12.3
An adaptation of the model of Bock and Levelt (1994) highlighting the four grammatical encoding operations: lexical selection, function assignment, constituent assembly, and inflection.

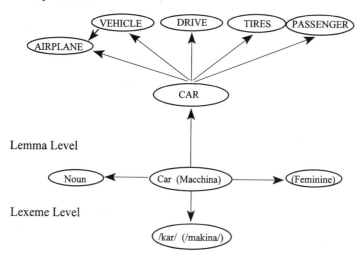

Figure 12.4
An abbreviated network of lexical concepts.

links; those associated with CAR and AIRPLANE, for example, are linked by virtue of their shared Noun property.

During function assignment, each lemma is linked to a grammatical function. If the intended message were "The gorilla pushed the car," for example, the lemmas for GORILLA and CAR would be assigned the subject and object functions, respectively.

Constituent Assembly and Inflection   As the next set of boxes in figure 12.3 shows, after lemmas have been selected and assigned a grammatical function, their order in the sentence is arranged and inflectional information is added. These processes of constituent assembly and inflection involve the retrieval of fragments of surface structure. These consist of maximal phrase structures, such as NP or VP, that are expanded down to the stem level of the head of the phrase. They indicate slots where the head stem and any associated freestanding closed-class morphemes are to be inserted. Inflections are already specified in the fragment. Examples are shown in (6) and (7).[5]

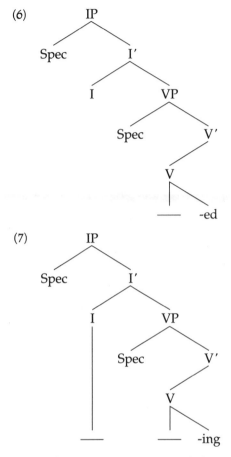

(6)

(7)

Fragments and freestanding closed-class morphemes are accessed from separate stores and then combined. Some authors working with models of this type have proposed that the items in each store are organized in rows and columns that are ordered according to morphosemantic complexity. For example, Lapointe (1985) proposed that in each V fragment store for English, a bare stem occupies the first row of the first column, V + s occupies the second row of the first column, I + V + ing occupies the first row of the second column, and so on. In the I store, auxiliary *is* appears in the first row of the third column, auxiliary *are* in the second row of the same column, and so on.

The locations of items in these fragment and closed-class morpheme stores are extremely important. Lapointe (1985) proposed that retrieval of the information in each of these stores involves moving along tracks defined by the top row and columns in the store. Retrieval can proceed only up or down a column track and only back and forth along the top row track. It is assumed that expenditure of resources is proportional to the distance traveled to retrieve information in the store. In speed-of-processing terms, the greater the distance to be traveled, the longer the time required

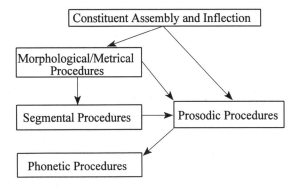

Figure 12.5
An adaptation of the model of Levelt (1989) highlighting phonological encoding procedures performed on the input provided by constituent assembly and inflection.

for successful retrieval. Therefore, retrieval of information in the upper left portion of the store will require less time than retrieval of information located deeper in the store. According to Lapointe, information from fragment stores is retrieved before information from freestanding closed-class morpheme stores.

Phonological Encoding   Phonological encoding constitutes the final level of operations shown in figure 12.3. A closer look at phonological encoding is provided in figure 12.5, adapted from Levelt (1989). Although this figure lacks some of the detail provided by Levelt, it conveys the information most relevant to the question at hand.

It is assumed that the surface structure resulting from constituent assembly and inflection serves as input for the generation of both the morphological form of each lemma and the prosody of the utterance. Input of the first type leads to the creation of each morphological form along with its "citation" metrical pattern, that is, the number of syllable peaks (if any) with marking on the peak that carries word accent. The segmental composition of each morphological form is then generated based on this information. Input of the second type (to the prosody generator) runs in parallel with the first and results in a phonetic plan for the utterance as a whole. These parallel paths are not wholly independent; when the citation metrical form of the word is generated, this information is passed to the prosody generator. Likewise, the segmental composition of the word is made known to the generator.

To examine the coordination of metrical and prosodic structure during phonological encoding more closely, we can borrow the framework of prosodic phonology (Nespor & Vogel, 1986; see also Hayes, 1989), which is fundamentally compatible with Levelt (1989). In the framework of prosodic phonology, there is a prosodic hierarchy, beginning with the syllable and extending to higher levels of organization. The prosodic categories beyond the syllable that are most relevant for larger units of speech include the phonological word, the clitic group, and the phonological phrase.

Phonological words are stems plus inflections, essentially the morphological forms that were generated from lemmas and those diacritical features that translate into inflections in the particular language. In a language such as English that permits bare stems, a phonological word may be either a bare stem (e.g., *take*) or a stem plus inflection (*takes*). In a language such as Italian, *prendo* (I take) and *prende* (he or she takes) are phonological words, but the impermissible *prend* is not.

The clitic group (CG) represents a higher level in the prosodic hierarchy. Here, phonological components interact with syntactic information beyond the level of inflection. A clitic group is an open-class word (which may be inflected) plus all adjacent and phonologically interacting closed-class morphemes. For example, the four-word utterance in (8) can be divided into two clitic groups of two words each, whereas the four-word utterance in (9) represents four distinct clitic groups.

(8)
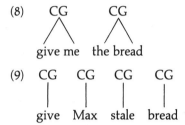

(9)
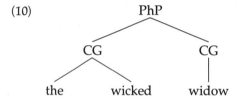

Clitic groups combine to form a phonological phrase (PhP). This prosodic category is made up of all clitic groups within a single syntactic phrase, up to and including the lexical head of the phrase. Thus, the noun phrase in (10) represents a single phonological phrase composed of two clitic groups, *the wicked* and *widow*.

(10)
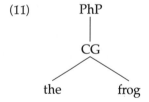

Despite the fact that phonological phrases constitute higher level categories, there are many contexts in which they can be quite short. For example, in (11), the phonological phrase consists of a single clitic group.

(11)

PhP
|
CG

the   frog

Yet there are contexts in which short phonological phrases can be merged with preceding phonological phrases, at least in languages such as English and Italian. The best studied context is one in which a phonological phrase consisting of a single clitic group, as in (11), is the complement of a preceding lexical head. An example is shown in (12).

(12)
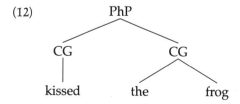

In this example, the phonological phrase *the frog* is a single clitic group that serves as the complement of the head V, *kissed*. As a result, it is subject to restructuring.

It is assumed that speech production phenomena such as substitutions and omissions may operate within each of these prosodic categories. Indeed, the work of Gerken (1991, 1994) nicely illustrates this point. Gerken's work began with the common observation that most of the weak syllables produced by young normally developing children appear in strong (S) syllable plus weak (W) syllable sequences. Thus, weak syllables are often produced in words such as *Tony* and *tiger*, but not in words such as *Jeanette* and *giraffe*. The latter are WS sequences.

Gerken hypothesized that young normally developing children would be more likely to produce or omit freestanding closed-class morphemes such as articles and pronouns as a function of their position in the phonological phrase. When the morpheme appears in utterance-initial position, and thus precedes rather than follows a strong syllable (e.g., "*He* kissed the bear," "*The* bear kissed him"), it should be more likely to be omitted. In contrast, when the morpheme appears immediately after a strong syllable, and thus has the potential to be grouped with the strong syllable in an SW pattern within the same phonological phrase (e.g., "The bear kissed *him*," "Pete kissed *the* bear"), it should be more likely to be produced. Children's responses to a sentence imitation task confirmed Gerken's hypothesis.

As children develop, they produce an increasing proportion of words and phrases that are free of the SW pattern. However, there is evidence to suggest that this developmental process is influenced by the children's available resources. Gerken, Landau, and Remez (1990) presented young children with several types of sentences to imitate. One type consisted of simple sentences (e.g., "Pete fix*es the* car") containing sentence-medial closed-class morphemes consisting of weak monosyllables, some of which occurred in positions that precluded organization into an SW pattern (e.g., *the* in the example above). Another type of sentence differed from the first in that weak nonsense syllables replaced the syllabic closed-class morphemes (e.g., "Pete push*eg le* truck"). Not surprisingly, Gerken et al. found that the weak syllables that could be organized as part of an SW sequence (e.g., *-es* and *-eg* above) were more likely to be included in the children's repetitions than the weak syllables that could not (e.g., *the* and *le*). Significantly, nonsense syllables were more likely to be included in the children's repetitions than were the closed-class morphemes. Both sentences with and without nonsense syllables required the children to produce weak syllables outside of an SW sequence. However, production of the closed-class morphemes required something more; these forms contained grammatical information involving number, definiteness, and the like that was absent from the nonsense equivalents. It seems likely that this additional information increased the processing demands of the task for the children.

The increase in processing could have taken two forms. First, each sentence presented to the child might have been reconstructed according to the child's own grammatical system before being produced. Although repetition responses sometimes exceed children's spontaneous productions in complexity, there is long-standing evidence of modifications by children that reflect this reconstruction process (e.g., Slobin & Welch, 1971). According to this possibility, then, the reconstruction of sentences containing closed-class morphemes would be more difficult because these morphemes would have to be retrieved from the fragment and closed-class morpheme stores before the morphological and segmental information was passed on to the prosody generator. A second possibility is that the children recognized the closed-class morphemes in the stimulus sentences, deemed them of little semantic importance, and deleted them from their response. The nonsense syllables, in contrast, were not recognized, and thus were retained in memory and repeated. Even here, the child must access the closed-class morphemes from the appropriate cells of fragment and closed-class morpheme stores, verify them as matching the items held in memory, and delete these forms from the morphological and segmental information that is transmitted to the prosody generator. These additional operations are not required in the case of the nonsense syllables.

Data from Children with SLI   At this point, it seems useful to consider how Bishop's (1994) proposal of slow processing speed in children with SLI might translate into problems at different points in the language production model. It would be reasonable to hypothesize that children with SLI will be slower than normal in each of the processes required in speech production. Thus, even tasks involving lexical selection should show this effect.

The available data are in fact consistent with this assumption. Studies by Anderson (1965), Kail and Leonard (1986), and Leonard, Nippold, Kail, and Hale (1983) provide evidence that the RTs of children with SLI are significantly slower than those of a group of age controls on simple picture-naming tasks. In these tasks, items are selected to ensure high levels of accuracy, and only the RTs for accurate responses are recorded.

The Kail and Leonard (1986) investigation employed a second naming task. In this task, the children heard an unfinished sentence that could be logically and grammatically completed by the name of the picture presented in the instant the final word should have been heard. According to the type of speech production model considered here, the sentence context leads to the activation of concepts associated with the picture name, and hence the target name receives some degree of activation even before the picture appears. Furthermore, the grammatical context immediately preceding the picture presentation promotes activation of grammatical information that is contained in the word's lemma. Thus, some of the work of retrieving the name of the picture is already accomplished due to the activation from the sentence.

Consequently, RTs should be faster than in the simple (picture-only) naming task. This proved true for both groups of children, though the RTs of the children with SLI were nevertheless slower than those of the age controls. But the differences, though highly significant, narrowed in the sentence plus picture task. The RTs of the children with SLI decreased by approximately 120 ms relative to the picture-only task. The corresponding decrease for the age controls was approximately 50 ms. This suggests that the slower RTs for the children with SLI were not attributable to a single component of the response (e.g., the act of uttering the word) but, rather, to several (and possibly all) components of the response. Put differently, the greater the amount of work that must be accomplished to generate a response—as in the picture-only relative to the sentence plus picture task—the slower the RTs of children with SLI will be relative to the norm.

Let's now consider how slower processing might play havoc with the production of grammatical elements. It can be recalled that the items in both the fragment and the freestanding closed-class morpheme stores vary in their location, and that items in more remote locations require more time to retrieve. This leads to the possibility that slower speed will result in the retrieval of an item from a shallower cell because deeper cells could not be reached in time without disrupting the utterance. More precisely, because the child must retrieve words, assign functions, and retrieve fragments and closed-class morphemes in real time, slow processing will lead to competition between completing retrieval of an item in a deeper cell of a store and proceeding with the rest of the sentence. If proceeding with the rest of the sentence wins out, bare stems and missing function words can result.

The assumption of slow processing within this type of speech production model, then, would lead to the expectation that when inappropriate items are produced, they constitute entries in shallower cells than the presumed targets. In general, this seems to hold rather well for children with SLI. For example, bare stems are often produced when an inflected verb is required. And, indeed, V is located in a more accessible cell than either $V + s$ or $V + ed$. In a sentence imitation task, Masterson and Kamhi (1992) found that children with SLI were more likely to have problems with grammatical inflections in sentences with complex (as opposed to simple) grammatical structure (although, as will be seen below, phonological complexity also played a role).

Another feature of the speech production model is that resources are allocated to fragment stores before they are allocated to freestanding closed-class morpheme stores. Thus, once a fragment is retrieved, there may be insufficient time for retrieval of the closed-class morpheme. Again, this works rather well for the available data on children with SLI. For example, fragments with missing freestanding closed-class morphemes are often seen in these children's speech, as in *Mommy eating*. A plausible account of such utterances is that once the fragment $I + V + ing$ is retrieved, there is insufficient time to retrieve *is* from the I store.

Proposals of this type can also be applied quite nicely to data from other languages. In Italian, verbs are always inflected, and thus a bare V does not appear in the fragment store. The most difficult verb inflection for Italian-speaking children with SLI is the third-person plural inflection; when children fail to use it in obligatory contexts, the third singular is used in its place. It is noteworthy that the third-person singular inflection is in a more accessible location in the fragment store in Lapointe's (1985) scheme.

Slow processing could also have adverse effects during phonological encoding. For example, longer and more grammatically complex sentences require a greater number of lemmas to be retrieved and assigned according to function, and a greater number of fragments and closed-class morphemes to be retrieved with these items located in deeper stores. Temporal demands for these processes might exceed the processing speed of children with SLI, and hence processes involved in phonological encoding might be left incomplete.

Evidence consistent with this outcome can be found in the literature. For some time, investigators have noted phonological errors coinciding with grammatical difficulties in children with SLI (e.g., Menyuk & Looney, 1972). Many of the studies conducted have employed sentence imitation tasks in which grammatical complexity and, often, phonological complexity are manipulated (Masterson & Kamhi, 1992; Panagos & Prelock, 1982; Prelock & Panagos, 1989; Schmauch, Panagos, & Klich, 1978). Results show that increases in grammatical complexity usually lead to phonological simplification. In this respect, the difference between children with SLI and control children is one of degree, not of kind. Control children also show greater phonological simplification when grammatical complexity increases, though not to the same extent.

In some studies, the reverse effect is also evident, where phonological complexity seems to result in grammatical simplification. This, too, can be handled by the type of speech production model under consideration here. For example, if the morphological, metrical, and segmental forms of the lemmas appearing early in the utterance are complex, they will require more time to be generated—a real problem for children whose processing is slow to begin with. This will place time pressures on the processes of constituent assembly and inflection for the remainder of the sentence, resulting in the retrieval of shallower fragments and closed-class morphemes in their respective stores. It might be pointed out in this regard that the deleterious effects of phonological complexity on grammatical inflections in particular seem to be less dramatic or even absent when time pressures are reduced, as when spontaneous utterances are the focus of analysis. Of course, here the child controls the complexity of the sentence as a whole. And, indeed, when grammatical inflections appear along with phonologically complex words, the utterances tend to have relatively simple syntactic structure.

The interaction among metrical, prosodic, and morphological information is also likely to place severe processing demands on children with SLI. There is a fair amount of evidence that children with SLI have great difficulty with weak syllables, especially when they precede strong syllables (e.g., D. Ingram, 1981). Chiat and Hirson (1987) described a child with SLI who often omitted weak syllables appearing before strong syllables in the same word. This production pattern was accompanied by a tendency to omit freestanding closed-class morphemes that preceded open-class words beginning with a strong syllable. In contexts in which closed-class morphemes appeared at the end of an utterance and thus could not precede a strong syllable, the child was much more likely to produce them.

McGregor and Leonard (1994) presented a sentence imitation task to a group of children with SLI and replicated Gerken's (1991) basic findings. The children were more likely to omit articles and pronouns when these were in sentence-initial position and could not unite with a preceding strong syllable to form an SW pattern. This bias toward an SW pattern for closed-class morphemes was also seen in the MLU controls who participated in the study. However, the overall percentage of omissions was significantly higher for the children with SLI.

Bortolini and Leonard (1996) made use of the prosodic phonology framework to examine the intragroup variability seen in two different groups of children with SLI, a group acquiring English and a group acquiring Italian. The subgroup of English-speaking children who showed the greatest tendency to omit articles and uncontractible copula forms were in general more restricted in the contexts in which they produced weak syllables. These children's use of weak syllables was much more likely to be limited to SW contexts within a word, clitic group, or phonological phrase. For example, closed-class morphemes that formed clitic groups with preceding open-class words (e.g., "fix them") did not seem especially difficult for the children, whereas those that formed clitic groups with a following open-class word were quite problematic (e.g., "to get"). Although the use of word-initial weak syllables was difficult for these children, the production of these syllables was more likely if the word followed a stressed open-class word in the same phonological phrase (e.g., "get another").

The findings for Italian were equally informative. As was found for English, the subgroup of Italian-speaking children with SLI who showed the greatest tendency to omit freestanding closed-class morphemes such as articles were more restricted in the contexts in which they used weak syllables in general. Relatively few weak syllables were produced that could not be placed in an SW sequence. This relative dependence on an SW pattern might have been responsible for these children's profile of inflection use. Most of the inflections examined by Bortolini and Leonard (1996) required the use of a weak syllable that immediately follows a strong syllable. For example, the first- and third-person singular inflected forms for *cantare* (to sing) are *cánto* and *cánta*, respectively. The plural forms of the nouns *líbro* (book) and *matíta* (pencil) are *líbri* and *matíte*. The third-person plural verb inflection differs in this respect. The inflection consists of two syllables (e.g., *-ano*), both weak. Thus, to produce third-person plural forms (e.g., *cántano*), children must make use of an SWW pattern. The children who were limited in their use of freestanding closed-class morphemes did not have special difficulties with the grammatical inflections under investigation except for the third-person plural inflection. Frequently, this inflection was replaced by the third-person singular inflection (e.g., *cánta*) in these children's speech, an inflection that allows for an SW sequence.

Within the framework of prosodic phonology, limitations in the ability to produce weak syllables outside of an SW sequence might also explain some of the cross-linguistic differences that seem to hold between English- and Italian-speaking children with SLI. The former do not appear to omit object pronouns frequently, whereas object clitics are often omitted by their Italian-speaking counterparts (Leonard, Sabbadini, Volterra, & Leonard, 1988). Both English object pronouns and Italian object clitics are weak syllables, but they differ in their typical locations in clitic groups, as shown in (13) and (14).

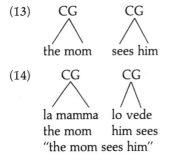

(13)   CG   CG

        the mom   sees him

(14)   CG   CG

        la mamma   lo vede
        the mom   him sees
        "the mom sees him"

The English object pronoun follows the verb and therefore can participate in an SW pattern. The Italian clitic usually precedes the verb; its use, then, requires production of a WS sequence.

Although these studies emphasize the important influence of prosodic factors on the degree and profile of grammatical morpheme use by children with SLI, they also contain evidence for effects that might be attributed to general speed of processing limitations. Despite the correlation between limitations in the use of weak-syllable closed-class morphemes that fall outside of an SW sequence and limitations on the use of word-initial weak syllables, omissions of the former are more likely. For example, in the Bortolini and Leonard (1996) study, the Italian-speaking children with SLI with the most limited use of word-initial weak syllables were twice as likely to produce them as they were to produce closed-class morphemes in similar prosodic contexts. It is likely that these differences were attributable to the fact that the closed-class morphemes had to be retrieved from fragment and closed-class morpheme stores before all metrical and segmental information could be transmitted to the prosody generator. The children with SLI may have proceeded with the rest of the utterance without performing these retrieval operations in an effort not to disrupt expression of the intended message. In the case of words with initial weak syllables, the metrical and segmental information could be transmitted without waiting for information retrieved from these stores.

The above sketch of possible sources of production difficulties does not take into consideration external discourse demands. Yet, surely these also play a role. For example, Evans (1996) discovered that omissions of grammatical inflections were more likely to occur in children with SLI (with production deficits only) when the discourse demands were the greatest. Specifically, omissions were most frequent when the children had to provide a topically related conversational turn within two seconds of the previous speaker's utterance. In such cases, time pressures are exerted at the outset, at the level of the message in figure 12.3. From that point on, the difficulties that might arise should be the same as those outlined above.

In summary, the notion of generalized slow processing within a speech production model seems viable as an account of performance limitations in children with SLI. This proposal has the advantage of breadth; performance glitches in a variety of linguistic domains can be accommodated. Future research will determine whether the pieces fall together in a sufficiently coherent manner to attribute all such performance-based production problems to a common factor.

Notes

1. Some studies have used passive constructions such as *The boy is being pushed*. In such cases, it becomes clear by the beginning of the word *being* that the sentence is not active. However, processing demands are not minimized, because nonpassive constructions such as *The boy is being bad* are still possible at this point.
2. Studies examining the effects of prosody on language learning also implicate rate effects. Weinert (1992) found that children with SLI did not benefit from prosodic cues in learning novel grammatical rules, showing no improvement over their learning of material presented in a monotone. However, overall durations of the material to be learned appeared to be longer in the monotone condition. The higher demands on rate of processing in the prosody condition might have offset some of the facilitating effects that the prosody provided.
3. And Fletcher (1983) did in fact find problems with inflections as well auxiliaries, upon reviewing data from one of his earlier collaborative studies (Fletcher & Peters, 1980).
4. This interpretation will be complicated if—as Clahsen (1989) has contended—word order errors are the result of problems with aspects of grammatical morphology. This possibility was explored in some detail in chapter 11.
5. For sake of consistency, the structures of fragments are presented in the form they are assumed to take in the principles and parameters framework. The work reviewed here employed earlier linguistic frameworks.

Chapter 13

SLI as a Processing Deficit in Specific Mechanisms

The accounts of SLI discussed in the last chapter share the assumption that the source of the problem is a general processing capacity limitation. Where areas of special weakness are seen in children with SLI, they are presumably the natural result of trying to learn, or produce, details of the language that are especially fragile under difficult conditions. The processing limitation serves to push these details over the edge much as, in average listeners, the auditory perceptual distinction between *fin* and *thin* in noisy conditions will suffer to a much greater degree than will other perceptual distinctions.

In contrast, the accounts discussed in this chapter assume that deficits in processing are more localized. The specific mechanism affected varies across these accounts; however, in each account, the consequences of the deficit are assumed to be widespread.

Phonological Memory Deficits

Gathercole and Baddeley (1990) have advanced the idea that part of the problem seen in SLI is attributable to deficits of phonological memory. According to these investigators, when learning new linguistic material, listeners first place the material in a separate phonological store in working memory (see Baddeley, 1986; Gathercole & Baddeley, 1993). The phonological representations in this store are assumed to fade quickly unless reactivated through rehearsal. Material adequately stored and rehearsed in the phonological store then enters long-term memory.

In a series of experiments, Gathercole and Baddeley (1990) found that children with SLI had greater difficulty repeating nonsense words and recalling lists of real words than a group of mental age controls and a group of younger controls matched according to vocabulary comprehension test scores. Apart from their poorer performance on these tasks, the children with SLI did not seem to differ from the controls in the way they responded. Like the controls, for example, the children with SLI had greater difficulty recalling lists of phonologically similar words than words that were quite different in their phonological composition. No differences between the children with SLI and the controls were seen on tasks assessing the discrimination of minimal pairs and articulation rate.

This pattern of findings prompted Gathercole and Baddeley (1990) to propose that the children with SLI exhibited a deficit in the storage of phonological information in working memory. They offered three possible explanations for this deficit. The first is that initial segmental analysis performed by these children was imprecise, leading to phonological representations that were less discriminable at retrieval. Another plausible explanation is that phonological traces decayed more rapidly in these children's phonological stores. Finally, it seemed possible that these children's phonological stores were limited in capacity, resulting in the storage of fewer items, or the same number of items stored but in degraded form. According to Gathercole and Baddeley, these children's problems with phonological memory could have been responsible for their below-age level vocabulary abilities, given that the acquisition of new words depends upon the availability of a stable and distinct phonological representation.

More recently, Gathercole and Baddeley (1993) have raised the possibility that deficits of phonological memory adversely affect the comprehension of grammar as well as lexical learning. They provide examples of sentences that are difficult to process on-line, such as semantically reversible sentences, sentences with passive constructions, and center-embedded sentences. Complete understanding of these types of sentences requires off-line analysis, and such analysis no doubt relies on a temporary phonological representation of the sentence. So, if phonological memory is limited, these sentences may be incompletely analyzed and comprehension will suffer. This, of course, is a case in which phonological memory directly influences the comprehension of grammar. An indirect effect is also possible, given the role of phonological memory in lexical learning and the important role that the lexicon—and especially the verb lexicon—plays in determining argument structure.

Support for the Gathercole and Baddeley (1990, 1993) proposal comes from several quarters. Findings of limitations in nonsense word repetition in children with SLI can be seen in studies by Kamhi and Catts (1986), and Kamhi, Catts, Mauer, Apel, and Gentry (1988). For example, in the Kamhi et al. (1988) investigation, children with SLI were poorer than mental age controls on the repetition of monosyllabic nonsense words (e.g., [dap]), strings of three monosyllabic nonsense words (e.g., [tab]–[gul]–[fɪv]), and multisyllabic nonsense syllables (e.g., [kəsæbən]), with the gap between the two groups widening across these three tasks.

Montgomery (1995b) presented both a nonsense word repetition task and a sentence comprehension task to children with SLI and a group of younger normally developing children matched according to language comprehension test scores. The nonsense words in the repetition task ranged from one to four syllables. The items in the sentence comprehension task were divided into longer and shorter sentences with similar syntactic structure (e.g., *The dirty little boy climbs the great big tall tree; The little boy climbs the tree*). For both types of sentences the child's task was to identify the picture that corresponded to the sentence from an array that consisted of the correct picture and three foils. Interestingly, comprehension of the added elements in the longer sentences (e.g., *dirty, great, big, tall* in the example above) was not required to identify the correct picture.

The children with SLI were less accurate than the control children in their repetition of the three- and four-syllable words, and in their comprehension of the longer sentences. Montgomery also found a significant, moderate positive correlation between the children's nonsense word repetition and sentence comprehension abilities. The results were interpreted as consistent with the proposals of Gathercole and Baddeley (1990, 1993), because both the longer nonsense words and the longer sentences (that involved additional, but nonessential information) were presumed to require greater phonological memory capacity.

Gillam, Cowan, and Day (1995) provided additional evidence on the possible nature of the phonological memory deficit in these children. These researchers employed a serial recall task in which digit lists were presented with or without a final nonsense word at the end of the list. Although the children serving as subjects were asked to ignore the nonsense word, it had (the expected) detrimental effects on the children's recall of the last few digits on the list. However, this effect was substantially greater for the school-age children with SLI participating in the study than for a group of age controls and a group of younger controls matched according to reading ability and digit span. Gillam et al. reasoned that for recall to be adversely affected by the additional nonsense item to such a degree, the memory trace of the digits must have been present to begin with (i.e., they were evidently present in the condition not employing nonsense words). Therefore, the drop in performance in the nonsense word condition might be attributable to a slower rate of coding information in working memory on the part of children with SLI. Because they had not yet coded the information, the phonetic information in the memory trace was vulnerable to interference from the nonsense word appearing at the end of the list.

Montgomery (1995a) studied the phonological memory skills of school-age children with SLI and younger controls matched according to language comprehension test scores. His findings were very similar to those of Gathercole and Baddeley (1990), except that the children with SLI also had greater difficulty than the control children on a discrimination task. For this task, Montgomery employed nonsense word pairs that ranged from one to four syllables. The clearest differences between the groups occurred with the four-syllable items. Although differences on tasks of this type ordinarily suggest discrimination problems, Montgomery pointed out that the longer lengths of these nonsense words might have caused them to decay in phonological memory before the children had an opportunity to complete their response.

James, van Steenbrugge, and Chiveralls (1994) compared children with SLI, age controls, and younger language test controls on the same range and types of tasks employed by Gathercole and Baddeley (1990). Like Gathercole and Baddeley, they found that children with SLI were more limited in their nonsense word repetition and word recall than both control groups. However, in contrast with Gathercole and Baddeley, the phoneme discrimination skills of the children with SLI were also poorer. Monosyllabic stimuli were used for the discrimination task, and therefore decay in phonological memory doesn't seem to be a likely explanation for this finding. Both real word minimal pairs and nonsense word minimal pairs were included. For the real words, the children with SLI scored worse than both comparison groups; for nonsense words, the children with SLI were comparable with the younger controls but worse than the age controls.

The results of several studies reviewed in chapter 3 are consistent with the phonological memory deficit account. Some of these investigations dealt with the word recall abilities of children with SLI. In each of these studies, the children with SLI were found to recall fewer words than control children (e.g., Kail, Hale, Leonard, & Nippold, 1984; Kirchner & Klatsky, 1985).

Studies of lexical learning in children with SLI also lend support to the Gathercole and Baddeley (1990, 1993) view. Haynes (1982) presented multisyllabic nonsense words serving as outer space referents in a story context to children with SLI, age controls, and receptive vocabulary age controls. The children's understanding of these nonsense words was then assessed in a task in which they had to identify the target from a series of nonsense words differing from the target by one, two, or three segments. The children with SLI showed lower accuracy than the other two groups in selecting the appropriate nonsense words, and when they erred, the errors were as likely to be selections differing from the target by three segments as by only one. Errors of this type represented a smaller percentage of the errant responses of each of the other two groups. Haynes interpreted her findings as suggesting that children with SLI have difficulties constructing phonological representations of unfamiliar words.

As was seen in chapter 3, not all findings from lexical learning studies show differences between children with SLI and younger vocabulary-matched controls. For example, Leonard, Schwartz, Chapman, Rowan, Prelock, Terrell, Weiss, and Messick (1982) found that children with SLI and younger controls matched for expressive vocabulary size were comparable in their comprehension of unfamiliar words introduced during play activities. However, in this study, the words were monosyllabic (e.g., *gauge*, *sheath*), and thus might not have taxed the children's phonological memory.

The lexical learning study by Rice, Oetting, Marquis, Bode, and Pae (1994) produced results that cannot be dismissed as easily. In that study, the words learned most easily by a group of children with SLI were multisyllabic (e.g., *aviate*, *crustacean*), whereas the most difficult were monosyllabic (e.g., *sprint*, *sphere*).

Perhaps the strongest evidence that clashes with the notion of phonological memory deficits in children with SLI comes from van der Lely and Howard (1993). These investigators studied a group of children with SLI and a group of younger normally developing children matched on a battery of three language tests, two of the tests tapping language production and one dealing with vocabulary comprehension. The children participated in three experiments involving the recall of lists of monosyllabic words or nonsense words. Variables examined in these experiments included whether the words in the list were semantically similar or phonologically similar, or whether the children responded by repeating the items in the list or by pointing to corresponding pictures. No significant differences between the two groups of children emerged from any of the experiments.[1]

It will be important to determine the reasons for the discrepancies between the word recall data of van der Lely and Howard (1993) and those of other investigators, because even if details of the phonological memory deficit account are incorrect, findings of relatively poor word recall performance by children with SLI have been a mainstay in the literature.

One factor that adds to the credibility of the phonological memory deficit account is that the notion of loss from memory is actually incorporated into other accounts. In particular, the surface account allows for the possibility that closed-class material that has been perceived may be lost from memory before morphological analysis is complete (see chapter 12). That is, because the operations of hypothesizing a grammatical function and entering the item in the proper cell of a paradigm are assumed to proceed slowly in these children, the perceived closed-class form can decay before these operations are completed. If such a possibility is viable as one scenario for incomplete processing in the surface account, it is surely a reasonable candidate as a major source of these children's problems, as the phonological memory deficit account holds.

In fact, there is reason to expect a relationship between performance on some of the tasks used to test phonological memory and the use of inflections by children with SLI. According to the evidence presented in several of the studies on phonological memory, children with SLI recall words more poorly than normally developing children but are affected in the same way by the phonological similarities among the words in the list to be recalled. Therefore, it seems that there should be a close relationship between these children's recall of phonologically similar words and their use of consonantal inflections. Words with inflections of this type (e.g., *jumps*, *jumped*) bear a phonological similarity to their bare-stem counterparts. Thus, children with SLI with the lowest degree of inflection use should also show especially poor word recall.

Thus far, the presumed adverse affects of phonological memory deficits on language have revolved around lexical acquisition and, to a lesser extent, syntactic structure. It also seems that an assumption of deficits of this type might be helpful in explaining some of the prosodic limitations observed in children with SLI. In particular, production accounts offer no explanation for why pronunciations such as [ɛfənt] for "elephant" occur more frequently than those such as [ɛlə]. According to approaches that assume a dominant strong–weak syllable pattern in the speech of these children, [ɛlə] should be more likely because [lə] is the weak syllable immediately following the initial strong syllable.[2] However, from the standpoint of phonological memory, the syllable [fənt] is more likely because, as a final syllable, it can benefit from recency effects.[3]

The findings of Montgomery (1995a) and James, van Steenbrugge, and Chiveralls (1994) suggest that the perceptual skills of children with SLI are part of the problem. The work by Tallal and her colleagues would seem to reinforce this idea. However, as we noted in the last chapter, some of the tasks used by Tallal had a memory as well as a perceptual requirement. In these tasks, the children had to hold in memory which of two stimuli was the target. If a child lost track of which stimulus required the panel to be pressed, an error was recorded even if (unknown to the examiner), the stimuli were distinguishable to the child. One way to look at the role of this memory component would be to score such tasks in terms of the percentage of stimulus changes (e.g., from [ba] to [pa] or vice versa) that were accompanied by a change in response (not pressing to pressing or vice versa). If this percentage is significantly greater than the percentage of adjacent identical stimuli that produced a change in response, it could be concluded that memory contributed to the children's performance.

In principle, phonological memory deficits can co-occur with other problems. In fact, this must be the case if the phonological memory deficit account is correct. There are too many other problems documented in children with SLI that seem to have no connection to phonological memory. Some of these might be related to limitations in working memory of a different sort. Specifically, as we saw in chapter 5, these children's problems with visual imagery are most apparent when the image must be held in the mind's eye for several seconds. In the model of Baddeley (1986), working memory of this type is controlled by a component called the "visual-spatial scratch pad." Although an assumption of deficits in two different components of a system is less parsimonious than an assumption of a single source of difficulty, there is currently no basis to rule out such a possibility. Of course, other nonlinguistic limitations, such as those associated with symbolic play, probably require an explanation that falls well outside the domain of memory, visual or phonological.

SLI as a Deficit of Temporal Processing

Among the most enduring findings in the literature on SLI is the finding that children with SLI perform quite poorly on tasks requiring the processing of brief stimuli and the processing of stimuli that are presented in rapid succession. In either case, demands are placed on the children that seem to have a more detrimental effect on their performance than is true for normally developing children. Most of this evidence was reviewed in chapter 6, and includes studies by Tallal and Piercy (1973a, 1973b, 1974, 1975), Tallal and Stark (1981), Alexander and Frost (1982), Robin, Tomblin, Kearney, and Hug (1989), and Leonard, McGregor, and Allen (1992).

Tallal and her colleagues have been the dominant researchers in this area, and have interpreted these data as reflecting a temporal processing deficit in children with SLI. This characterization of the deficit is no doubt based on the fact that stimulus duration and rate of presentation influence the time within which processing must take place. However, much of the evidence is not evidence of temporal processing. As pointed out by Studdert-Kennedy and Mody (1995), difficulty in distinguishing, for example, [ba] and [da] when each has a 40-ms formant transition does not constitute difficulty with temporal processing. These two stimuli are identical in duration and rate of frequency change. They differ in the loci and directions of their frequency trajectories, which is a spectral contrast.

The processing deficit described by Tallal and her colleagues is assumed to be malleable. This has been demonstrated in a follow-up study of children with SLI conducted by Bernstein and Stark (1985), a study by Robin, Tomblin, Kearney, and Hug (1989) in which children with SLI were given practice with difficult stimuli, and training studies using stimuli with extended durations by Alexander and Frost (1982) and, most recently, by Merzenich, Jenkins, Johnston, Schreiner, Miller, and Tallal (1996) and Tallal, Miller, Bedi, Byma, Wang, Nagarajan, Schreiner, Jenkins, and Merzenich (1996). In the latter two studies, very large gains were seen as a result of concentrated practice over a period of 20 days.

In the Tallal, Miller, Bedi, Byma, Wang, Nagarajan, Schreiner, Jenkins, and Merzenich (1996) investigation, the dependent measures included scores on language comprehension tests. The remarkable gains in these test scores constituted the strongest evidence thus far that problems with brief material play a causal role in SLI. Accordingly, details of this important work will be considered here.

The children with SLI participating in the experimental condition in the Tallal, Miller, Bedi, Byma, Wang, Nagarajan, Schreiner, Jenkins, and Merzenich (1996) study (age five to ten years) heard speech that was prolonged by 50% and selectively amplified such that fast transitional elements were boosted by up to 20 dB. The children with SLI serving in the control condition heard the same material in unmodified form. The same two groups of children were the participants in the Merzenich, Jenkins, Johnston, Schreiner, Miller, and Tallal (1996) study. The children in the experimental condition of this study received practice hearing and responding to nonverbal and verbal stimulus pairs integrated into computer games. Initially, the two sounds in each pair were distinguishable by cues of relatively long duration. Amplitude was also greater for the transitional information in the verbal stimuli. As training progressed, these cues were reduced in duration and amplitude. The children in the control condition participated in video games that apparently had no auditory perceptual component.

The decision not to have the control children respond to stimuli identical to those used in the experimental condition but with shorter durations may have been made out of concern that the children would fail and experience frustration. No other reason for this decision is obvious; in the companion study by Tallal, Miller, Bedi, Byma,

Wang, Nagarajan, Schreiner, Jenkins, and Merzenich (1996), the control condition involved the use of unmodified speech to serve as a contrast to the altered speech employed in the experimental condition. The work of Alexander and Frost (1982) reviewed in chapter 6 suggests that had brief stimuli been used in the control condition of the Merzenich, Jenkins, Johnston, Schreiner, Miller, and Tallal study (1996), the children in that condition would have improved, though not to the same extent as the children in the experimental condition. More important, because the control children apparently did not receive auditory stimuli in the Merzenich et al. study, they received less auditory perceptual stimulation in total than the children in the experimental condition. This might have contributed to the differences between the two groups on the language comprehension test scores reported by Tallal et al. It should be mentioned, of course, that this criticism can be of comfort only to detractors who believe in the first place that practice with auditory stimuli of brief duration can have a positive impact on language comprehension test scores.

Merzenich, Jenkins, Johnston, Schreiner, Miller, and Tallal (1996) noted that the impressive gains made by the children in the experimental condition argue for the assumption that the learning machinery of children with SLI is probably intact. The atypical findings reported for children with SLI on neuroanatomical and neurophysiological measures, these investigators reasoned, are probably the result of these children's history of learning with processing limitations. If these children are given treatment that alters the course of this learning, structural and physiological changes for the better should also be seen.

The children with SLI who participated in the training activities were young enough to make major gains in language learning once barriers to this process were removed. To the extent that work on age of second language learning is any indication (Newport, 1991), these children were within an age range that should have allowed them to attain native abilities if learning was not delayed further. Nevertheless, the gains seen in these children's test scores were extraordinary. When interpreting their findings, Tallal, Miller, Bedi, Byma, Wang, Nagarajan, Schreiner, Jenkins, and Merzenich (1996) made the following point:

> It seems unlikely that these children learned the equivalent of approximately 2 years of language in 1 month. Rather, it appears that they had already developed considerably more language competence than they were able to demonstrate or use "on line" under normal listening and speaking conditions.   (p. 83)

This point makes a great deal of sense but raises an important question: If the children's processing problems were the reason for their language learning difficulties, how did they acquire the language competence that this training regimen allowed them to reveal? This leads to the necessary conclusion that the processing deficit did not hinder their language development to the extent generally assumed, because at each point when the children had been tested, up until the time they were recruited for this study, their on-line performance underestimated their true language abilities. However, even if the effects of this training are interpreted as strengthening these children's language performance rather than altering their language competence, the results are extremely important. Any means of helping children to perform consistently at their optimal level would be a great benefit. Furthermore, this interpretation of the findings does not alter the interpretation of the investigators that processing limitations contribute significantly to these children's difficulties.

The work of Tallal and Merzenich will no doubt trigger a great deal of new research on this question. For example, in the prolonged speech used in the Tallal, Miller, Bedi, Byma, Wang, Nagarajan, Schreiner, Jenkins, and Merzenich study (1996), brief transitional information was also selectively amplified. Unlike the case for the nonverbal and verbal stimuli in the Merzenich, Jenkins, Johnston, Schreiner, Miller, and Tallal (1996) study where amplitude was used to enhance salience early and was then removed, the amplification of the altered speech used by Tallal et al. was never removed. Researchers will probably pursue the question of whether manipulations of duration alone are sufficient to create changes.

The origins of the presumed processing deficit are at this point unknown. Merzenich (1995) has offered some possibilities. One of these is that children with SLI may have had an atypically high number of bouts with otitis media with effusion (OME), or longer periods with this condition than most children. It seems that the type of hearing impairment that can result from OME could cause difficulties with details of the speech stream that are brief in duration. Dobie and Berlin (1979) found that sentences that had been reduced by 20 dB possessed significantly degraded information for inflections and function words. For example, the attenuated version of the utterance *Where are Jack's gloves to be placed?* resembled *Where Jack glove be place?* Unfortunately, as noted in chapter 1, the literature on the relationship between OME and language development does not present a clear picture, nor is it evident that children with SLI are more susceptible to this condition than are other children.

Merzenich (1995) also raised the possibility that these processing limitations might develop because children with SLI happen to focus on available but less ideal (and presumably less temporally sensitive) acoustic cues as they begin to listen actively to the ambient language. The failure to process cues of brief duration at the outset might make this task onerous when the children try to focus on these cues some months later. Although Merzenich did not elaborate, some possible candidates are the co-occurring cues seen in English, such as aspiration and (long lag) voice onset time of word-initial consonants, and vowel duration and the voicing of word-final consonants. In chapter 3, it was noted that the phonology of some children with SLI reflects evidence of using one cue without the ordinarily co-occurring cue. Of course, in those studies, it might have been the case that the children's production of one cue without the other had a production basis (e.g., difficulty pronouncing final voiced consonants). Furthermore, this list is clearly too short to serve as a sufficient inventory of possible ways to be diverted from cues of brief duration. Future research may uncover additional cues that meet the necessary criteria.

SLI as Grammatical Analysis by Unsuitable Mechanisms

Locke (1992, 1993, 1994) has offered an account of SLI that is based on a wide range of facts about children with this disorder. According to Locke, children with SLI exhibit a neuromaturational delay. This delay is seen in the late appearance of several cognitive and motor milestones (see chapter 1). Lexical development, of course, is also slow. In Locke's account, a certain amount of lexical material must be acquired before a presumably time-locked grammatical analysis mechanism is activated. The optimal period of functioning of this mechanism is approximately 20 to 36 months of age, with a decline in functioning thereafter.

The later acquisition of the requisite amount of lexical material by children with SLI means that activation of the grammatical analysis mechanism will be postponed. As a result, there will be a reduced period of optimal functioning of the mechanism. Grammatical analysis of the language input will therefore be incomplete at the end of the optimal period.

Locke proposed that when the grammatical analysis mechanism comes to the end of its period of optimal functioning, less efficient, compensatory mechanisms will be pressed into service for additional grammatical learning. Because these mechanisms are ill-equipped for such learning, the development of grammar in these children will remain impaired. Grammatical development will continue, but at such a slow pace that normal linguistic functioning will probably not be the end result:

> Given enough speaking activity over a sufficient interval, this new allocation of neural resources may produce a passable linguistic product, although subtle residual signs are likely to remain and literacy skills may be hard won. (Locke, 1994, p. 613)

This account makes use of an interesting combination of factors. It assumes that potential problems begin with the presence of a global delay but the most serious difficulties arise because modular components of language functioning are most affected by this general delay. The delay involves the late emergence of abilities, but apparently, once these abilities appear, their subsequent rate of development is not protracted. This seems to apply to lexical ability as well. Once lexical items begin to be comprehended, lexical development proceeds at a near-normal rate. Of course, because of the late start in the acquisition of lexical items, the modular grammatical analysis mechanism—whose period of functioning is unalterable—has less time to do its work and must leave part of the job to mechanisms that are not up to the task.

There are some appealing features in Locke's account. Other accounts have the responsibility of trying to explain how subtle nonlinguistic deficits seen in children with SLI are related to the language problem. In Locke's account, no relationship has to be assumed because these deficits are thought to reflect a general neuromaturational delay. Later on, the conspicuous difficulty with language can be attributed to the short period of operation of the grammatical analysis mechanism. The subtle deficits in other higher cognitive functions that can be observed in adolescence and beyond might be attributed to the fact that these individuals cannot make full use of language mediation to supplement nonlinguistic reasoning, as discussed in chapter 5.

Locke's account also contains an explanation for reports of neuroanatomical and neurophysiological differences between children with SLI and normally developing children. The task of grammatical learning must be assumed by mechanisms that are not designed for this purpose. This change in brain activity will result in structural changes in the brain. For example, Locke speculated that compensatory hypertrophy of right hemisphere mechanisms can result when more suitable left hemispheric mechanisms are not activated. In chapter 7, we saw that there is some debate about this interpretation of brain development, though the evidence of compensatory brain activity in other primates reported by Merzenich and others (e.g., Merzenich, Nelson, Stryker, Cynader, Schoppmann, & Zook, 1984), and the Merzenich, Jenkins, Johnston, Schreiner, Miller, and Tallal (1996) interpretation of their findings of improvement in auditory processing by children with SLI, are consistent with Locke's view on the matter.

Another nice feature of Locke's account is that it is highly testable. Clarke and Leonard (1996) evaluated one aspect of Locke's account, its assumptions about lexical development. It is assumed that once lexical comprehension begins, lexical development proceeds at a close to typical rate. This means that children with SLI should not fall further behind their peers in lexical ability across time. On the other hand, these children's lexical abilities shouldn't catch up either, at least during the early childhood years, when vocabulary continues to accumulate at a rapid rate.

Clarke and Leonard (1996) performed two types of analyses. First, they examined the vocabulary comprehension abilities of children with SLI whose speech was limited to single-word utterances. Because the absence of multi-word utterances suggested that the grammatical analysis mechanism had not begun to do its work, Clarke and Leonard reasoned that these children would have to have had vocabulary comprehension abilities that were below age level. This expectation was confirmed by the data.

Another analysis dealt with whether the vocabulary comprehension of children with SLI remained below age level once grammatical analysis was clearly under way. Clarke and Leonard (1996) examined the vocabulary comprehension scores of a group of children with SLI whose MLUs ranged from 2.4 to 4.2 morphemes. The assumption of continuing lexical limitations was not confirmed; the majority of the children showed vocabulary scores that were in the age-appropriate range. Such a finding suggests that if these children experienced a delay in emergence of vocabulary comprehension, subsequent development was faster than normal, allowing the children to catch up to their peers in this area.

Of the two assumptions tested by Clarke and Leonard (1996), the one that was not confirmed seems less crucial to Locke's account. Specifically, once enough lexical material is learned to activate the grammatical analysis mechanism, the rate of lexical development thereafter is less essential. Though an acceleration in the rate of lexical development from this point on cannot be readily explained within this account, this is not a fatal flaw. Most of the accounts of SLI discussed in this book must make recourse to extratheoretical factors to handle some of the details seen in these children's linguistic or nonlinguistic behavior. The true test of this account will come when other essential assumptions are examined. The finding by Clarke and Leonard that children with SLI in the single-word period are behind in vocabulary comprehension constitutes the clearing of only the first hurdle in the process of evaluating Locke's intriguing account.

Notes

1. In a subsequent interchange, Gathercole and Baddeley (1995) stressed that methodological differences between the Gathercole and Baddeley (1990) and van der Lely and Howard (1993) studies were too great to assume a failure to replicate. Howard and van der Lely (1995) defended their tasks as appropriate for the assessment of phonological memory.

2. Of course, there is the possibility that /l/ is problematic for these children and therefore avoided; as was seen in chapter 3, phonological avoidance is fairly common. However, the production of a glide as a substitute—hence [w]—is also a reasonable alternative, in which case [ɛwə] would be expected. Most of the available examples (e.g., [tɛfon] instead of [tɛlə]) pose the same interpretation problems.

3. A perceptual explanation of this production is that the final syllable is lengthened and hence is more salient than the second syllable.

PART VI

Conclusions

Chapter 14

Why Study SLI Revisited

In chapter 1, we discussed several theoretical and clinical reasons to study SLI. In this chapter, we review some of these, in light of evidence presented in this volume.

SLI and Learnability Theories

The Contribution of SLI to Learnability Theories

Two arguments were made for the inclusion of children with SLI in learnability theories. The first was one of principle: If learnability theories are designed to explain how "all normal" children acquire language, then a group of children cannot be excluded from consideration if their chief abnormality is the way they acquire language. The second argument was that the discrepancy between linguistic and nonlinguistic abilities in children with SLI seemed to offer an ideal means of evaluating the relative adequacy of learnability theories. According to theories that assume modularity, large gaps between linguistic and nonlinguistic abilities should be possible. According to theories that assume that both types of abilities rely on the same cognitive operations, these gaps must be rather small. We shall revisit both of these arguments here.

Nothing presented in these pages seems to jeopardize the argument that children with SLI should in principle be included in theories of language learnability. Some investigators contend that at least certain types of children with SLI show deficits only in language; obviously, these children are strong candidates for consideration. However, even for the many children who show subtle deficits in areas beyond language, a logical argument for exclusion is not easily made. Although these children's nonlinguistic abilities may be deficient, their linguistic abilities are even more limited. A gap therefore exists, consistent with modularity accounts. Furthermore, proponents of modularity accounts would hardly urge the exclusion of such children on the grounds that these children exhibit subtle limitations in nonlinguistic abilities such as mental imagery, for in these accounts, abilities of this type are not supposed to be linked to language.

Indeed, assumptions of modularity need not be abandoned even if a nonlinguistic factor is suspected of being the source of the problem. Locke (1994) assumed that a general neuromaturational delay was responsible for children's slow lexical development and that a sufficient number of lexical items had to be acquired to trigger a separate, time-locked grammatical analysis mechanism. Because a sufficient number of lexical items was not acquired until well into the optimal period of functioning of this mechanism, insufficient time was available for this mechanism to do its work.

The account of Tallal and her colleagues (e.g., Tallal, Miller, Bedi, Byma, Wang, Nagarajan, Schreiner, Jenkins, & Merzenich, 1996), though not expressly modular, can certainly coexist with modularity. According to this account, children with SLI have special difficulty with auditory information that is rapid in nature. The difficulty is not limited to linguistic material. However, this difficulty can be minimized through training and significant language gains can result. According to Tallal et al., such findings can occur only if the language-learning machinery is intact from the beginning. The language mechanism plays the hand it is dealt, as it were.

It appears that the available evidence from children with SLI—true to its billing—can be interpreted as compatible with learnability theories that assume modularity, even evidence from children with deficits in nonlinguistic areas. But are there provisions that such theories must contain, given the data from these children?

One possibility, based on the proposals of Locke (1994), is that grammatical development is not simply on a biological clock, working with whatever lexical information is available once it begins. Time may well march on, but a minimum number of lexical items may be required as a wake-up call. It is not clear that theories assuming maturation of grammatical principles have accorded such an important role to the lexicon.

An account such as that of Tallal and her colleagues might appear to offer relatively little in the way of implications for learnability theories. At best, it might provide a basis for comparing equally viable theories; the one that can also explain the manner of language development when there are distortions in the input would win out. However, as Merzenich (1995) has pointed out, there are several plausible causes for the improper processing of rapid information in these children, and given the malleability of these processing limitations, a broken system does not seem to be one of them. One of the remaining possibilities, according to Merzenich, is that early in life these children may have focused on some of the wrong cues among sets of co-occurring phonetic cues in the input. If this is true, children with SLI would play more than a tiebreaking role in deciding between two viable learnability theories. If certain language-learning strategies adopted by otherwise normal children lead to language deficits, then the identification and explication of these strategies should be a central concern of these theories.

The fact that modularity theories cannot be ruled out, given the findings from children with SLI, does not mean that nonmodularity theories are out of the running. The findings of deficits in nonlinguistic areas allow for the possibility that the language deficits are related to a more general problem. However, theories of this type are faced with the challenge of accounting for the observed gaps between linguistic and nonlinguistic abilities, even if the latter are found wanting.

One reasonable solution to this problem comes in the form of speed of processing limitations observed in children with SLI. Considerable evidence suggests that these children show slower RTs than normally developing children on a wide range of tasks. Theories assuming speed of processing limitations as the source of the problem in SLI might be on the right path if future work confirms the impression given in chapter 12 that (1) those operations requiring the greatest speed of processing are the most deficient in children with SLI and (2) language involves a greater proportion of speed-dependent operations than do other domains.

The evidence of the important role of speed of processing has implications for learnability theories. Nonmodularity theories assume that many different types of information are brought to bear in learning the lexicon and grammar of the ambient language. Such information includes the position of a word in a sentence, the phonetic similarities between final consonants or vowels in adjacent words, and the form–function transparency and reliability, among others. However, the speed at which this information is registered and analyzed receives relatively little attention. A separate literature contains much evidence that speed of processing on a wide range of tasks increases with age, but this evidence is rarely put to use in learnability theories. Perhaps the findings from children with SLI will promote this application.

Evaluating Accounts of SLI According to Learnability Assumptions
Learnability theories can assist in the evaluation of accounts of SLI just as the study of SLI may benefit the evaluation of learnability theories. This is not circular reasoning, wherein a favored account of SLI has the privilege of dictating new provisions for learnability theories, whereas an account out of favor is rejected because it doesn't conform to every detail of these theories. Rather, alternative accounts of SLI might be compared in terms of whether they make sense from the standpoint of what is assumed about the learning and organization of language in general. If two accounts of SLI can accommodate the available data but one account must assume language-learning operations that have not been attested in normally developing children, it should probably be viewed as a weaker account. This is not to say that children with SLI necessarily acquire language in the same way as typical children do. However, as a first step, the explanation of their language development should be sought within the bounds of known operations.

Some of the competing accounts of SLI meet this standard; others do not. The extended optional infinitive account assumes that children with SLI are late in realizing some fact about the ambient language that younger normally developing children also fail to grasp at the outset. The narrow rule account assumes that linguistic rules of children with SLI are too restricted, which is also an argument that has been made for the early stages of normal development. Some general processing limitation accounts place emphasis on measures that are known to distinguish normally developing children of different ages, such as speed of processing. Others emphasize factors such as morphological richness that contribute to the differences in the rate of grammatical morpheme development seen among normally developing children acquiring different languages.

On the other hand, some accounts assume a type of deficit that does not bear a close resemblance to anything seen in normal language development. One obvious example is the implicit rule deficit account. Another example is the temporal processing deficit account; there does not seem to be any theory of normal language acquisition that treats developing auditory skills as a driving force. Because the presumed sources of the problem in these accounts are not already incorporated into learnability theories, these accounts have extra work to do. They not only must document that problems exist in the areas assumed but also, without the benefit of normal development as a compass, explain how these problems can lead to the patterns of language deficits observed in children with SLI.

The Clinical and Educational Ramifications

The Data for SLI Suggest Directions for Assessment and Treatment
The findings reported for children with SLI have important clinical and educational implications, though in many cases, additional data are needed before these applica-

tions are formalized. The problems of many children with SLI are long-standing. Therefore, it seems prudent to identify these children as early as possible. Intervention produces gains that are larger than otherwise expected. For this reason, the earlier these children are identified, the earlier intervention can take place and the rate of their development can be accelerated. This does not ensure that these children will ultimately achieve age-appropriate functioning. Nevertheless, the earlier this process begins, the longer the children will be in a position to benefit from treatment.

Unfortunately, even though children with SLI are slow in developing words and word combinations, it is difficult to pick them out from late talkers who will be age-appropriate in their language skills by the time they reach kindergarten. Methods are needed for more accurate identification at a young age. One possible aid to this identification is family history information; children with SLI are much more likely to have parents and siblings with a history of language problems.

Intervention programs for preschoolers with SLI should have a component that permits the tracking of those children who by kindergarten or first grade appear to have overcome their language difficulties. Children with SLI are at risk for reading difficulties, and it is possible that these reading problems are exacerbated, if not caused, by deficits in language that only appeared to go away during preschool.

Intervention should emphasize language abilities. However, the fact that many children with SLI perform below the level of their peers on certain nonlinguistic tasks suggests that an expanded curriculum might be appropriate. Furthermore, given the results of treatment studies, there is no reason that intervention programs designed for children with SLI should exclude children with similar language difficulties whose nonverbal intelligence test scores fall slightly under 85.

It is not yet possible to identify subgroups of children with SLI in a valid and reliable manner. Professionals must therefore make decisions without a firm basis for knowing whether the differences seen among children with SLI require only slight modifications of the same treatment approach or a wholesale change in the approach adopted. However, recent evidence lends support to the traditional clinical distinction between production deficits only and deficits in both comprehension and production. This distinction should probably be viewed as two ends of a continuum. Nevertheless, children with problems limited to production appear to differ from other children in factors such as speed of processing and family history of language problems, as well as in comprehension ability.

Many children with SLI show areas of special difficulty in addition to moderate language deficits that are more broadly based. Both of these facts need to be considered during clinical management. For example, when attending to children's especially weak inflectional morphology, the materials devised for treatment—such as the lexical items to serve as the stems to be inflected—must be chosen carefully. Even though the children's lexical abilities may exceed their ability with inflections, their knowledge of lexical items may be below age-level expectations. Conversely, efforts designed to assist children with part of their broader problem must take into account the obstacles that areas of special difficulty might create. For example, particular speech acts or discourse devices targeted for teaching might, on close inspection, require morphosyntactic constructions that are beyond the children's abilities.

The crosslinguistic differences seen in the profiles of children with SLI should serve as a caution that weaknesses can be based on factors such as frequency and transparency as much as on the concepts or grammatical functions that the problematic details of language convey. This means that efforts aimed at helping children acquire these details should place as much emphasis on how the material is presented as on what is presented.

Accounts of SLI Matter

The competing accounts of SLI have important things to say about clinical work. In general, treatment can be thought of as presenting the child with an ideal input, in which the material to be learned appears more frequently in a more interpretable manner. However, the way "more interpretable" is defined depends on the nature of the problem assumed. In chapter 2, we saw how treatment was set up differently, depending on whether the problem was assumed to be based on processing factors (specifically, problems with grammatical morphemes of brief duration) or on limitations of linguistic knowledge (that verbs must be finite in main clauses). Ideal input based on the first of these assumptions required placing the forms to be learned in utterance positions (e.g., utterance-final position) that lead to a natural increase in their durations. Ideal input based on the second assumption required contrasts for finite and nonfinite to appear, as well as some explicit cue to indicate that nonfinite forms cannot be used in place of finite forms.

Other examples are easy to identify. An account that assumes a time-locked grammatical analysis mechanism is tantamount to an urgent plea for early intervention. Young children with SLI must be assisted in acquiring as many lexical items as possible. Otherwise, when the optimal period of functioning of the grammatical analysis mechanism arrives, there won't be a sufficient number of lexical items to activate it.

According to morphological richness accounts, children with SLI acquiring a language such as English will be assisted if during treatment the input is heavily weighted toward the inflectional material that does exist in the language. The missing agreement deficit account calls for treatment of English-speaking children to include work on number agreement between nouns and demonstratives (e.g., *this book, these books; that book, those books*) as well as subject–verb agreement. The narrow rule account suggests that a stricter criterion should be established before assuming that a child has acquired a rule to an adequate degree. It is not enough to observe application of the newly learned pattern to new exemplars. These exemplars need to differ considerably from those used in treatment, to ensure that the rule is sufficiently broad in scope.

Clearly, different accounts of SLI suggest different areas of emphasis for treatment. However, given that we do not yet know which account is closest to the true state of affairs, practitioners are faced with some difficult decisions. One decision might be to tailor intervention to the assumptions of one of these accounts until or unless the account is found to be unsustainable. Even if the account proves faulty, the children are still likely to receive some benefit from treatment because the provisions of any single account are likely to overlap with at least a few of the requirements of another account, owing to the fact that most accounts are grounded in empirical evidence. The accounts may espouse different reasons for a problem with some area of language, but they are likely to agree on the fact that this area of language requires clinical attention. However, at this point, it seems risky to follow the tenets of one account without regard for compatible aspects of other accounts. For example, presenting grammatical morphemes in contexts in which they are lengthened may assist children, but the children's gains may be only partial if grammatical contrasts or production activities are not included.

A second alternative is to wait until a winner emerges from the ranks of the contending accounts. In the meantime, treatment could be based on sensible principles that might coincide with those of certain accounts, though not by design. There is risk in this approach as well. For example, we might attempt to assist children in their grammatical learning before they have a large enough lexicon. Problematic features of grammar might be presented with sufficient frequency but not in salient or contrastive contexts. Grammatical morphemes such as -ing, in, and on might be the focus of treatment because they are ordinarily the first to be acquired by young normally developing children, even though they are not at the heart of the children's difficulty with grammatical morphology.

Until the competing accounts have been thoroughly evaluated, and the accumulating evidence tips in favor of one of them, practitioners would be wise to emphasize in their intervention plan those elements that are compatible with most or all accounts. For example, verb forms can be presented frequently, in utterance-final position (where they are longer) as well as utterance-medial position. In both positions, the examples used can illustrate a contrast in tense, agreement, and finiteness. These provisions are compatible with the extended optional infinitive account, the missing agreement account, the surface account, the morphological richness account, and the temporal processing deficit account. By adding activities that give the children an opportunity to produce the verb forms in the above contexts, an element compatible with a more general speed of processing account has been added. No single account calls for each of these details, of course; but none of these details does violence to the assumptions of any of these accounts, either. Although this tack might be characterized as a compromise, it is at the same time empirically based and sensitive to phenomena that could prove to be determining factors in SLI.

Accounts of SLI, then, do matter, as do the data on which they are based. The work of interpreting findings from children with SLI provides important clues as to how these children might be more effectively assisted in their language learning.

References

Albertini, J. (1980). The acquisition of five grammatical morphemes: Deviance or delay? *Proceedings of the Symposium on Research in Child Language Disorders, 1*, 94–111. Madison: University of Wisconsin, Madison.

Alexander, D., & Frost, B. (1982). Decelerated synthesized speech as a means of shaping speed of auditory processing of children with delayed language. *Perceptual and Motor Skills, 55*, 783–792.

Altenberg, B. (1987). *Prosodic patterns in spoken English: Studies in the correlation between prosody and grammar for text-to-speech conversion.* Lund, Sweden: Lund University Press.

American Psychiatric Association (1994). *Diagnostic and statistical manual of mental disorders IV.* Washington, DC: American Psychiatric Association.

Anderson, J. (1965). Initiatory delay in congenital aphasoid conditions. *Cerebral Palsy Journal, 26*, 9–12.

Aram, D., Ekelman, B., & Nation, J. (1984). Preschoolers with language disorders: 10 years later. *Journal of Speech and Hearing Research, 27*, 232–244.

Aram, D., Hack, M., Hawkins, S., Weissman, G., & Borawski-Clark, E. (1991). Very-low-birthweight children and speech and language development. *Journal of Speech and Hearing Research, 34*, 1169–1179.

Aram, D., Morris, R., & Hall, N. (1993). Clinical and research congruence in identifying children with specific language impairment. *Journal of Speech and Hearing Research, 36*, 580–591.

Aram, D., & Nation, J. (1975). Patterns of language behavior in children with developmental language disorders. *Journal of Speech and Hearing Research, 18*, 229–241.

Aram, D., & Nation, J. (1980). Preschool language disorders and subsequent language and academic difficulties. *Journal of Communication Disorders, 13*, 159–170.

Aram, D., & Nation, J. (1982). *Child language disorders.* St. Louis: Mosby.

Arnold, G. (1961). The genetic background of developmenatal language disorders. *Folia Phoniatrica, 13*, 246–254.

Baddeley, A. (1966). Short-term memory for word sequences as a function of acoustic, semantic and formal similarity. *Quarterly Journal of Experimental Psychology, 18*, 362–365.

Baddeley, A. (1986). *Working memory.* Oxford: Oxford University Press.

Bain, B., & Olswang, L. (1995). Examining readiness for learning two-word utterances by children with specific expressive language impairment: Dynamic assessment. *American Journal of Speech-Language Pathology and Audiology, 4*, 81–91.

Baker, L., & Cantwell, D. (1982). Psychiatric disorders in children with different types of communication disorders. *Journal of Communication Disorders, 15*, 113–126.

Baker, L., Cantwell, D., & Mattison, R. (1980). Behavior problems in children with pure speech disorders and in children with combined speech and language disorders. *Journal of Abnormal Child Psychology, 8*, 245–256.

Ball, J., & Cross, T. (1981). Formal and pragmatic factors in childhood autism and aphasia. Paper presented at the Symposium on Research in Child Language Disorders, University of Wisconsin, Madison.

Ball, J., Cross, T., & Horsborough, K. (1981). A comparative study of the linguistic abilities of autistic, dysphasic, and normal children. Paper presented at the International Congress on the Study of Child Language, Vancouver, Canada.

Barbeito, C. (1972). A comparison of two methods for teaching prepositions to language handicapped preschool children. Unpublished doctoral dissertation, University of Denver.

Bartke, S. (1994). Dissociations in SLI children's inflectional morphology: New evidence from agreement inflections and noun plurals in German. Paper presented at the Meeting of the European Group for Child Language Disorders, Garderen, The Netherlands.

Bastian, H. (1880). *The brain as an organ of mind.* New York: Appleton.

Bates, E. (1976). *Language in context.* New York: Academic Press.

Bates, E., Camaioni, L., & Volterra, V. (1975). The acquisition of performatives prior to speech. *Merrill-Palmer Quarterly, 21,* 205–226.

Bates, E., & MacWhinney, B. (1987). Competition, variation, and language learning. In B. MacWhinney (Ed.), *Mechanisms of language acquisition* (pp. 157–193). Hillsdale, NJ: Lawrence Erlbaum.

Bayley, N. (1969). *Bayley scales of infant development.* New York: Psychological Corporation.

Beastrom, S., & Rice, M. (1986). Comprehension and production of the articles *a* and *the.* Paper presented at the Convention of the American Speech-Language-Hearing Association, Detroit.

Beckman, M., & Edwards, J. (1990). Lengthening and shortening and the nature of prosodic constituency. In J. Kingston & M. Beckman (Eds.), *Papers in laboratory phonology, 1, Between grammar and the physics of speech* (pp. 152–178). Cambridge: Cambridge University Press.

Bedrosian, J., & Willis, T. (1987). Effects of treatment on the topic performance of a school-age child. *Language, Speech, and Hearing Services in Schools, 18,* 158–167.

Beebe, H. (1946). Sigmatismus nasalis. *Journal of Speech Disorders, 11,* 35–37.

Beers, M. (1992). Phonological processes in Dutch language impaired children. *Scandinavian Journal of Logopedics and Phoniatrics, 17,* 9–16.

Beitchman, J., Wilson, B., Brownlie, E., Walters, H., & Lancee, W. (1996). Long-term consistency in speech/language profiles: I. Developmental and academic outcomes. *Journal of the American Academy of Child and Adolescent Psychiatry, 35,* 804–814.

Bellaire, S., Plante, E., & Swisher, L. (1994). Bound-morpheme skills in the oral language of school-age, language-impaired children. *Journal of Communication Disorders, 27,* 265–279.

Benasich, A., & Tallal, P. (1993). Assessing auditory temporal processing in 5- to 9-month-old infants. Paper presented at the Meeting of the Society for Research in Child Development, New Orleans.

Bender, J. (1940). A case of delayed speech. *Journal of Speech Disorders, 5,* 363.

Benedikt, M. (1865). Ueber Aphasie, Agraphie und verwandte pathologische Zustände. *Wiener Medizinische Presse, 6,* 1189–1190.

Benton, A. (1964). Developmental aphasia and brain damage. *Cortex, 1,* 40–52.

Berko, J. (1958). The child's learning of English morphology. *Word, 14,* 150–177.

Berkovits, R. (1993). Progressive utterance-final lengthening in syllables with final fricatives. *Language and Speech, 36,* 89–98.

Bernhardt, B. (1992). Developmental implications of nonlinear phonological theory. *Clinical Linguistics and Phonetics, 6,* 259–281.

Bernhardt, B., & Gilbert, J. (1992). Applying linguistic theory to speech-language pathology: The case for nonlinear phonology. *Clinical Linguistics and Phonetics, 6,* 123–145.

Bernhardt, B., & Stoel-Gammon, C. (1994). Nonlinear phonology: Introduction and clinical application. *Journal of Speech and Hearing Research, 37,* 123–143.

Bernstein, L., & Stark, R. (1985). Speech perception development in language-impaired children: A four-year follow-up study. *Journal of Speech and Hearing Disorders, 50,* 21–30.

Bever, T. (1970). The cognitive basis for linguistic structure. In J. Hayes (Ed.), *Cognition and the development of language* (pp. 279–362). New York: Wiley.

Billard, C., Loisel Dufour, M., Gillet, P., & Ballanger, M. (1989). Evolution du langage oral et du langage écrit dans une population de dysphasie de développement de forme expressive. *ANAE, 1,* 16–22.

Bird, D., Bishop, D., & Freeman, N. (1995). Phonological awareness and literacy development in children with expressive phonological impairments. *Journal of Speech and Hearing Research, 38,* 446–462.

Bishop, D. (1979). Comprehension in developmental language disorders. *Developmental Medicine and Child Neurology, 21,* 225–238.

Bishop, D. (1982). Comprehension of spoken, written and signed sentences in childhood language disorders. *Journal of Child Psychology and Psychiatry, 23,* 1–20.

Bishop, D. (1989). Autism, Asperger's syndrome and semantic-pragmatic disorder: Where are the boundaries? *British Journal of Disorders of Communication, 24,* 107–121.

Bishop, D. (1990). Handedness, clumsiness and developmental language disorders. *Neuropsychologia, 28,* 681–690.

Bishop, D. (1992a). The biological basis of specific language impairment. In P. Fletcher & D. Hall (Eds.), *Specific speech and language disorders in children* (pp. 2–17). London: Whurr.

Bishop, D. (1992b). The underlying nature of specific language impairment. *Journal of Child Psychology and Psychiatry, 33,* 3–66.

Bishop, D. (1994). Grammatical errors in specific language impairment: Competence or performance limitations? *Applied Psycholinguistics, 15,* 507–550.

Bishop, D., & Adams, C. (1990). A prospective study of the relationship between specific language impairment, phonological disorders and reading retardation. *Journal of Child Psychology and Psychiatry, 31,* 1027–1050.

Bishop, D., & Adams, C. (1991). What do referential communication tasks measure? A study of children with specific language impairment. *Applied Psycholinguistics, 12,* 199–215.

Bishop, D., & Adams, C. (1992). Comprehension problems in children with specific language impairment: Literal and inferential meaning. *Journal of Speech and Hearing Research, 35,* 119–129.

Bishop, D., & Edmundson, A. (1986). Is otitis media a major cause of specific developmental language disorders? *British Journal of Disorders of Communication, 21,* 321–338.

Bishop, D., & Edmundson, A. (1987a). Language-impaired 4-year-olds: Distinguishing transient from persistent impairment. *Journal of Speech and Hearing Disorders, 52,* 156–173.

Bishop, D., & Edmundson, A. (1987b). Specific language impairment as a maturational lag: Evidence from longitudinal data on language and motor development. *Developmental Medicine and Child Neurology, 29,* 442–459.

Bishop, D., & Rosenbloom, L. (1987). Classification of childhood language disorders. In W. Yule & M. Rutter (Eds.), *Language development and disorders* (pp. 16–41). London: Mac Keith Press.

Blackwell, A., & Bates, E. (1995). Inducing agrammatic profiles in normals: Evidence for the selective vulnerability of morphology under cognitive resource limitation. *Journal of Cognitive Neuroscience, 7,* 228–257.

Blank, M., Gessner, M., & Esposito, A. (1979). Language without communication: A case study. *Journal of Child Language, 6,* 329–352.

Bliss, L. (1989). Selected syntactic usage of language-impaired children. *Journal of Communication Disorders, 22,* 277–289.

Bloom, L. (1967). A comment on Lee's "Developmental sentence types: A method for comparing normal and deviant syntactic development." *Journal of Speech and Hearing Disorders, 32,* 294–296.

Bloom, L., & Lahey, M. (1978). *Language development and language disorders.* New York: Wiley.

Bock, K., & Levelt, W. (1994). Grammatical encoding. In M. Gernsbacher (Ed.), *Handbook of psycholinguistics* (pp. 945–984). San Diego: Academic Press.

Bol, G., & de Jong, J. (1992). Auxiliary verbs in Dutch SLI children. *Scandinavian Journal of Logopedics and Phoniatrics, 17,* 17–21.

Bol, G., & Kuiken, F. (1988). Grammaticale analyse van taalontwikkelingsstoornissen. Unpublished manuscript, University of Amsterdam.

Bond, Z., & Wilson, H. (1980). Acquisition of the voicing contrast by language-delayed and normal-speaking children. *Journal of Speech and Hearing Research, 23,* 152–161.

Bondurant, J., Romeo, D., & Kretschmer, R. (1983). Language behaviors of mothers of children with normal and delayed language. *Language, Speech, and Hearing Services in Schools, 14,* 233–242.

Bortolini, U. (1995). I disturbi fonologici. In G. Sabbadini (Ed.), *Manuale di neuropsicologia dell'età evolutiva* (pp. 342–357). Bologna: Zanichelli.

Bortolini, U., & Leonard, L. (1991). The speech of phonologically disordered children acquiring Italian. *Clinical Linguistics and Phonetics, 5,* 1–12.

Bortolini, U., & Leonard, L. (1996). Phonology and grammatical morphology in specific language impairment: Accounting for individual variation in English and Italian. *Applied Psycholinguistics, 17,* 85–104.

Bortolini, U., Leonard, L., & McGregor, K. (1992). Overregularization in the grammatical morphology of children with specific language impairment. Poster presented at the Symposium on Research in Child Language Disorders, University of Wisconsin, Madison.

Bosch-Galceran, L., & Serra-Raventós, M. (1994). Grammatical morphology deficits in Spanish SLI children. Paper presented at the Meeting of the European Group for Language Disorders, Garderen, The Netherlands.

Bottari, P., Cipriani, P., & Chilosi, A. (1994). Dissociations in the acquisition of clitic pronouns by dysphasic children: A case study from Italian. Working paper, Scientific Institute Stella Maris, Pisa, Italy.

Bottari, P., Cipriani, P., & Chilosi, A. (1995). Past participle agreement and root infinitives in Italian children. Paper presented at the Boston University Conference on Language Development, Boston.

Bouillion, K. (1973). The comparative efficacy of nondirective group play therapy with preschool speech or language delayed children. Unpublished doctoral dissertation, Texas Tech University.

Boyd, R. (1980). Language intervention for grade one children. *Language, Speech, and Hearing Services in Schools, 11,* 30–40.

Braine, M. (1976). Children's first word combinations. *Monographs of the Society for Research in Child Development 41* (Serial no. 164).

Brinton, B., & Fujiki, M. (1982). A comparison of request—response sequences in the discourse of normal and language-disordered children. *Journal of Speech and Hearing Disorders, 47,* 57—62.

Brinton, B., & Fujiki, M. (1995). Conversational intervention with children with language impairment. In M. Fey, J. Windsor, & S. Warren (Eds.), *Language intervention: Preschool through the primary school years* (pp. 183—212). Baltimore: Paul H. Brookes.

Brinton, B., Fujiki, M., Winkler, E., & Loeb, D. (1986). Responses to requests for clarification in linguistically normal and language-impaired children. *Journal of Speech and Hearing Disorders, 51,* 370—378.

Broadbent, W. (1872). On the cerebral mechanism of speech and thought. *Medico-Chirurgical Transactions, 55,* 145—194.

Brooks, A., & Benjamin, B. (1989). The use of structured role play therapy in the remediation of grammatical deficits in language delayed children: Three case studies. *Journal of Childhood Communication Disorders, 12,* 171—186.

Brown, J., Redmond, A., Bass, K., Liebergott, J., & Swope, S. (1975). Symbolic play in normal and language-impaired children. Paper presented at the Convention of the American Speech-Language-Hearing Association, Washington, DC.

Brown, L., Sherbenou, R., & Johnsen, S. (1982). *Test of nonverbal intelligence.* Austin, TX: Pro-Ed.

Brown, R. (1973). *A first language.* Cambridge, MA: Harvard University Press.

Bruck, M., & Ruckenstein, S. (1978). Teachers talk to language delayed children. Paper presented at the Boston University Conference on Language Development, Boston.

Burgemeister, B., Blum, H., & Lorge, I. (1972). *The Columbia mental maturity scale.* New York: Psychological Corporation.

Byrne, B., Willerman, L., & Ashmore, L. (1974). Severe and moderate language impairment: Evidence for distinctive etiologies. *Behavior Genetics, 4,* 331—345.

Camarata, S., & Gandour, J. (1984). On describing idiosyncratic phonologic systems. *Journal of Speech and Hearing Disorders, 50,* 4—45.

Camarata, S., & Nelson, K. E. (1992). Treatment efficiency as a function of target selection in the remediation of child language disorders. *Clinical Linguistics and Phonetics, 6,* 167—178.

Camarata, S., Nelson, K. E., & Camarata, M. (1994). Comparison of conversational-recasting and imitative procedures for training grammatical structures in children with specific language impairment. *Journal of Speech and Hearing Research, 37,* 1414—1423.

Camarata, S., Newhoff, M., & Rugg, B. (1981). Perspective taking in normal and language disordered children. *Proceedings of the Symposium on Research in Child Language Disorders, 2,* 81—88. Madison: University of Wisconsin, Madison.

Camarata, S., & Swisher, L. (1990). A note on intelligence assessment within studies of specific language impairment. *Journal of Speech and Hearing Research, 33,* 205—207.

Candler, A., & Hildreth, B. (1990). Characteristics of language disorders in learning disabled students. *Academic Therapy, 25,* 333—343.

Capreol, K. (1994). Symbolic play training: Who profits? Master's thesis, University of British Columbia, Vancouver.

Casby, M. (1992). An intervention approach for naming problems in children. *American Journal of Speech-Language Pathology and Audiology, 1,* 35—42.

Catts, H. (1989). Defining dyslexia as a developmental language disorder. *Annals of Dyslexia, 39,* 50—64.

Catts, H. (1991). Early identification of dyslexia: Evidence from a follow-up study of speech-language impaired children. *Annals of Dyslexia, 41,* 163—177.

Catts, H. (1993). The relationship between speech-language impairments and reading disabilities. *Journal of Speech and Hearing Research, 36,* 948—958.

Catts, H., & Jensen, P. (1983). Speech timing of phonologically disordered children: Voicing contrasts of initial and final stop consonants. *Journal of Speech and Hearing Research, 26,* 501—510.

Catts, H., Swank, L., McIntosh, S., & Stewart, L. (1990). Precursors of reading disorders in language-impaired children. *Working Papers in Language Development, 5,* 44—54.

Ceci, S. (1983). Automatic and purposive semantic processing characteristics of normal and language/learning-disabled children. *Developmental Psychology, 19,* 427—439.

Chapman, K., Leonard, L., Rowan, L., & Weiss, A. (1983). Inappropriate word extensions in the speech of young language-disordered children. *Journal of Speech and Hearing Disorders, 48,* 55—62.

Chapman, R. (1978). Comprehension strategies in young children. In J. Kavanaugh & W. Strange (Eds.), *Speech and language in the laboratory, school, and clinic* (pp. 308—327). Cambridge, MA: MIT Press.

Chevrie-Muller, C. (1996). Troubles spécifiques du développement du langage "dysphasies de développement." In C. Chevrie-Muller & J. Narbona (Eds.), *Le langage de l'enfant: Aspects normaux et pathologiques* (pp. 255–281). Paris: Masson.

Chiat, S. (1983). Why *Mikey's* right and *my key's* wrong: The significance of stress and word boundaries in a child's output system. *Cognition, 14,* 275–300.

Chiat, S. (1989). The relation between prosodic structure, syllabification and segmental realization: Evidence from a child with fricative stopping. *Clinical Linguistics and Phonetics, 5,* 329–337.

Chiat, S., & Hirson, A. (1987). From conceptual intention to utterance: A study of impaired language output in a child with developmental dysphasia. *British Journal of Disorders of Communication, 22,* 37–64.

Chilosi, A., & Cipriani, P. (1991). *Il bambino disfasico.* Pisa: Edizioni del Cerro.

Chilosi, A., Cipriani, P., Giorgi, A., & Pfanner, L. (1993). Problemi di classificazione dei disordini specifici del linguaggio in età evolutiva. In S. Frasson, L. Lena, & P. Zottis (Eds.), *Diagnosi precoce e prevenzione dei disturbi del linguaggio e della communicazione* (pp. 145–161). Tirrenia: Edizione del Cerro.

Chin, S., & Dinnsen, D. (1991). Feature geometry in disordered phonologies. *Clinical Linguistics and Phonetics, 5,* 329–337.

Chomsky, N. (1981). *Lectures on government and binding.* Dordrecht, The Netherlands: Foris.

Chomsky, N. (1986). *Barriers.* Cambridge, MA: MIT Press.

Chomsky, N. (1993). A minimalist program for linguistic theory. In K. Hale & S. Keyser (Eds.), *The view from Building 20: Essays in linguistics in honor of Sylvain Bromberger* (pp. 1–52). Cambridge, MA: MIT Press.

Chomsky, N., & Halle, M. (1968). *The sound pattern of English.* New York: Harper & Row.

Cipriani, P., Chilosi, A., & Bottari, P. (1995). Language acquisition and language recovery in developmental dysphasia and acquired childhood aphasia. In K. E. Nelson & Z. Réger (Eds.), *Children's language, 8,* 245–273. Hillsdale, NJ: Lawrence Erlbaum.

Cipriani, P., Chilosi, A., Bottari, P., & Pfanner, L. (1993). *L'acquisizione della morfosintassi in italiano.* Padua: Unipress.

Cipriani, P., Chilosi, A., Bottari, P., Pfanner, L., Poli, P., & Sarno, S. (1991). L'uso della morfologia grammaticale nella disfasia congenita. *Giornale Italiano di Psicologia, 18,* 765–779.

Clahsen, H. (1989). The grammatical characterization of developmental dysphasia. *Linguistics, 27,* 897–920.

Clahsen, H. (1991). *Child language and developmental dysphasia.* Amsterdam: John Benjamins.

Clahsen, H. (1993). Linguistic perspectives on specific language impairment. *Working Papers Series "Theorie des Lexikons," 37.*

Clahsen, H., & Hansen, D. (1993). The missing agreement account of specific language impairment: Evidence from therapy experiments. *Essex Research Reports in Linguistics, 2,* 1–37.

Clahsen, H., & Rothweiler, M. (1992). Inflectional rules in children's grammars: Evidence from German participles. In G. Booij & J. van Marle (Eds.), *Yearbook of Morphology 1992* (pp. 1–34). Dordrecht, The Netherlands: Kluwer.

Clahsen, H., Rothweiler, M., Woest, A., & Marcus, G. (1992). Regular and irregular inflection in the acquisition of German noun plurals. *Cognition, 45,* 225–255.

Clark, M., & Plante, E. (1995). Morphology in the inferior frontal gyrus in developmentally language-disordered adults. Paper presented at the Conference on Cognitive Neuroscience, San Francisco.

Clarke, M., & Leonard, L. (1996). Lexical comprehension and grammatical deficits in children with specific language impairment. *Journal of Communication Disorders, 29,* 95–105.

Clarus, A. (1874). Ueber Aphasie bei Kindern. *Jahrbuch für Kinderheilkunde und Physische Erziehung, 7,* 369–400.

Cleave, P., & Rice, M. (1995). Acquisition of BE: A detailed analysis. Poster presented at the Convention of the American Speech-Language-Hearing Association, Orlando, FL.

Clifford, J., Reilly, J., & Wulfeck, B. (1995). Narratives from children with language impairment: An exploration in language and cognition. Technical Report CND-9509. San Diego: Center for Research in Language, University of California at San Diego.

Coën, R. (1886). *Pathologie und Therapie der Sprachanomalien.* Vienna: Urban & Schwarzenberg.

Cohen, H., Gelinas, C., Lassonde, M., & Geoffroy, G. (1991). Auditory lateralization for speech in language-impaired children. *Brain and Language, 41,* 395–401.

Cohen, M., Campbell, R., & Yaghmai, F. (1989). Neuropathological abnormalities in developmental dysphasia. *Annals of Neurology, 25,* 567–570.

Coker, C., Umeda, N., & Browman, P. (1973). Automatic synthesis from ordinary English text. *IEEE Transactions on Audio and Electroacoustics, AU-21,* 293–298.

Cole, K., & Dale, P. (1986). Direct language instruction and interactive language instruction with language delayed preschool children: A comparison study. *Journal of Speech and Hearing Research, 29*, 209–217.

Cole, K., Dale, P., & Mills, P. (1990). Defining language delay in young children by cognitive referencing: Are we saying more than we know? *Applied Psycholinguistics, 11*, 291–302.

Compton, A. (1970). Generative studies of children's phonological disorders. *Journal of Speech and Hearing Disorders, 35*, 315–339.

Compton, A. (1976). Generative studies of children's phonological disorders: Clinical ramifications. In D. Morehead & A. Morehead (Eds.), *Normal and deficient child language* (pp. 61–96). Baltimore: University Park Press.

Connell, P. (1980a). An experimental analysis of expansion training. Unpublished paper, University of Wisconsin–Milwaukee.

Connell, P. (1980b). Topic comment analysis of disordered child language. Unpublished paper, University of Wisconsin–Milwaukee.

Connell, P. (1986a). Acquisition of semantic role by language-disordered children: Differences between production and comprehension. *Journal of Speech and Hearing Research, 29*, 366–374.

Connell, P. (1986b). Teaching subjecthood to language-disordered children. *Journal of Speech and Hearing Research, 29*, 481–492.

Connell, P. (1987). An effect of modeling and imitation teaching procedures on children with and without specific language impairment. *Journal of Speech and Hearing Research, 30*, 105–113.

Connell, P., Gardner-Gletty, D., Dejewski, J., & Parks-Reinick, L. (1981). Response to Courtright and Courtright. *Journal of Speech and Hearing Research, 24*, 146–148.

Connell, P., & Stone, C. (1992). Morpheme learning of children with specific language impairment under controlled instructional conditions. *Journal of Speech and Hearing Research, 35*, 844–852.

Conti-Ramsden, G. (1989). Proper name usage: Mother–child interaction with language-impaired and non-language-impaired children. *First Language, 9*, 271–285.

Conti-Ramsden, G. (1990). Maternal recasts and other contingent replies to language-impaired children. *Journal of Speech and Hearing Disorders, 55*, 262–274.

Conti-Ramsden, G., & Dykins, J. (1991). Mother–child interactions with language-impaired children and their siblings. *British Journal of Disorders of Communication, 26*, 337–354.

Conti-Ramsden, G., & Friel-Patti, S. (1983). Mothers' discourse adjustments to language-impaired and non-language-impaired children. *Journal of Speech and Hearing Disorders, 48*, 360–367.

Conti-Ramsden, G., & Friel-Patti, S. (1984). Mother–child dialogues: A comparison of normal and language-impaired children. *Journal of Communication Disorders, 17*, 19–35.

Conti-Ramsden, G., Hutcheson, G., & Grove, J. (1995). Contingency and breakdown: Children with SLI and their conversations with mothers and fathers. *Journal of Speech and Hearing Research, 38*, 1290–1302.

Cooper, J., Moodley, M., & Reynell, J. (1978). *Helping language development.* New York: St. Martin's Press.

Cooper, J., Moodley, M., & Reynell, J. (1979). The Developmental Language Programme: Results from a five year study. *British Journal of Disorders of Communication, 14*, 57–69.

Courchesne, E., Lincoln, A., Yeung-Courchesne, R., Elmasian, R., & Grillon, C. (1989). Pathophysiologic findings in nonretarded autism and receptive developmental language disorder. *Journal of Autism and Developmental Disorders, 19*, 1–17.

Courtright, J., & Courtright, I. (1976). Imitative modeling as a theoretical base for instructing language-disordered children. *Journal of Speech and Hearing Research, 19*, 655–663.

Courtright, J., & Courtright, I. (1979). Imitative modeling as a language intervention strategy: The effects of two mediational variables. *Journal of Speech and Hearing Research, 22*, 389–402.

Cousins, A. (1979). Grammatical morpheme development in an aphasic child. Paper presented at the Boston University Conference on Language Development, Boston.

Cowell, P., Jernigan, T., Denenberg, V., & Tallal, P. (1994). Language- and learning-impairment and prenatal risk: An MRI study of the corpus callosum and cerebral volume. Unpublished manuscript.

Crago, M., & Allen, S. (1994). Morphemes gone askew: Linguistic impairment in Inuktitut. *McGill Working Papers in Linguistics, 10*, 206–215.

Crago, M., & Gopnik, M. (1994). From families to phenotypes: Theoretical and clinical implications of research into the genetic basis of specific language impairment. In R. Watkins & M. Rice (Eds.), *Specific language impairments in children* (pp. 35–51). Baltimore: Paul H. Brookes.

Craig, H. (1993). Social skills of children with specific language impairment. *Language, Speech and Hearing Services in Schools, 24*, 206–215.

Craig, H., & Evans, J. (1989). Turn exchange characteristics of SLI children's simultaneous and non-simultaneous speech. *Journal of Speech and Hearing Disorders, 54*, 334–347.

Craig, H., & Evans, J. (1993). Pragmatics and SLI: Within-group variation in discourse behaviors. *Journal of Speech and Hearing Research, 36*, 777–789.

Craig, H., & Gallagher, T. (1986). Interactive play: The frequency of related verbal responses. *Journal of Speech and Hearing Research, 29*, 375–383.

Craig, H., & Koenigsknecht, R. (1973). A group remediation program for children with atypical syntax utilizing operant and programmed procedures. Paper presented at the Convention of the American Speech-Language-Hearing Association, Detroit.

Craig, H., & Washington, J. (1993). The access behaviors of children with specific language impairment. *Journal of Speech and Hearing Research, 36*, 322–337.

Crais, E., & Chapman, R. (1987). Story recall and inferencing skills in language/learning-disabled and non-disabled children. *Journal of Speech and Hearing Disorders, 52*, 50–55.

Crais, E. (1988). Language/learning disabled children's storytelling compared with same-age and younger peers. Paper presented at the Convention of the American Speech-Language-Hearing Association, Boston.

Cromer, R. (1978). The basis of childhood dysphasia: A linguistic approach. In M. Wyke (Ed.), *Developmental dysphasia* (pp. 85–134). London: Academic Press.

Cromer, R. (1983). Hierarchical planning disability in the drawings and constructions of a special group of severely aphasic children. *Brain and Cognition, 2*, 144–164.

Cross, T. (1981). The linguistic experience of slow learners. In A. Nesdale, C. Pratt, R. Grieve, J. Field, D. Illingsworth, & J. Hogben (Eds.), *Advances in child development: Theory and research* (pp. 110–121). Nedlands: University of Western Australia.

Cross, T., Nienhuys, T., & Kirkman, M. (1985). Parent–child interaction with receptively disabled children: Some determinants of maternal speech style. In K. Nelson (Ed.), *Children' language, 5*, 247–290. Hillsdale, NJ: Lawrence Erlbaum.

Crystal, D., Fletcher, P., & Garman, M. (1976). *The grammatical analysis of language disability*. London: Edward Arnold.

Culatta, B., & Horn, D. (1982). A program for achieving generalization of grammatical rules to spontaneous discourse. *Journal of Speech and Hearing Disorders, 47*, 174–180.

Cunningham, C., Siegel, L., van der Spuy, H., Clark, M., & Bow, S. (1985). The behavioral and linguistic interactions of specifically language-delayed and normal boys with their mothers. *Child Development, 56*, 1389–1403.

Curtiss, S., & Tallal, P. (1991). On the nature of the impairment in language impaired children. In J. Miller (Ed.), *Research in child language disorders: A decade of progress* (pp. 189–211). Austin, TX: Pro-Ed.

Dalalakis, J. (1994). Developmental language impairment in Greek. *McGill Working Papers in Linguistics, 10*, 216–227.

Dalby, M. (1977). Aetiological studies in language retarded children. *Neuropaediatrie* (supplement), *8*, 499–500.

Darley, F., & Moll, K. (1960). Reliability of language measures and size of language samples. *Journal of Speech and Hearing Research, 3*, 166–173.

deAjuriaguerra, J., Jaeggi, A., Guignard, F., Kocher, F., Maquard, M., Roth, S., & Schmid, E. (1965). Évolution et prognostic de la dysphasie chez l'enfant. *La Psychiatrie de l'Enfant, 8*, 291–352.

de Jong, J., Fletcher, P., & Ingham, R. (1994). Verb argument structure in specifically language impaired children: Preliminary findings. Paper presented at the Meeting of the European Group for Child Language Disorders, Garderen, The Netherlands.

de Villiers, J., & de Villiers, P. (1973). A cross-sectional study of the acquisition of grammatical morphemes in child speech. *Journal of Psycholinguistic Research, 2*, 267–278.

Dinnsen, D. (1984). Methods and empirical issues in analyzing functional misarticulation. *ASHA Monographs*, no. 22, 5–17.

Dinnsen, D., Chin, S., Elbert, M., & Powell, T. (1990). Some constraints on functionally disordered phonologies: Phonetic inventories and phonotactics. *Journal of Speech and Hearing Research, 33*, 28–37.

Dobie, R., & Berlin, C. (1979). Influence of otitis media on hearing and development. *Annals of Otology, Rhinology and Laryngology, 88* (supplement 60), 48–53.

Doehring, D. (1960). Visual spatial memory in aphasic children. *Journal of Speech and Hearing Research, 3*, 138–149.

Dollaghan, C. (1987). Fast mapping of normal and language-impaired children. *Journal of Speech and Hearing Disorders, 52,* 218–222.

Dollaghan, C. (1995). Perception of "lexical" and "grammatical" phonemes by children with specific language impairment. Paper presented at the Symposium on Research in Child Language Disorders, University of Wisconsin, Madison.

Dollaghan, C., & Kaston, N. (1986). A comprehension monitoring program for language-impaired children. *Journal of Speech and Hearing Disorders, 51,* 264–271.

Dromi, E., & Berman, R. (1982). A morphemic measure of early language development: Data from modern Hebrew. *Journal of Child Language, 9,* 403–424.

Dromi, E., Leonard, L., & Shteiman, M. (1993). The grammatical morphology of Hebrew-speaking children with specific language impairment: Some competing hypotheses. *Journal of Speech and Hearing Research, 36,* 760–771.

Dukes, P. (1974). An exploratory study of the comparative effectiveness of two language intervention programs in teaching normal and language deviant preschool children. Unpublished doctoral dissertation, Kent State University.

Dunn, L., & Dunn, L. (1981). *Peabody picture vocabulary test-revised.* Circle Pines, MN: American Guidance Service.

Dunn, M., Flax, J., Sliwinski, M., & Aram, D. (1996). The use of spontaneous language measures as criteria for identifying children with specific language impairment: An attempt to reconcile clinical and research incongruence. *Journal of Speech and Hearing Research, 39,* 643–654.

Edwards, J., & Lahey, M. (1996). Auditory lexical decisions of children with specific language impairment. *Journal of Speech and Hearing Research, 39,* 1263–1273.

Edwards, M. L. (1980). The use of "favorite sounds" by children with phonological disorders. Paper presented at the Boston University Conference on Language Development, Boston.

Edwards, M. L., & Bernhardt, B. (1973). Phonological analyses of the speech of four children with language disorders. Unpublished paper, Stanford University.

Eisenson, J. (1966). Perceptual disturbances in children with central nervous system dysfunctions and implications for language development. *British Journal of Disorders of Communication, 1,* 21–32.

Eisenson, J. (1968). Developmental aphasia: A speculative view with therapeutic implications. *Journal of Speech and Hearing Disorders, 33,* 3–13.

Eisenson, J. (1972). *Aphasia in children.* New York: Harper & Row.

Elliott, L., & Hammer, M. (1988). Longitudinal changes in auditory discrimination in normal children and children with language-learning problems. *Journal of Speech and Hearing Disorders, 53,* 467–474.

Elliott, L., & Hammer, M. (1993). Fine-grained auditory discrimination: Factor structures. *Journal of Speech and Hearing Research, 36,* 396–409.

Elliott, L., Hammer, M., & Scholl, M. (1989). Fine-grained auditory discrimination in normal children and children with language-learning problems. *Journal of Speech and Hearing Research, 32,* 112–119.

Ellis Weismer, S. (1985). Constructive comprehension abilities exhibited by language-disordered children. *Journal of Speech and Hearing Research, 28,* 175–184.

Ellis Weismer, S. (1991). Hypothesis-testing abilities of language-impaired children. *Journal of Speech and Hearing Research, 34,* 1329–1338.

Ellis Weismer, S., & Hesketh, L. (1993). The influence of prosodic and gestural cues on novel word acquisition by children with specific language impairment. *Journal of Speech and Hearing Research, 36,* 1013–1025.

Ellis Weismer, S., & Hesketh, L. (1996). Lexical learning by children with specific language impairment: Effects of linguistic input presented at varying speaking rates. *Journal of Speech and Hearing Research, 39,* 177–190.

Ellis Weismer, S., & Murray-Branch, J. (1989). Modeling versus modeling plus evoked production training: A comparison of two language intervention methods. *Journal of Speech and Hearing Disorders, 54,* 269–281.

Ellis Weismer, S., Murray-Branch, J., & Miller, J. (1994). A prospective longitudinal study of language development in late talkers. *Journal of Speech and Hearing Research, 37,* 852–867.

Elman, J. (1993). Learning and development in neural networks: The importance of starting small. *Cognition, 48,* 71–99.

Evans, J. (1996). SLI subgroups: Interaction between discourse constraints and morphosyntactic deficits. *Journal of Speech and Hearing Research, 39,* 655–660.

Evans, M. A., & Schmidt, F. (1991). Repeated book reading with two children: Language-normal and language-impaired. *First Language, 11,* 269–287.

Evesham, M. (1977). Teaching language skills to children. *British Journal of Disorders of Communication, 12,* 23–29.

Ewing, A. (1930). *Aphasia in children.* New York: Oxford University Press.

Eyer, J., & Leonard, L. (1994). Learning past tense morphology with specific language impairment: A case study. *Child Language Teaching and Therapy, 10,* 127–138.

Eyer, J., & Leonard, L. (1995). Functional categories and specific language impairment: A case study. *Language Acquisition, 4,* 177–203.

Fant, G., Kruckenberg, A., & Nord, L. (1991). Durational correlates of stress in Swedish, French, and English. *Journal of Phonetics, 19,* 351–365.

Farmer, A., & Florance, K. (1977). Segmental duration differences: Language disordered and normal children. In J. Andrews & M. Burns (Eds.), *Selected papers in language and phonology, 2, Language remediation.* Evanston, IL: Institute for Continuing Professional Education.

Farnetani, E., & Fori, S. (1982). Lexical stress in spoken sentences: A study of duration and vowel formant pattern. *Quaderni del Centro di Studio per le Ricerche di Fonetica, 1,* 104–133.

Farwell, C. (1972). A note on the production of fricatives in linguistically deviant children. *Papers and Reports on Child Language Development, 4,* 93–101.

Fazio, B. (1994). The counting abilities of children with specific language impairment: A comparison of oral and gestural tasks. *Journal of Speech and Hearing Research, 37,* 358–368.

Fazio, B. (1996). Mathematical abilities of children with specific language impairment: A 2-year follow-up. *Journal of Speech and Hearing Research, 39,* 839–849.

Fee, E. J. (1995). The phonological system of a specifically language-impaired population. *Clinical Linguistics and Phonetics, 9,* 189–209.

Fellbaum, C., Miller, S., Curtiss, S., & Tallal, P. (1995). An auditory processing deficit as a possible source of SLI. In D. MacLaughlin & S. McEwen (Eds.), *Proceedings of the 19th Annual Boston University Conference on Language Development, 1,* 204–215. Somerville, MA: Cascadilla Press.

Fenson, L., Dale, P., Reznick, S., Thal, D., Bates, E., Hartung, J., Pethick, S., & Reilly, J. (1993). *The MacArthur communicative development inventories.* San Diego: Singular Publishing.

Fey, M. (1985). Articulation and phonology: Inextricable constructs in speech pathology. *Human Communication, 9,* 7–16.

Fey, M. (1986). *Language intervention with young children.* San Diego: College-Hill Press.

Fey, M., Cleave, P., Long, S., & Hughes, D. (1993). Two approaches to the facilitation of grammar in children with language impairment: An experimental evaluation. *Journal of Speech and Hearing Research, 36,* 141–157.

Fey, M., & Leonard, L. (1983). Pragmatic skills of children with specific language impairment. In T. Gallagher & C. Prutting (Eds.), *Pragmatic assessment and intervention issues in language* (pp. 65–82). San Diego: College-Hill Press.

Fey, M., & Leonard, L. (1984). Partner age as a variable in the conversational performance of specifically language-impaired children and normal-language children. *Journal of Speech and Hearing Research, 27,* 413–423.

Fey, M., Leonard, L., Fey, S., & O'Connor, K. (1978). The intent to communicate in language-impaired children. Paper presented at the Boston University Conference on Language Development, Boston.

Fey, M., Leonard, L., & Wilcox, K. (1981). Speech-style modifications of language-impaired children. *Journal of Speech and Hearing Disorders, 46,* 91–97.

Fey, M., Long, S., & Cleave, P. (1994). Reconsideration of IQ criteria in the definition of specific language impairment. In R. Watkins & M. Rice (Eds.), *Specific language impairments in children* (pp. 161–178). Baltimore: Paul H. Brookes.

Fischel, J., Whitehurst, G., Caulfield, M., & DeBaryshe, B. (1989). Language growth in children with expressive language delay. *Pediatrics, 82,* 218–227.

Fletcher, P. (1983). From sound to syntax: A learner's guide. *Proceedings from the Wisconsin Symposium on Research in Child Language Disorders, 4,* 1–31. Madison: University of Wisconsin, Madison.

Fletcher, P. (1991). Evidence from syntax for language impairment. In J. Miller (Ed.), *Research on child language disorders* (pp. 169–187). Austin, TX: Pro-Ed.

Fletcher, P. (1992). Subgroups in school-age language-impaired children. In P. Fletcher & D. Hall (Eds.), *Specific speech and language disorders in children* (pp. 152–163). London: Whurr.

Fletcher, P., & Garman, M. (1988). Normal language development and language impairment: Syntax and beyond. *Clinical Linguistics and Phonetics, 2,* 97–113.

Fletcher, P., & Peters, J. (1980). Verb-forms in normal and language-impaired children. Paper presented at the American Speech-Language-Hearing Convention, Los Angeles.

Fletcher, P., & Peters, J. (1984). Characterizing language impairment in children: An exploratory study. *Language Testing, 1,* 33–49.

Foley, M., Schwartz, D., & Shamow, N. (1976). The use of verbal instructions and a fading procedure for training temporal relations to a language impaired child. *Working Papers in Experimental Speech Pathology and Audiology, 5,* 32–40.

Folger, M. K., & Leonard, L. (1978). Language and sensorimotor development during the early period of referential speech. *Journal of Speech and Hearing Research, 21,* 519–527.

Forrest, K., & Rockman, B. (1988). Acoustic and perceptual analysis of word-initial stop consonants in phonologically disordered children. *Journal of Speech and Hearing Research, 31,* 449–459.

Forrest, K., Weismer, G., Hodge, M., Dinnsen, D., & Elbert, M. (1990). Statistical analysis of word-initial /k/ and /t/ produced by normal and phonologically disordered children. *Clinical Linguistics and Phonetics, 4,* 327–340.

Freedman, P., & Carpenter, R. (1976). Semantic relations used by normal and language-impaired children at Stage I. *Journal of Speech and Hearing Research, 19,* 784–795.

Friedman, P., & Friedman, K. (1980). Accounting for individual differences when comparing the effectiveness of remedial language teaching methods. *Applied Psycholinguistics, 1,* 151–170.

Fried-Oken, M. (1981). What's that? Teachers' interrogatives to language delayed and normal children. Paper presented at the Symposium on Research in Child Language Disorders, University of Wisconsin, Madison.

Fried-Oken, M. (1984). The development of naming skills in normal and language deficient children. Doctoral dissertation, Boston University.

Friel-Patti, S. (1976). Good-looking: An analysis of verbal and nonverbal behaviors in a group of language disordered children. Paper presented at the Convention of the American Speech-Language-Hearing Association, Houston, TX.

Friel-Patti, S. (1978). The interface of selected verbal and nonverbal behaviors in mother–child dyadic interactions with normal and language disordered children. Doctoral dissertation, Purdue University.

Friel-Patti, S., & Finitzo, T. (1990). Language learning in a prospective study of otitis media with effusion in the first two years of life. *Journal of Speech and Hearing Research, 33,* 188–194.

Fröschels, E. (1918). *Kindersprache und Aphasie.* Berlin: Karger.

Frumkin, B., & Rapin, I. (1980). Perception of vowels and consonant-vowels of varying duration in language impaired children. *Neuropsychologia, 18,* 443–454.

Fujiki, M., & Brinton, B. (1991). The verbal noncommunicator: A case study. *Language, Speech and Hearing Services in Schools, 22,* 322–333.

Fujiki, M., Brinton, B., & Sonnenberg, E. (1990). Repair of overlapping speech in the conversations of specifically language-impaired and normally developing children. *Applied Psycholinguistics, 11,* 201–215.

Fujiki, M., Brinton, B., & Todd, C. (1996). Social skills with specific language impairment. *Language, Speech, and Hearing Services in Schools, 27,* 195–202.

Fukuda, S., & Fukuda, S. (1994). Developmental language impairment in Japanese: A linguistic investigation. *McGill Working Papers in Linguistics, 10,* 150–177.

Fygetakis, L., & Gray, B. (1970). Programmed conditioning of linguistic competence. *Behavior Research and Therapy, 8,* 455–460.

Galaburda, A., Sherman, G., Rosen, G., Aboitiz, F., & Geschwind, N. (1985). Developmental dyslexia: Four consecutive patients with cortical anomalies. *Annals of Neurology, 18,* 222–233.

Gale, D., Liebergott, J., & Griffin, S. (1981). Getting it: Children's requests for clarification. Paper presented at the Convention of the American Speech-Language-Hearing Association, Los Angeles.

Gall, F. (1835). *The function of the brain and each of its parts. 5, Organology.* Boston: Marsh, Capen, & Lyon.

Gallagher, T., & Craig, H. (1984). Pragmatic assessment: Analysis of a highly frequent repeated utterance. *Journal of Speech and Hearing Disorders, 49,* 368–377.

Gallagher, T., & Darnton, B. (1978). Conversational aspects of the speech of language disordered children: Revision behaviors. *Journal of Speech and Hearing Research, 21,* 118–135.

Gandour, J. (1981). The nondeviant nature of deviant phonological systems. *Journal of Communication Disorders, 14,* 11–29.

Gardner, H. (1983). *Frames of mind: The theory of multiple intelligences.* New York: Basic Books.

Garvey, M., & Gordon, N. (1973). A follow-up study of children with disorders of speech development. *British Journal of Disorders of Communication, 8*, 17–28.

Gathercole, S., & Baddeley, A. (1990). Phonological memory deficits in language disordered children: Is there a causal connection? *Journal of Memory and Language, 29*, 336–360.

Gathercole, S., & Baddeley, A. (1993). *Working memory and language*. Hillsdale, NJ: Lawrence Erlbaum.

Gathercole, S., & Baddeley, A. (1995). Short-term memory may yet be deficient in children with language impairments: A comment on van der Lely & Howard (1993). *Journal of Speech and Hearing Research, 38*, 463–466.

Gavin, W., Klee, T., & Membrino, I. (1993). Differentiating specific language impairment from normal language development using grammatical analysis. *Clinical Linguistics and Phonetics, 7*, 191–206.

Gentner, D. (1982). Why nouns are learned before verbs: Linguistic relativity versus natural partitioning. In S. Kuczaj (Ed.), *Language development, 2, Language, thought, and culture* (pp. 301–334). Hillsdale, NJ: Lawrence Erlbaum.

Gérard, C. (1991). *L'enfant dysphasique*. Paris: Editions Universitaires.

Gerken, L. A. (1991). The metrical basis for children's subjectless sentences. *Journal of Memory and Language, 30*, 431–451.

Gerken, L. A. (1994). Young children's representation of prosodic phonology: Evidence from English-speakers' weak syllable productions. *Journal of Memory and Language, 33*, 19–38.

Gerken, L. A., Landau, B., & Remez, R. (1990). Function morphemes in young children's speech perception and production. *Developmental Psychology, 27*, 204–216.

German, D. (1987). Spontaneous language profiles of children with word-finding problems. *Language, Speech, and Hearing Services in Schools, 18*, 217–230.

Gertner, B., Rice, M., & Hadley, P. (1994). The influence of communicative competence on peer preferences in a preschool classroom. *Journal of Speech and Hearing Research, 37*, 913–923.

Geschwind, N., & Levitsky, W. (1968). Human brain: Asymmetries in the temporal speech region. *Science, 161*, 186–187.

Gesell, A., & Amatruda, C. (1947). *Developmental diagnosis* (Second edition). New York: Hoeber.

Giattino, J., Pollack, E., & Silliman, E. (1978). Adult input to language impaired children. Paper presented at the Convention of the American Speech-Language-Hearing Association, San Francisco.

Gierut, J., & Dinnsen, D. (1986). On word-initial voicing: Converging sources of evidence in phonologically disordered speech. *Language and Speech, 29*, 97–114.

Gierut, J., Simmerman, C., & Neumann, H. (1994). Phonemic structures of delayed phonological systems. *Journal of Child Language, 21*, 291–316.

Gilbert, J. (1977). A voice onset time analysis of apical stop production in 3-year-olds. *Journal of Child Language, 4*, 103–110.

Gilger, J., Borecki, I., Smith, S., DeFries, J., & Pennington, B. (1996). The etiology of extreme scores for complex phenotypes. In C. Chase, G. Rosen, & G. Sherman (Eds.), *Developmental dyslexia: Neural and cognitive mechanisms underlying speech, language, and reading* (pp. 63–85). Baltimore: York Press.

Gillam, R., Cowan, N., & Day, L. (1995). Sequential memory in children with and without language impairment. *Journal of Speech and Hearing Research, 38*, 393–402.

Gillam, R., & Johnston, J. (1985). Development of print awareness in language-disordered preschoolers. *Journal of Speech and Hearing Research, 28*, 521–526.

Gillam, R., & Johnston, J. (1992). Spoken and written language relationships in language/learning-impaired and normally achieving school-age children. *Journal of Speech and Hearing Research, 35*, 1303–1315.

Gleitman, L. (1990). The structural sources of verb meanings. *Language Acquisition, 1*, 3–55.

Gleitman, L., & Gleitman, H. (1992). A picture is worth a thousand words, but that's the problem: The role of syntax in vocabulary acquisition. *Current Directions in Psychological Science, 1*, 31–35.

Gleitman, L., Gleitman, H., Landau, B., & Wanner, E. (1988). Where langauge begins: Initial representations for language learning. In F. Newmeyer (Ed.), *Linguistics: The Cambridge survey, 3*, 150–193. Cambridge, England: Cambridge University Press.

Goad, H., & Gopnik, M. (1994). Phoneme discrimination in familial language impairment. *McGill Working Papers in Linguistics, 10*, 10–15.

Goad, H., & Rebellati, C. (1994). Pluralization in familial language impairment: Affixation or compounding? *McGill Working Papers in Linguistics, 10*, 24–40.

Goldowsky, B., & Newport, E. (1993). Modeling the effects of processing limitations on the acquisition of morphology: The less is more hypothesis. In E. Clark (Ed.), *The proceedings of the twenty-fourth child language research forum* (pp. 124–138). Stanford, CA: Center for the Study of Language and Information.

Gopnik, M. (1990a). Feature-blind grammar and dysphasia. *Nature, 344,* 715.

Gopnik, M. (1990b). Feature blindness: A case study. *Language Acquisition, 1,* 139–164.

Gopnik, M. (1994a). The family. *McGill Working Papers in Linguistics, 10,* 1–4.

Gopnik, M. (1994b). Prologue. *McGill Working Papers in Linguistics, 10,* vii-x.

Gopnik, M., & Crago, M. (1991). Familial aggregation of a developmental language disorder. *Cognition, 39,* 1–50.

Gottsleben, R., Tyack, D., & Buschini, G. (1974). Three case studies in language training: Applied linguistics. *Journal of Speech and Hearing Disorders, 39,* 213–241.

Gray, B., & Fygetakis, L. (1968). The development of language as a function of programmed conditioning. *Behavior Research and Therapy, 6,* 455–460.

Gray, B., & Ryan, B. (1973). *A language program for the nonlanguage child.* Champaign, IL: Research Press.

Graybeal, C. (1981). Memory for stories in language-impaired children. *Applied Psycholinguistics, 2,* 269–283.

Greenfield, P., & Schneider, L. (1977). Building a tree structure: The development of hierarchical complexity and interrupted strategies in children' construction activity. *Developmental Psychology, 13,* 299–313.

Grela, B., & Leonard, L. (1997). The use of subject arguments by children with specific language impairment. *Clinical Linguistics and Phonetics, 11,* 443–453.

Griffin, S. (1979). Requests for clarification made by normal and language impaired children. Master's thesis, Emerson College.

Griffiths, C. (1969). A follow-up study of children with disorders of speech. *British Journal of Disorders of Communication, 4,* 46–56.

Grimm, H. (1983). Kognitions- und interaktionspsychologische Aspekte der Entwicklungs-dysphasie. *Sprache und Kognition, 3,* 169–186.

Grimm, H. (1984). Zur Frage der sprachlichen Wissenskonstruktion. Oder: Erwerben dysphasische Kinder die Sprache anders? In E. Oksaar (Ed.), *Spracherwerb, Sprachkontakt, Sprachkonflikt* (pp. 30–53). Berlin: de Gruyter.

Grimm, H. (1986). Ontogenese der Sprache als Fortsetzung nicht-sprachlichen Handelns. In H. Bosshardt (Ed.), *Perspektiven auf Sprache. Interdisziplinäre Beiträge zum Gedenken an Hans Hörmann* (pp. 166–184). Berlin: de Gruyter.

Grimm, H. (1987). Developmental dysphasia: New theoretical perspectives and empirical results. *German Journal of Psychology, 11,* 8–22.

Grimm, H. (1991). Entwicklungskritische Dialogmerkmale in Mutter-Kind-Dyaden mit dysphasisch sprachgestörten und sprachunauffälligen Kindern. Paper presented at the Annual Meeting of Developmental Psychology, Cologne, Germany.

Grimm, H. (1993). Syntax and morphological difficulties in German-speaking children with specific language impairment: Implications for diagnosis and intervention. In H. Grimm & H. Skowronek (Eds.), *Language acquisition problems and reading disorders: Aspects of diagnosis and intervention* (pp. 25–63). Berlin: Walter de Gruyter.

Grimm, H., & Weinert, S. (1990). Is the syntax development of dysphasic children deviant and why? New findings to an old question. *Journal of Speech and Hearing Research, 33,* 220–228.

Grunwell, P. (1981). *The nature of phonological disability in children.* London: Edward Arnold.

Grunwell, P. (1992). Assessment of child phonology in the clinical context. In C. Ferguson, L. Menn, & C. Stoel-Gammon (Eds.), *Phonological development: Models, research, implications* (pp. 457–483). Timonium, MD: York Press.

Grunwell, P., & Russell, J. (1990). A phonological disorder in an English-speaking child. *Clinical Linguistics and Phonetics, 4,* 29–38.

Guilfoyle, E., Allen, S., & Moss, S. (1991). Specific language impairment and the maturation of functional categories. Paper presented at the Boston University Conference on Language Development, Boston.

Guilfoyle, E., & Noonan, M. (1992). Functional categories and language acquisition. *Canadian Journal of Linguistics, 37,* 241–272.

Gulotta, E., Becciu, M., Mazzoncini, B., & Sechi, E. (1991). La comprensione sintattica nei bambini con disturbo specifico di linguaggio. *I Care, 4,* 115–118.

Günther, H. (1981). Untersuchungen zum Sprachverhalten agrammatischer Kinder mit Ziel- und Modellsatzmethode. In G. Kegel & H. Günther (Eds.), *Psycholinguistische Untersuchungen zum kindlichen Agrammatismus* (pp. 35–59). Munich: Institut für Phonetische und Sprachliche Kommunikation.

Gutzmann, H. (1894). *Des Kindes Sprache und Sprachfehler.* Leipzig: Weber.

Haarmann, H., Just, M., & Carpenter, P. (1997). Aphasic sentence comprehension as a resource deficit: A computational approach. *Brain and Language.*

Haber, L. (1982). An analysis of linguistic deviance. In K. E. Nelson (Ed.), *Children's language, 3,* 247–285. Hillsdale, NJ: Lawrence Erlbaum.

Hadley, P., & Rice, M. (1991). Conversational responsiveness of speech- and language-impaired preschoolers. *Journal of Speech and Hearing Research, 34,* 1308–1317.

Hadley, P., & Rice, M. (1996). Emergent uses of BE and DO: Evidence from children with specific language impairment. *Language Acquisition, 5,* 209–243.

Haggerty, R., & Stamm, J. (1978). Dichotic auditory fusion levels in children with learning disabilities. *Neuropsychologia, 16,* 349–360.

Håkansson, G., & Nettelbladt, U. (1996). Similarities betweeen SLI and L2 children: Evidence from the acquisition of Swedish word order. In C. Johnson & J. Gilbert (Eds.), *Children's language, 9,* 135–151. Hillsdale, NJ: Lawrence Erlbaum.

Hall, P., & Tomblin, J. B. (1975). Case study: Therapy procedures for remediation of a nasal lisp. *Language, Speech, and Hearing Services in Schools, 6,* 29–32.

Hall, P., & Tomblin, J. B. (1978). A follow-up study of children with articulation and language disorders. *Journal of Speech and Hearing Disorders, 43,* 227–241.

Hansson, K. (1992). Swedish verb morphology and problems with its acquisition in language impaired children. *Scandinavian Journal of Logopedics and Phoniatrics, 17,* 23–29.

Hansson, K. (1997). Patterns of verb usage in Swedish children with SLI: An application of recent theories. *First Language, 17,* 195–217.

Hansson, K., & Nettelbladt, U. (1990). Comparison between Swedish children with normal and disordered language development regarding word order patterns and use of grammatical markers. Paper presented at the Meeting of the European Group for Language Disorders, Røros, Norway.

Hansson, K., & Nettelbladt, U. (1995). Grammatical characteristics of Swedish children with SLI. *Journal of Speech and Hearing Research, 38,* 589–598.

Hardy, J. (1965). On language disorders in young children: A reorganization of thinking. *Journal of Speech and Hearing Disorders, 30,* 3–16.

Hargrove, P. (1982). Misarticulated vowels: A case study. *Language, Speech, and Hearing Services in Schools, 13,* 86–95.

Hargrove, P., Holmberg, C., & Zeigler, M. (1986). Changes in spontaneous speech associated with therapy hiatus: A retrospective study. *Child Language Teaching and Therapy, 2,* 266–280.

Hayes, B. (1989). The prosodic hierarchy in meter. In P. Kiparsky & G. Youmans (Eds.), *Phonetics and phonology: Rhythm and meter* (pp. 201–260). San Diego: Academic Press.

Haynes, C. (1982). Vocabulary acquisition problems in language disordered children. Masters thesis, Guys Hospital Medical School, University of London.

Haynes, C. (1992). A longitudinal study of language-impaired children from a residential school. In P. Fletcher & D. Hall (Eds.), *Specific speech and language disorders in children* (pp. 166–182). London: Whurr.

Haynes, C., & Naidoo, S. (1991). *Children with specific speech and language impairment.* London: Mac Keith Press.

Hegde, M. (1980). An experimental-clinical analysis of grammatical and behavioral distinctions between verbal auxiliary and copula. *Journal of Speech and Hearing Research, 23,* 864–877.

Hegde, M., & Gierut, J. (1979). The operant training and generalization of pronouns and a verb form in a language delayed child. *Journal of Communication Disorders, 12,* 23–34.

Hegde, M., Noll, M., & Pecora, R. (1979). A study of some factors affecting generalization of language training. *Journal of Speech and Hearing Disorders, 44,* 301–320.

Henderson, B. (1978). Older language impaired children's processing of rapidly changing acoustic signals. Paper presented at the Convention of the American Speech-Language-Hearing Association, San Francisco.

Hester, P., & Hendrickson, J. (1977). Training functional expressive language: The acquisition and generalization of five-element syntactic responses. *Journal of Applied Behavior Analysis, 10,* 316.

Hill, M., & Clark, M. (1984). Mothers' intervention strategies in a structured question–answer dialogue. Paper presented at the Convention of the American Speech-Language-Hearing Association, San Francisco.

Hinckley, A. (1915). A case of retarded speech development. *Pediatric Seminary, 22,* 121–146.

Hoar, N. (1977). Paraphrase capabilities of language impaired children. Paper presented at the Boston University Conference on Language Development, Boston.

Hochman, R., Thal, D., & Maxon, A. (1977). Temporal integration in dysphasic children. *Journal of the Acoustical Society of America, 62*, Supplement no. 1, S97.

Hodson, B., & Paden, E. (1981). Phonological processes which characterize unintelligible and intelligible speech in early childhood. *Journal of Speech and Hearing Disorders, 46*, 369–373.

Hoeffner, J., & McClelland, J. (1993). Can a perceptual processing deficit explain the impairment of inflectional morphology in developmental dysphasia? A computational investigation. Paper presented at the Stanford Child Language Research Forum, Stanford, CA.

Hoff-Ginsberg, E. (1986). Function and structure in maternal speech: Their relation to the child's development of syntax. *Developmental Psychology, 22*, 155–163.

Hoff-Ginsberg, E., Kelly, D., & Buhr, J. (1995). Syntactic bootstrapping by children with specific language impairment: Implications for a theory of specific language impairment. Paper presented at the Boston University Conference on Language Development, Boston.

Hoffer, P., & Bliss, L. (1990). Maternal verbal responsiveness with language-impaired, stage-matched, and age-matched normal children. *Journal of Applied Developmental Psychology, 11*, 305–319.

Holtz, A. (1988). Untersuchungen zur Entwicklung der Pluralmorphologie bei sprachbehinderten Kindern. *Ulmer Publikationen zur Sprachbehindertenpädagogik, 5*, 15–45.

Horsborough, K., Cross, T., & Ball, J. (1985). Conversational interaction between mothers and their autistic, dysphasic, and normal children. In T. Cross & L. Riach (Eds.), *Issues in research development* (pp. 470–476). Victoria, Australia: Institute of Early Childhood Development.

Hoskins, B. (1979). A study of hypothesis testing behavior in language disordered children. Paper presented at the Convention of the American Speech-Language-Hearing Association, Atlanta.

Howard, D., & van der Lely, H. (1995). Specific language impairment in children is *not* due to a short-term memory deficit: Response to Gathercole & Baddeley. *Journal of Speech and Hearing Research, 38*, 466–472.

Huer, M. (1989). Acoustic tracking of articulation errors: [r]. *Journal of Speech and Hearing Disorders, 54*, 530–534.

Hughes, D., & Carpenter, R. (1989). Short-term syntax learning and generalization effects. *Child Language Teaching and Therapy, 5*, 266–280.

Hughes, M., & Sussman, H. (1983). An assessment of cerebral dominance in language-disordered children via a time-sharing paradigm. *Brain and Language, 19*, 48–64.

Hulme, C., & Tordoff, V. (1989). Working memory development: The effects of speech rate, word length, and acoustic similarity on serial recall. *Journal of Experimental Child Psychology, 47*, 72–87.

Huntley, R., Holt, K., Butterfill, A., & Latham, C. (1988). A follow-up study of a language intervention programme. *British Journal of Disorders of Communication, 23*, 127–140.

Hurst, J., Baraitser, M., Auger, E., Graham, F., & Norell, S. (1990). An extended family with a dominantly inherited speech disorder. *Neurology, 32*, 347–355.

Hyams, N. (1987). The setting of the null subject parameter: A reanalysis. Paper presented at the Boston University Conference on Language Development, Boston.

Ingram, D. (1972a). The acquisition of questions and its relation to cognitive development in normal and linguistically deviant children: A pilot study. *Papers and Reports on Child Language Development, 4*, 13–18.

Ingram, D. (1972b). The acquisition of the English verbal auxiliary and copula in normal and linguistically deviant children. *Papers and Reports on Child Language Development, 4*, 79–91.

Ingram, D. (1976). *Phonological disability in children.* London: Edward Arnold.

Ingram, D. (1979). Early patterns of grammatical development. Paper presented at the Conference on Language Behavior in Infancy and Early Childhood, Santa Barbara, CA.

Ingram, D. (1981). *Procedures for the phonological analysis of children's language.* Baltimore: University Park Press.

Ingram, D. (1987a). Categories of phonological disorders. In *Proceedings of the first international symposium on specific speech and language disorders in children* (pp. 88–99). Brentford, UK: Association for All Speech Impaired Children.

Ingram, D. (1987b). Phonological impairment in children. Paper presented at the International Symposium on Language Acquisition and Language Impairment in Children, Salsomaggiore, Italy.

Ingram, D. (1990). The acquisition of the feature [voice] in normal and phonologically delayed English children. Paper presented at the Convention of the American Speech-Language-Hearing Association, Seattle.

Ingram, D., & Carr, L. (1994). When morphology ability exceeds syntactic ability: A case study. Paper presented at the Convention of the American Speech-Language-Hearing Association, New Orleans.

Ingram, D., & Terselic, B. (1983). Final ingression: A case of deviant child phonology. *Topics in Language Disorders, 3,* 45–50.

Ingram, T. T. S. (1959). Specific developmental disorders of speech in childhood. *Brain, 82,* 450–467.

Ingram, T. T. S., Mason, A., & Blackburn, I. (1970). A retrospective study of 82 children with reading disability. *Developmental Medicine and Child Neurology, 12,* 271–283.

Ingram, T. T. S., & Reid, J. (1956). Developmental aphasia observed in a department of child psychiatry. *Archives of Disorders in Childhood, 31,* 162–172.

Inhelder, B. (1963). Observations sur les aspects opératifs et figuratifs de la pensée chez des enfants dysphasiques. *Problèmes de Psycholinguistique, 6,* 143–153.

Jackson, T., & Plante, E. (1997). Gyral morphology in the posterior sylvian region in families affected by developmental language disorders. *Brain Imaging and Behavior.*

Jacobs, T. (1981). Verbal dominance, complexity, and quantity of speech in pairs of language-disabled and normal children. Doctoral dissertation, University of Southern California.

James, D., van Steenbrugge, W., & Chiveralls, K. (1994). Underlying deficits in language-disordered children with central auditory processing difficulties. *Applied Psycholinguistics, 15,* 311–328.

Jernigan, T., Hesselink, J., Sowell, E., & Tallal, P. (1991). Cerebral structure on magnetic resonance imaging in language- and learning-impaired children. *Archives of Neurology, 48,* 539–545.

Johnson, C., & Sutter, J. (1984). Past time language in expressively delayed children. Paper presented at the Convention of the American Speech-Language-Hearing Association, San Francisco.

Johnson, C., & Sutter, J. (1985). Parental time talk to children with expressive language delays. *Proceedings of the Symposium on Research in Child Language Disorders, 6,* 126–136 Madison: University of Wisconsin, Madison.

Johnston, J. (1982). Interpreting the Leiter IQ: Performance profiles of young normal and language-disordered children. *Journal of Speech and Hearing Research, 25,* 291–296.

Johnston, J. (1988). Specific language disorders in the child. In N. Lass, L. McReynolds, J. Northern, & D. Yoder (Eds.), *Handbook of speech-language pathology and audiology* (pp. 685–715). Toronto: Decker.

Johnston, J. (1991). Questions about cognition in children with specific language impairment. In J. Miller (Ed.), *Research on child language disorders* (pp. 299–307). Austin, TX: Pro-Ed.

Johnston, J. (1994). Cognitive abilities of children with language impairment. In R. Watkins & M. Rice (Eds.), *Specific language impairments in children* (pp. 107–121). Baltimore: Paul H. Brookes.

Johnston, J. (1995). Comments on research design. Panel Presentation at the Conference on SLI and Williams Syndrome, University of New Mexico, Albuquerque.

Johnston, J., Blatchley, M., & Olness, G. (1990). Miniature language system acquisition by children with different learning proficiencies. *Journal of Speech and Hearing Research, 33,* 335–342.

Johnston, J., & Ellis Weismer, S. (1983). Mental rotation abilities in language-disordered children. *Journal of Speech and Hearing Research, 26,* 397–403.

Johnston, J., & Kamhi, A. (1984). Syntactic and semantic aspects of the utterances of language-impaired children. The same can be less. *Merrill-Palmer Quarterly, 30,* 65–85.

Johnston, J., Miller, J., Curtiss, S., & Tallal, P. (1993). Conversations with children who are language impaired: Asking questions. *Journal of Speech and Hearing Research, 36,* 973–978.

Johnston, J., Miller, J., Tallal, P., & Curtiss, S. (1994). Past tense morphology in children with specific language impairment: Longitudinal data. Poster presented at the Convention of the American Speech-Language-Hearing Association, New Orleans.

Johnston, J., & Ramstad, V. (1983). Cognitive development in preadolescent language impaired children. *British Journal of Disorders of Communication, 18,* 49–55.

Johnston, J., & Schery, T. (1976). The use of grammatical morphemes by children with communicative disorders. In D. Morehead & A. Morehead (Eds.), *Normal and deficient child language* (pp. 239–258). Baltimore: University Park Press.

Johnston, J., & Smith, L. (1989). Dimensional thinking in language impaired children. *Journal of Speech and Hearing Research, 32,* 33–38.

Johnston, J., Smith, L., & Box, P. (1988). Six ways to skin a cat: Communication strategies used by language impaired preschoolers. Paper presented at the Symposium on Research in Child Language Disorders, University of Wisconsin, Madison.

Johnston, R., Stark, R., Mellits, E., & Tallal, P. (1981). Neurological status of language-impaired and normal children. *Annals of Neurology, 10,* 159–163.

Kail, R. (1991). Processing time declines exponentially during childhood and adolescence. *Developmental Psychology, 27,* 259–266.

Kail, R. (1994). A method of studying the generalized slowing hypothesis in children with specific language impairment. *Journal of Speech and Hearing Research, 37,* 418–421.

Kail, R., Hale, C., Leonard, L., & Nippold, M. (1984). Lexical storage and retrieval in language-impaired children. *Applied Psycholinguistics, 5,* 37–49.

Kail, R., & Leonard, L. (1986). Word-finding abilities in language-impaired children. *ASHA Monographs,* no. 25.

Kail, R., & Salthouse, T. (1994). Processing speed as a mental capacity. *Acta Psychologica, 86,* 199–225.

Kaiser, A., Yoder, P., & Keetz, A. (1992). Evaluating milieu teaching. In S. Warren & J. Reichle (Eds.), *Causes and effects in communication and language intervention* (pp. 9–47). Baltimore: Paul H. Brookes.

Kaltenbacher, E., & Kany, W. (1985). Kognitive Verarbeitungsstrategien und Syntaxerwerb bei dysphasischen und sprachunauffälligen Kindern. In I. Füssenich & B. Gläss (Eds.), *Dysgrammatismus: Theoretische und praktische Probleme bei der interdisziplinärem Beschreibung gestörter Kindersprache* (pp. 180–219). Heidelberg: HVA Edition Schiefele.

Kaltenbacher, E., & Lindner, K. (1990). Some aspects of delayed and deviant development in language-impaired children learning German. Paper presented at the Meeting of the European Group for Child Language Disorders, Røros, Norway.

Kamhi, A. (1981). Nonlinguistic symbolic and conceptual abilities of language-impaired and normally-developing children. *Journal of Speech and Hearing Research, 24,* 446–453.

Kamhi, A., & Catts, H. (1986). Toward an understanding of developmental language and reading disorders. *Journal of Speech and Hearing Disorders, 51,* 337–347.

Kamhi, A., & Catts, H. (Eds.). (1989). *Reading disabilities: A developmental language perspective.* Boston: Little Brown.

Kamhi, A., Catts, H., Koenig, L., & Lewis, B. (1984). Hypothesis-testing and nonlinguistic symbolic abilities in language-impaired children. *Journal of Speech and Hearing Disorders, 49,* 169–176.

Kamhi, A., Catts, H., Mauer, D., Apel, K., & Gentry, B. (1988). Phonological and spatial processing abilities in language- and reading-impaired children. *Journal of Speech and Hearing Disorders, 53,* 316–327.

Kamhi, A., Gentry, B., Mauer, D., & Gholson, B. (1990). Analogical reasoning and transfer in language-impaired children. *Journal of Speech and Hearing Disorders, 55,* 140–148.

Kamhi, A., & Johnston, J. (1992). Semantic assessment: Determining propositional complexity. In W. Secord & J. Damico (Eds.), *Best practices in school speech-language pathology: Descriptive/nonstandardized language assessment* (pp. 99–105). San Antonio, TX: Psychological Corporation.

Kamhi, A., & Koenig, L. (1985). Metalinguistic awareness in normal and language-disordered children. *Language, Speech, and Hearing Services in Schools, 16,* 199–210.

Kamhi, A., Minor, J., & Mauer, D. (1990). Content and intratest performance profiles on the Columbia and the TONI. *Journal of Speech and Hearing Research, 33,* 375–379.

Kamhi, A., Nelson, L., Lee, R., & Gholson, B. (1985). The ability of language-disordered children to use and modify hypotheses in discrimination learning. *Applied Psycholinguistics, 6,* 435–451.

Kamhi, A., Ward, M., & Mills, E. (1995). Hierarchical planning abilities in children with specific language impairments. *Journal of Speech and Hearing Research, 38,* 1108–1116.

Karlin, I. (1954). Aphasia in children. *American Journal of Disabilities in Children, 87,* 752–767.

Katz, W., Curtiss, S., & Tallal, P. (1992). Rapid automatized naming and gesture by normal and language-impaired children. *Brain and Language, 43,* 623–641.

Kaufman, A., & Kaufman, N. (1983). *Kaufman assessment battery for children.* Circle Pines, MN: American Guidance Service.

Kegel, G. (1981). Zum Einfluß von Syntax und Semantik auf die Nachsprechleistungen agrammatischer Kinder. In G. Kegel & H. Günther (Eds.), *Psycholinguistische Untersuchungen zum kindlichen Agrammatismus* (pp. 61–80). Munich: Institut für Phonetische und Sprachliche Kommunikation.

Kelly, D., & Rice, M. (1994). Preferences for verb interpretation in children with specific language impairment. *Journal of Speech and Hearing Research, 37,* 182–192.

Kerr, J. (1917). Congenital or developmental aphasia. *Journal of Delinquency, 2,* 6.

Kerschensteiner, M., & Huber, W. (1975). Grammatical impairment in developmental aphasia. *Cortex, 11,* 264–282.

Kessler, C. (1975). Postsemantic processes in delayed child language related to first and second language learning. In D. Dato (Ed.), *Georgetown University roundtable on language and linguistics* (pp. 159–178). Washington, DC: Georgetown University Press.

Kewley-Port, D., & Preston, M. (1974). Early apical stop production: A voice onset time analysis. *Journal of Phonetics, 2,* 195–210.

Khan, L., & James, S. (1983). Grammatical morpheme development in three language disordered children. *Journal of Childhood Communication Disorders, 6,* 85–100.

Kilborn, K. (1991). Selective impairment of grammatical morphology due to induced stress in normal listeners: Implications for aphasia. *Brain and Language, 41,* 275–288.

King, G., & Fletcher, P. (1993). Grammatical problems in school-age children with specific language impairment. *Clinical Linguistics and Phonetics, 7,* 339–352.

King, G., Schelletter, I., Sinka, P., Fletcher, P., & Ingham, R. (1995). Are English-speaking SLI children with morpho-syntactic deficits impaired in their use of locative-contact and causative alternating verbs? *University of Reading Working Papers in Linguistics, 2,* 45–65.

King, R., Jones, C., & Lasky, E. (1982). In retrospect: A fifteen year follow-up report of speech-language-disordered children. *Language, Speech, and Hearing Services in Schools, 13,* 24–32.

Kirchner, D., & Klatsky, R. (1985). Verbal rehearsal and memory in language-disordered children. *Journal of Speech and Hearing Research, 28,* 556–565.

Klatt, D. (1975). Vowel lengthening is syntactically determined in a connected discourse. *Journal of Phonetics, 3,* 129–140.

Klecan-Aker, J., & Kelty, K. (1990). An investigation of the oral narratives of normal and language-learning disabled children. *Journal of Childhood Communication Disorders, 13,* 207–216.

Klee, T., & Fitzgerald, M. (1985). The relation between grammatical development and mean length of utterance in morphemes. *Journal of Child Language, 12,* 251–268.

Klee, T., Schaffer, M., May, S., Membrino, I., & Mougey, K. (1989). A comparison of the age–MLU relation in normal and specifically language-impaired preschool children. *Journal of Speech and Hearing Disorders, 54,* 226–233.

Kolvin, I., Fundudis, T., George, G., Wrate, R., & Scarth, L. (1979). Predictive importance-behaviour. In T. Fundudis, I. Kolvin, & R. Garside (Eds.), *Speech retarded and deaf children: Their psychological development* (pp. 67–77). London: Academic Press.

Korkman, M., & Häkkinen-Rihu, P. (1994). A new classification of developmental language disorders. *Brain and Language, 47,* 96–116.

Kouri, T., Lewis, M., & Schlosser, M. (1992). Children's action word comprehension as a function of syntactic context. Paper presented at the Convention of the American Speech-Language-Hearing Association, San Antonio, TX.

Kracke, I. (1975). Perception of rhythmic sequences by receptive aphasic and deaf children. *British Journal of Disorders of Communication, 10,* 43–51.

Kutas, M., & Hillyard, S. (1980). Reading senseless sentences: Brain potentials reflect semantic incongruity. *Science, 207,* 203–205.

Laferriere, D., & Cirrin, F. (1984). Functions of mothers' questions to their language disabled children. Paper presented at the Convention of the American Speech-Language-Hearing Association, San Francisco.

Lahey, M. (1988). *Language disorders and language development.* New York: Macmillan.

Lahey, M. (1990). Who shall be called language disordered? Some reflections and one perspective. *Journal of Speech and Hearing Disorders, 55,* 612–620.

Lahey, M., & Edwards, J. (1995). Specific language impairment: Preliminary investigation of factors associated with family history and with patterns of language performance. *Journal of Speech and Hearing Research, 38,* 643–657.

Lahey, M., & Edwards, J. (1996). Why do children with specific language impairment name pictures more slowly than their peers? *Journal of Speech and Hearing Research, 39,* 1081–1098.

Lahey, M., Liebergott, J., Chesnick, M., Menyuk, P., & Adams, J. (1992). Variability in the use of grammatical morphemes: Implications for understanding language impairment. *Applied Psycholinguistics, 13,* 373–398.

Lamesch, B. (1982). Language acquisition by a child in an institutional environment. In F. Lowenthal, F. Vandamme, & J. Cordier (Eds.), *Language and language acquisition* (pp. 303–308). New York: Plenum.

Landau, W., Goldstein, R., & Kleffner, F. (1960). Congenital aphasia: A clinicopathologic study. *Neurology, 10,* 915–921.

Landau, W., & Kleffner, F. (1957). Syndrome of acquired aphasia with convulsive disorder in children. *Neurology, 7,* 523–530.

Lapointe, S. (1985). A theory of verb form use in the speech of agrammatic aphasics. *Brain and Language, 24,* 100–155.

Lasky, E., & Klopp, K. (1982). Parent–child interactions in normal and language-disordered children. *Journal of Speech and Hearing Disorders, 47*, 7–18.

Lavrand, M. (1897). Mutité chez des entendants. *Revue Internationale de Rhinologie, Otologie et Laryngologie, 3*, 95–97.

Lea, J. (1975). An investigation into the association between rhythmic ability and language ability in a group of children with severe speech and language disorders. Master's thesis, University of London, Guy's Hospital Medical School.

Lee, L. (1966). Developmental sentence types: A method for comparing normal and deviant syntactic development. *Journal of Speech and Hearing Disorders, 31*, 311–330.

Lee, L. (1974). *Developmental sentence analysis.* Evanston, IL: Northwestern University Press.

Lee, L. (1976). Normal and atypical semantic development. Manuscript, Northwestern University, Evanston, IL.

Lee, L., & Canter, S. (1971). Developmental sentence scoring: A clinical procedure for estimating syntactic development in children's spontaneous speech. *Journal of Speech and Hearing Disorders, 36*, 315–340.

Lee, L., Koenigsknecht, R., & Mulhern, S. (1975). *Interactive language development teaching.* Evanston, IL: Northwestern University Press.

Leemans, G. (1994). The acquisition of verb placement in Dutch SLI children. Paper presented at the Meeting of the European Group for Child Language Disorders, Garderen, The Netherlands.

Leiter international performance scale (1979). Chicago: Stoelting.

Le Normand, M. T., & Chevrie-Muller, C. (1989). Exploration de la production lexicale chez six enfants dysphasiques. *Rééducation Orthophonique, 27*, 345–361.

Le Normand, M. T., & Chevrie-Muller, C. (1991). Individual differences in the production of word classes in eight specific language-impaired preschoolers. *Journal of Communication Disorders, 24*, 331–351.

Le Normand, M. T., Leonard, L., & McGregor, K. (1993). A cross-linguistic study of article use by children with specific language impairment. *European Journal of Disorders of Communication, 28*, 153–163.

Leonard, C. M., Voeller, K., Lombardino, L., Morris, M., Hynd, G., Alexander, A., Anderson, H., Garofalakis, M., Honeyman, J., Mao, J., Agee, O., & Staab, E. (1993). Anomalous cerebral structure in dyslexia revealed with magnetic resonance imaging. *Archives of Neurology, 50*, 461–469.

Leonard, L. (1972). What is deviant language? *Journal of Speech and Hearing Disorders, 37*, 427–446.

Leonard, L. (1973). The nature of deviant articulation. *Journal of Speech and Hearing Disorders, 38*, 156–161.

Leonard, L. (1974). A preliminary view of generalization in language training. *Journal of Speech and Hearing Disorders, 39*, 429–436.

Leonard, L. (1975). Developmental considerations in the management of language disabled children. *Journal of Learning Disabilities, 8*, 232–237.

Leonard, L. (1979). Language impairment in children. *Merrill-Palmer Quarterly, 25*, 205–232.

Leonard, L. (1981). Facilitating linguistic skills in children with specific language impairment. *Applied Psycholinguistics, 2*, 89–118.

Leonard, L. (1982a). The nature of specific language impairment in children. In S. Rosenberg (Ed.), *Handbook of applied psycholinguistics* (pp. 295–327). Hillsdale, NJ: Lawrence Erlbaum.

Leonard, L. (1982b). Phonological deficits in children with developmental language impairment. *Brain and Language, 16*, 73–86.

Leonard, L. (1984). Semantic considerations in early language training. In K. Ruder & M. Smith (Eds.), *Developmental language intervention* (pp. 141–169). Baltimore: University Park Press.

Leonard, L. (1985). Unusual and subtle phonological behavior in the speech of phonologically disordered children. *Journal of Speech and Hearing Disorders, 50*, 4–13.

Leonard, L. (1986). Conversational replies of children with specific language impairment. *Journal of Speech and Hearing Research, 29*, 114–119.

Leonard, L. (1987). Is specific language impairment a useful construct? In S. Rosenberg (Ed.), *Advances in applied psycholinguistics, 1, Disorders of first-language development* (pp. 1–39). New York: Cambridge University Press.

Leonard, L. (1989). Language learnability and specific language impairment in children. *Applied Psycholinguistics, 10*, 179–202.

Leonard, L. (1991). Specific language impairment as a clinical category. *Language, Speech, and Hearing Services in Schools, 22*, 66–68.

Leonard, L. (1992a). Models of phonological development and children with phonological disorders. In C. Ferguson, L. Menn, & C. Stoel-Gammon (Eds.), *Phonological development: Models, research, implications* (pp. 495–507). Timonium, MD: York Press.

Leonard, L. (1992b). The use of morphology by children with specific language impairment: Evidence from three languages. In R. Chapman (Ed.), *Processes in language acquisition and disorders* (pp. 186–201). St. Louis: Mosby-Yearbook.

Leonard, L. (1994). Some problems facing accounts of morphological deficits in children with specific language impairments. In R. Watkins & M. Rice (Eds.), *Specific language impairments in children* (pp. 91–105). Baltimore: Paul H. Brookes.

Leonard, L. (1995). Functional categories in the grammars of children with specific language impairment. *Journal of Speech and Hearing Research, 38,* 1270–1283.

Leonard, L., Bolders, J., & Miller, J. (1976). An examination of the semantic relations reflected in the language usage of normal and language handicapped children. *Journal of Speech and Hearing Research, 19,* 371–392.

Leonard, L., Bortolini, U., Caselli, M. C., McGregor, K., & Sabbadini, L. (1992). Morphological deficits in children with specific language impairment: The status of features in the underlying grammar. *Language Acquisition, 2,* 151–179.

Leonard, L., Bortolini, U., Caselli, M. C., & Sabbadini, L. (1993). The use of articles by Italian-speaking children with specific language impairment. *Clinical Linguistics and Phonetics, 7,* 19–27.

Leonard, L., & Brown, B. (1984). The nature and boundaries of phonologic categories: A case study of an unusual phonologic pattern in a language-impaired child. *Journal of Speech and Hearing Disorders, 49,* 419–428.

Leonard, L., Camarata, S., Rowan, L., & Chapman, K. (1982). The communicative functions of lexical usage by language impaired children. *Applied Psycholinguistics, 3,* 109–125.

Leonard, L., Camarata, S., Schwartz, R., Chapman, K., & Messick, C. (1985). Homonymy and the voiced–voiceless distinction in the speech of children with specific language impairment. *Journal of Speech and Hearing Research, 28,* 215–224.

Leonard, L., Devescovi, A., & Ossella, T. (1987). Context-sensitive phonological patterns in children with poor intelligibility. *Child Language Teaching and Therapy, 3,* 125–132.

Leonard, L., & Dromi, E. (1994). The use of Hebrew verb morphology by children with specific language impairment and children developing language normally. *First Language, 14,* 283–304.

Leonard, L., & Eyer, J. (1996). Surface properties of grammatical morphology and morphological deficits in children with specific language impairment. In J. Morgan & K. Demuth (Eds.), *Signal to syntax* (pp. 233–247). Hillsdale, NJ: Lawrence Erlbaum.

Leonard, L., Eyer, J., Bedore, L., & Grela, B. (1997). Three accounts of the grammatical morpheme difficulties of English-speaking children with specific language impairment. *Journal of Speech and Hearing Research.*

Leonard, L., & Fey, M. (1991). Facilitating grammatical development: The contribution of pragmatics. In T. Gallagher (Ed.), *Pragmatics of language* (pp. 333–355). San Diego: Singular Publishing.

Leonard, L., & Leonard, J. (1985). The contribution of phonetic context to an unusual phonological pattern: A case study. *Language, Speech, and Hearing Services in Schools, 16,* 110–118.

Leonard, L., McGregor, K., & Allen, G. (1992). Grammatical morphology and speech perception in children with specific language impairment. *Journal of Speech and Hearing Research, 35,* 1076–1085.

Leonard, L., Nippold, M., Kail, R., & Hale, C. (1983). Picture naming in language-impaired children. *Journal of Speech and Hearing Research, 26,* 609–615.

Leonard, L., Sabbadini, L., Leonard, J., & Volterra, V. (1987). Specific language impairment in children: A cross-linguistic study. *Brain and Language, 32,* 233–252.

Leonard, L., Sabbadini, L., Volterra, V., & Leonard, J. (1988). Some influences on the grammar of English- and Italian-speaking children with specific language impairment. *Applied Psycholinguistics, 9,* 39–57.

Leonard, L., Schwartz, R., Allen, G., Swanson, L., & Loeb, D. (1989). Unusual phonological behavior and the avoidance of homonymy in children. *Journal of Speech and Hearing Research, 32,* 583–590.

Leonard, L., Schwartz, R., Chapman, K., Rowan, L., Prelock, P., Terrell, B., Weiss, A., & Messick, C. (1982). Early lexical acquisition in children with specific language impairment. *Journal of Speech and Hearing Research, 25,* 554–564.

Leonard, L., Schwartz, R., Swanson, L., & Loeb, D. (1987). Some conditions that promote unusual phonological behavior in children. *Clinical Linguistics and Phonetics, 1,* 23–34.

Leonard, L., Steckol, K., & Panther, K. (1983). Returning meaning to semantic relations: Some clinical applications. *Journal of Speech and Hearing Disorders, 48,* 25–36.

Leonard, L., Steckol, K., & Schwartz, R. (1978). Semantic relations and utterance length in child language. In F. Peng & W. von Raffler-Engel (Eds.), *Language acquisition and developmental kinesics* (pp. 93–105). Tokyo: Bunka Hyoron.

Levelt, W. (1989). *Speaking: From intention to articulation*. Cambridge, MA: MIT Press.

Levi, G. (1972). Prestazioni sintattiche nel linguaggio disfasico: Contributo neurolinguistico. *Neuropsichiatria Infantile, 138*, 991–1010.

Levi, G., Capozzi, F., Fabrizi, A., & Sechi, E. (1982). Language disorders and prognosis for reading disabilities in developmental age. *Perceptual and Motor Skills, 54*, 1119–1122.

Levi, G., Fabrizi, A., La Barba, A., & Stievano, P. (1991). Strategie atipiche nella comprensione verbale dei bambini con disturbo specifico di linguaggio. *Psichiatria dell'infanzia e dell'adolescenza, 58*, 483–493.

Lewis, B. (1992). Pedigree analysis of children with phonology disorders. *Journal of Learning Disabilities, 25*, 586–597.

Lewis, B., & Freebairn, L. (1992). Residual effects of preschool phonological disorders in grade school, adolescence, and adulthood. *Journal of Speech and Hearing Research, 35*, 818–831.

Lewis, B., & Thompson, L. (1992). A study of developmental speech and language disorders in twins. *Journal of Speech and Hearing Research, 35*, 1086–1094.

Ley, J. (1929). Un cas d'audi-mutité idiopathique (aphasie congénitale) chez des jumeaux monozygotiques. *L'Encéphale, 24*, 121–165.

Liebmann, A. (1898). *Vorlesungen über Sprachstörungen, 3, Hörstummheit*. Berlin: Coblentz.

Liles, B. (1985a). Cohesion in the narratives of normal and language disordered children. *Journal of Speech and Hearing Research, 28*, 123–133.

Liles, B. (1985b). Production and comprehension of narrative discourse in normal and language disordered children. *Journal of Communication Disorders, 18*, 409–427.

Liles, B. (1987). Episode organization and cohesive conjunctives in narratives of children with and without language disorder. *Journal of Speech and Hearing Research, 30*, 185–196.

Liles, B., Duffy, R., Merritt, D., & Purcell, S. (1995). Measurement of narrative discourse ability in children with language disorders. *Journal of Speech and Hearing Research, 38*, 415–425.

Liles, B., & Purcell, S. (1987). Departures in the spoken narratives of normal and language-disordered children. *Applied Psycholinguistics, 8*, 185–202.

Liles, B., Schulman, M., & Bartlett, S. (1977). Judgments of grammaticality in normal and language-disordered children. *Journal of Speech and Hearing Disorders, 42*, 199–209.

Lincoln, A., Courchesne, E., Harms, L., & Allen, M. (1995). Sensory modulation of auditory stimuli in children with autism and receptive developmental language disorder: Event-related brain potential evidence. *Journal of Autism and Developmental Disorders, 25*, 521–539.

Lincoln, A., Dickstein, P., Courchesne, E., Elmasian, R., & Tallal, P. (1992). Auditory processing abilities in non-retarded adolescents and young adults with developmental receptive language disorder and autism. *Brain and Language, 43*, 613–622.

Lindner, K., & Johnston, J. (1992). Grammatical morphology in language-impaired children acquiring English or German as their first language: A functional perspective. *Applied Psycholinguistics, 13*, 115–129.

Lindner, K., Stoll, S., & Täubner, K. (1994). Aktionsarten and verb inflection in German: A comparison of normal children and children with SLI. Paper presented at the Meeting of the European Group for Child Language Disorders, Garderen, The Netherlands.

Ljubešić, M., & Kovačević, M. (1992). Some insights into specific language impairment in Croatian. *Scandinavian Journal of Logopedics and Phoniatrics, 17*, 37–43.

Locke, J. (1992). Learning spoken language in time to speak it well. Paper presented at the Convention of the American Speech-Language-Hearing Association, San Antonio, TX.

Locke, J. (1993). *The child's path to spoken language*. Cambridge, MA: Harvard University Press.

Locke, J. (1994). Gradual emergence of developmental language disorders. *Journal of Speech and Hearing Disorders, 37*, 608–616.

Loeb, D. (1994). Pronoun case errors of children with and without specific language impairment: Evidence from a longitudinal elicited imitation task. Paper presented at the Stanford Child Language Research Forum, Stanford, CA.

Loeb, D., & Leonard, L. (1988). Specific language impairment and parameter theory. *Clinical Linguistics and Phonetics, 2*, 317–327.

Loeb, D., & Leonard, L. (1991). Subject case marking and verb morphology in normally developing and specifically language-impaired children. *Journal of Speech and Hearing Research, 34*, 340–346.

Loeb, D., & Mikesic, E. (1992). Pronominal acquisition in language-impaired and normally developing children. *Kansas University Working Papers in Language Development, 6*, 285–303.

Loeb, D., Pye, C., Redmond, S., & Richardson, L. (1994). Verb alternations of children with specific language impairment. Poster presented at the Symposium on Research in Child Language Disorders, University of Wisconsin, Madison.

Lorentz, J. (1974). A deviant phonological system of English. *Papers and Reports on Child Language Development, 8,* 55—64.

Lorentz, J. (1976). An analysis of some deviant phonological rules of English. In D. Morehead & A. Morehead (Eds.), *Normal and deficient child language* (pp. 29—59). Baltimore: University Park Press.

Loucks, Y. (1987). Dispute patterns in children. Doctoral dissertation, University of Michigan, Ann Arbor.

Lovell, K., Hoyle, H., & Siddall, M. (1968). A study of some aspects of the play and language of young children with delayed speech. *Journal of Child Psychology and Psychiatry, 9,* 41—50.

Lowe, A., & Campbell, R. (1965). Temporal discrimination in aphasoid and normal children. *Journal of Speech and Hearing Research, 8,* 313—314.

Ludlow, C., Cudahy, E., Bassich, C., & Brown, G. (1983). The auditory processing skills of hyperactive, language impaired and reading disabled boys. In J. Katz & E. Lasky (Eds.), *Central auditory processing disorders: Problems of speech, language, and learning* (pp. 163—185). Baltimore: University Park Press.

Luschinger, R. (1970). Inheritance of speech deficits. *Folia Phoniatrica, 22,* 216—230.

MacKenzie, J., Newhoff, M., & Marinkovich, G. (1981). Normal children address language disordered peers: Dimensions of communicative competence. Paper presented at the Convention of the American Speech-Language-Hearing Association, Los Angeles.

Mackworth, N., Grandstaff, N., & Pribram, K. (1973). Orientation to pictorial novelty by speech-disordered children. *Neuropsychologia, 11,* 443—450.

MacLachlan, B., & Chapman, R. (1988). Communication breakdowns in normal and language learning-disabled children's conversation and narration. *Journal of Speech and Hearing Disorders, 53,* 2—7.

Macpherson, C., & Weber-Olson, M. (1980). Mother speech input to deficient and language normal children. Paper presented at the Symposium on Research in Child Language Disorders, University of Wisconsin, Madison.

Magnusson, E. (1983). *The phonology of language disordered children: Production, perception, and awareness.* Lund: Gleerup.

Magnusson, E., & Nauclér, K. (1990). Reading and spelling in language-disordered children. Linguistic and metalinguistic prerequisites: Report on a longitudinal study. *Clinical Linguistics and Phonetics, 4,* 49—61.

Manhardt, J., Hansen, I., & Rescorla, L. (1995). Narrative competency outcomes of specific expressive language impairment (SLI-E) at ages six, seven, and eight. Paper presented at the Boston University Conference on Language Development, Boston.

Marchman, V. (1993). Constraints on plasticity in a connectionist model of the English past tense. *Journal of Cognitive Neuroscience, 5,* 215—234.

Marchman, V., & Ellis Weismer, S. (1994). Patterns of productivity in children with SLI and NL: A study of the English past tense. Paper presented at the Symposium for Research in Child Language Disorders, University of Wisconsin, Madison.

Marchman, V., Wulfeck, B., & Ellis Weismer, S. (1995). Productive use of English past tense morphology in children with SLI and normal language. Technical Report CND-9514. Center for Research in Language, University of California at San Diego.

Marcus, G., Pinker, S., Ullman, M., Hollander, M., Rosen, T. J., & Xu, F. (1992). Overregularization in language acquisition. *Monographs of the Society for Research in Child Development, 57* (4, serial no. 228).

Marinkovich, G., Newhoff, M., & MacKenzie, J. (1980). Why can't you talk? Peer input to language disordered children. Paper presented at the Convention of the American Speech-Language-Hearing Association, Detroit.

Martin, S. (1991). Input training in phonological disorder: A case discussion. In M. Yavas (Ed.), *Phonological disorders in children: Theory, research and practice* (pp. 152—172). London: Routledge.

Masland, M., & Case, L. (1968). Limitation of auditory memory as a factor in delayed language development. *British Journal of Disorders of Communication, 3,* 139—142.

Masterson, J. (1993). The performance of children with language-learning disabilities on two types of cognitive tasks. *Journal of Speech and Hearing Research, 36,* 1026—1036.

Masterson, J., Evans, L., & Aloia, M. (1993). Verbal analogical reasoning in children with language-learning disabilities. *Journal of Speech and Hearing Research, 36,* 76—82.

Masterson, J., & Kamhi, A. (1992). Linguistic trade-offs in school-age children with and without language disorders. *Journal of Speech and Hearing Research, 35,* 1064—1075.

Maxwell, E. (1979). Competing analyses of a deviant phonology. *Glossa, 13,* 181–214.

Maxwell, E., & Weismer, G. (1982). The contribution of phonological, acoustic, and perceptual techniques to the characteristics of a misarticulating child's voice contrast for stops. *Applied Psycholinguistics, 3,* 29–44.

McCall, E. (1911). Two cases of congenital aphasia in children. *British Medical Journal, 1, 2628,* 1105; *2632,* 1407.

McCauley, R., & Swisher, L. (1984). Use and misuse of norm-referenced tests in clinical assessment: A hypothetical case. *Journal of Speech and Hearing Disorders, 49,* 338–348.

McCroskey, R., & Kidder, H. (1980). Auditory fusion among learning disabled, reading disabled, and normal children. *Journal of Learning Disabilities, 13,* 69–76.

McCroskey, R., & Thompson, N. (1973). Comprehension of rate controlled speech by children with specific learning disabilities. *Journal of Learning Disabilities, 6,* 29–35.

McGregor, K. (1994). Use of phonological information in a word-finding treatment for children. *Journal of Speech and Hearing Research, 37,* 1381–1393.

McGregor, K., & Leonard, L. (1989). Facilitating word-finding skills of language-impaired children. *Journal of Speech and Hearing Disorders, 54,* 141–147.

McGregor, K., & Leonard, L. (1994). Subject pronoun and article omissions in the speech of children with specific language impairment: A phonological interpretation. *Journal of Speech and Hearing Research, 37,* 171–181.

McGregor, K., & Leonard, L. (1995). Intervention for word-finding deficits in children. In M. Fey, J. Windsor, & S. Warren (Eds.), *Language intervention: Preschool through the elementary years* (pp. 85–105). Baltimore: Paul H. Brookes.

McGregor, K., & Schwartz, R. (1992). Converging evidence for underlying phonological representation in a child who misarticulates. *Journal of Speech and Hearing Research, 35,* 596–603.

McGregor, K., & Waxman, S. (1995). Multiple level naming abilities of children with word-finding deficits. Paper presented at the Boston University Conference on Language Development, Boston.

McReynolds, L. (1966). Operant conditioning for investigating speech sound discrimination in aphasic children. *Journal of Speech and Hearing Research, 9,* 519–528.

McReynolds, L., & Huston, K. (1971). A distinctive feature analysis of children's misarticulations. *Journal of Speech and Hearing Disorders, 36,* 155–166.

Meline, T. (1978). Referential communication by normal- and deficient-language children. Paper presented at the Convention of the American Speech-Language-Hearing Association, San Francisco.

Meline, T. (1986). Referential communication skills of learning disabled/language impaired children. *Applied Psycholinguistics, 7,* 129–140.

Meline, T. (1988). The encoding of novel referents by language-impaired children. *Language, Speech, and Hearing Services in Schools, 19,* 119–127.

Meline, T., & Meline, N. (1983). Facing a communicative obstacle: Pragmatics of language-impaired children. *Perceptual and Motor Skills, 56,* 469–470.

Menyuk, P. (1964). Comparison of grammar of children with functionally deviant and normal speech. *Journal of Speech and Hearing Research, 7,* 109–121.

Menyuk, P. (1968). The role of distinctive features in children's acquisition of phonology. *Journal of Speech and Hearing Research, 11,* 138–146.

Menyuk, P. (1969). *Sentences children use.* Cambridge, MA: MIT Press.

Menyuk, P. (1975). Children with language problems: What's the problem? In D. Dato (Ed.), *Georgetown University roundtable on language and linguistics* (pp. 129–144). Washington, DC: Georgetown University Press.

Menyuk, P. (1978). Linguistic problems in children with developmental dysphasia. In M. Wyke (Ed.), *Developmental dysphasia* (pp. 135–158). London: Academic Press.

Menyuk, P., & Looney, P. (1972). Relationships among components of the grammar in language disorder. *Journal of Speech and Hearing Research, 15,* 395–406.

Merino, B. (1983). Language development in normal and language handicapped Spanish-speaking children. *Hispanic Journal of Behavioral Sciences, 5,* 379–400.

Merritt, D., & Liles, B. (1987). Story grammar ability in children with and without language disorder: Story generation, story retelling, and story comprehension. *Journal of Speech and Hearing Research, 30,* 539–552.

Merzenich, M. (1995). Cortical learning mechanisms: Some implications for neurobehavioral development and rehabilitation. Keynote address at the Convention of the American Speech-Language-Hearing Association, Orlando, FL.

Merzenich, M., Jenkins, W., Johnston, P., Schreiner, C., Miller, S., & Tallal, P. (1996). Temporal processing deficits of language-learning impaired children ameliorated by training. *Science, 271,* 77–81.

Merzenich, M., Nelson, R., Stryker, M., Cynader, M., Schoppmann, A., & Zook, J. (1984). Somatosensory cortical map changes following digit amputation in adult monkeys. *Journal of Comparative Neurology, 224,* 591–605.

Messick, C., & Newhoff, M. (1979). Request form: Does the language-impaired child consider the listener? Paper presented at the Convention of the American Speech-Language-Hearing Association, Atlanta.

Messick, C., & Prelock, P. (1981). Successful communication: Mothers of language-impaired children vs. mothers of language-normal children. Paper presented at the Convention of the American Speech-Language-Hearing Association, Los Angeles.

Methé, S., & Crago, M. (1996). Verb morphology in French children with language impairment. Paper presented at the Symposium on Research in Child Language Disorders, University of Wisconsin, Madison.

Miller, J. (1996). The search for the phenotype of disordered language performance. In M. Rice (Ed.), *Toward a genetics of language* (pp. 297–314). Hillsdale, NJ: Lawrence Erlbaum.

Miller, J., & Chapman, R. (1981). The relation between age and mean length of utterance in morphemes. *Journal of Speech and Hearing Research, 24,* 154–161.

Millet, A., & Newhoff, M. (1978). Language disordered children: Language disordered mothers. Paper presented at the Convention of the American Speech-Language-Hearing Association, San Francisco.

Mills, A., Pulles, A., & Witten, F. (1992). Semantic and pragmatic problems in specifically language impaired children: One or two problems. *Scandinavian Journal of Logopedics and Phoniatrics, 17,* 51–57.

Mills, D., Thal, D., Di Iulio, L., Castaneda, C., Coffey-Corina, S., & Neville, H. (1995). Auditory sensory processing and language abilities in late talkers: An ERP study. Technical Report CND-9508. Center for Research in Language, University of California at San Diego.

Minifie, F., Darley, F., & Sherman, D. (1963). Temporal reliability of seven language measures. *Journal of Speech and Hearing Research, 6,* 139–149.

Mody, M. (1993). Bases of reading impairment in speech perception: A deficit in rate of auditory processing or in phonological encoding? Doctoral dissertation, City University of New York.

Monsees, E. (1961). Aphasia in children. *Journal of Speech and Hearing Disorders, 26,* 83–86.

Monsees, E. (1968). Temporal sequence and expressive language disorders. *Exceptional Children, 35,* 141–147.

Montgomery, J. (1993). Haptic recognition of children with specific language impairment: Effects of response modality. *Journal of Speech and Hearing Research, 36,* 98–104.

Montgomery, J. (1995a). Examination of phonological working memory in specifically language-impaired children. *Applied Psycholinguistics, 16,* 355–378.

Montgomery, J. (1995b). Sentence comprehension in children with specific language impairment: The role of phonological working memory. *Journal of Speech and Hearing Research, 38,* 187–199.

Montgomery, J., Scudder, R., & Moore, C. (1990). Language-impaired children's real-time comprehension of spoken language. *Applied Psycholinguistics, 11,* 273–290.

Moore, M. (1995). Error analysis of pronouns by normal and language-impaired children. *Journal of Communication Disorders, 28,* 57–72.

Moore, M., & Johnston, J. (1993). Expressions of past time by normal and language-impaired children. *Applied Psycholinguistics, 14,* 515–534.

Morehead, D. (1972). Early grammatical and semantic relations: Some implications for a general representational deficit in linguistically deviant children. *Papers and Reports in Child Language Development, 4,* 1–12.

Morehead, D., & Ingram, D. (1970). The development of base syntax in normal and linguistically deviant children. *Papers and Reports on Child Language Development, 2,* 55–75.

Morehead, D., & Ingram, D. (1973). The development of base syntax in normal and linguistically deviant children. *Journal of Speech and Hearing Research, 16,* 330–352.

Morley, M., Court, D., Miller, H., & Garside, R. (1955). Delayed speech and developmental aphasia, *British Medical Journal, 2, 4937,* 463–467.

Moseley, M. (1990). Mother–child interaction with preschool language-delayed children: Structuring conversations. *Journal of Communication Disorders, 23,* 187–203.

Moyer, H. (1898). Dumbness or congenital aphasia of a family type without deafness or obvious mental defect. *Chicago Medical Recorder, 15,* 305–309.

Mulac, A., & Tomlinson, C. (1977). Generalization of an operant remediation program for syntax with language-delayed children. *Journal of Communication Disorders, 10,* 231–244.

Muma, J. (1971). Language intervention: Ten techniques. *Language, Speech, and Hearing Services in Schools, 5,* 7–17.

Muma, J. (1986). *Language acquisition: A functionalist perspective.* Austin, TX: Pro-Ed.

Murphy, V. (1978). A comparison of four measures of visual imagery in normal and language-disordered children. Master's thesis, Northern Illinois University.

Myklebust, H. (1971). Childhood aphasia: An evolving concept. In L. E. Travis (Ed.), *Handbook of speech-language pathology and audiology* (pp. 1181–1202). New York: Appleton-Century-Crofts.

Nakamura, P., & Newhoff, M. (1982). Clinician speech adjustments to normal and language disordered children. Paper presented at the Convention of the American Speech-Language-Hearing Association, Toronto.

Neils, J., & Aram, D. (1986). Family history of children with developmental language disorders. *Perceptual and Motor Skills, 63,* 655–658.

Nelson, K. E., Camarata, S., Welsh, J., Butkovsky, L., & Camarata, M. (1996). Effects of imitative and conversational recasting treatment on the acquisition of grammar in children with specific language impairment and younger language-normal children. *Journal of Speech and Hearing Research, 39,* 850–859.

Nelson, K. E., Welsh, J., Camarata, S., Butkovsky, L., & Camarata, M. (1995). Available input for language-impaired children and younger children of matched language levels. *First Language, 43,* 1–18.

Nelson, L., Kamhi, A., & Apel, K. (1987). Cognitive strengths and weaknesses in language-impaired children: One more look. *Journal of Speech and Hearing Disorders, 52,* 36–43.

Nespor, M., & Vogel, I. (1986). *Prosodic phonology.* Dordrecht, The Netherlands: Foris.

Nettelbladt, U. (1983). *Developmental studies of dysphonology in children.* Lund: Liber.

Nettelbladt, U., & Hansson, K. (1990). Dialogues with language disordered children. How does the interactional style of the conversational partner influence the child? Paper presented at the Meeting of the European Group for Child Language Disorders, Røros, Norway.

Nettelbladt, U., & Hansson, K. (1993). Parents, peers and professionals in interaction with language impaired children. Paper presented at the Child Language Seminar, Plymouth, UK.

Nettelbladt, U., Sahlén, B., Ors, M., & Johannesson, P. (1989). A multidisciplinary assessment of children with severe language disorder. *Clinical Linguistics and Phonetics, 3,* 313–346.

Neville, H., Coffey, S., Holcomb, P., & Tallal, P. (1993). The neurobiology of sensory and language processing in language-impaired children. *Journal of Cognitive Neuroscience, 5,* 235–253.

Newcomer, P., Barenbaum, E., & Nodine, B. (1988). Comparison of the story production of learning disabled, normal achieving and low achieving children under two modes of production. *Learning Disabilities Quarterly, 11,* 82–96.

Newcomer, P., & Hammill, D. (1991). *Test of language development—primary 2.* Austin, TX: Pro-Ed.

Newhoff, M. (1977). Maternal linguistic behavior in relation to the linguistic and developmental ages of children. Doctoral dissertation, Memphis State University.

Newport, E. (1990). Maturational constraints on learning. *Cognitive Science, 14,* 11–28.

Newport, E. (1991). Contrasting concepts of the critical period of language. In S. Carey & R. Gelman (Eds.), *The epigenesis of mind* (pp. 111–130). Hillsdale, NJ: Lawrence Erlbaum.

Nice, M. (1925). A child who would not talk. *Pedagogical Seminary, 32,* 105–144.

Nichols, S., Townsend, J., & Wulfeck, B. (1995). Covert visual attention in language-impaired children. Technical Report CND-9502. Center for Research in Language, University of California at San Diego.

Nippold, M., Erskine, B., & Freed, D. (1988). Proportional and functional analogical reasoning in normal and language-impaired children. *Journal of Speech and Hearing Disorders, 53,* 440–448.

Nippold, M., & Fey, S. (1983). Metaphoric understanding in preadolescents having a history of language acquisition difficulties. *Language, Speech, and Hearing Services in Schools, 14,* 171–180.

Noterdaeme, M., Amorosa, H., Ploog, M., & Scheimann, G. (1988). Quantitative and qualitative aspects of associated movements in children with specific developmental speech and language disorders and in normal pre-school children. *Journal of Human Movement Studies, 15,* 151–169.

Nye, C., Foster, S., & Seaman, D. (1987). Effectiveness of language intervention with the language/learning disabled. *Journal of Speech and Hearing Disorders, 52,* 348–357.

Oetting, J., & Horohov, J. (1997). Past tense marking by children with and without specific language impairment. *Journal of Speech and Hearing Research 40,* 62–74.

Oetting, J., & Rice, M. (1993). Plural acquisition in children with specific language impairment. *Journal of Speech and Hearing Research, 36,* 1241–1253.

Oetting, J., Rice, M., & Swank, L. (1995). Quick incidental learning (QUIL) of words by school-age children with and without SLI. *Journal of Speech and Hearing Research, 38,* 434–445.

Ogiela, D. (1995). Pronoun case errors in normally developing and specifically language impaired children. Master's thesis, Purdue University.

O'Hara, M., & Johnston, J. (1997). Syntactic bootstrapping in children with specific language impairment. *European Journal of Disorders of Communication.*

Oller, D. K. (1973). Regularities in abnormal child phonology. *Journal of Speech and Hearing Disorders, 38,* 36–47.

Olswang, L., Bain, B., Dunn, C., & Cooper, J. (1983). The effects of stimulus variation on lexical learning. *Journal of Speech and Hearing Disorders, 48,* 192–201.

Olswang, L., Bain, B., & Johnson, G. (1992). The zone of proximal development: Dynamic assessment of language disordered children. In S. Warren & J. Reichle (Eds.), *Perspectives on communication and language intervention: Development, assessment, and remediation* (pp. 187–216). Baltimore: Paul H. Brookes.

Olswang, L., Bain, B., Rosendahl, P., Oblak, S., & Smith, A. (1986). Language learning: Moving performance from a context-dependent to -independent state. *Child Language Teaching and Therapy, 2,* 180–210.

Olswang, L., & Carpenter, R. (1978). Elicitor effects on the language obtained from young language-impaired children. *Journal of Speech and Hearing Disorders, 43,* 76–88.

Olswang, L., & Coggins, T. (1984). The effects of adult behavior on increasing language delayed children's production of early relational meanings. *British Journal of Disorders of Communication, 19,* 15–34.

Orton, S. (1937). *Reading, writing, and speech problems in children.* New York: Norton.

Padget, S. (1988). Speech- and language-impaired three and four year olds: A five-year follow-up study. In R. Masland & M. Masland (Eds.), *Pre-school prevention of reading failure* (pp. 52–77). Timonium, MD: York Press.

Paluszek, S., & Feintuch, F. (1979). Comparing imitation and comprehension training in two language impaired children. *Working Papers in Experimental Speech-Language-Pathology and Audiology, 8,* 72–91.

Panagos, J., & Prelock, P. (1982). Phonological constraints on the sentence productions of language-disordered children. *Journal of Speech and Hearing Research, 25,* 171–177.

Panther, K., & Steckol, K. (1981). Training symbolic play skills in language impaired children. Paper presented at the Convention of the American Speech-Language-Hearing Association, Los Angeles.

Paul, R. (1991a). Assessing communication skills in toddlers. *Clinics in Communication Disorders: Infant Assessment, 1* (2), 7–23.

Paul, R. (1991b). Profiles of toddlers with slow expressive language development. *Topics in Language Disorders, 11* (4), 1–13.

Paul, R. (1993). Patterns of development in late talkers: Preschool years. *Journal of Childhood Communication Disorders, 15,* 7–14.

Paul, R., & Cohen, D. (1984). Outcome of severe disorders of language acquisition. *Journal of Autism and Developmental Disorders, 14,* 405–421.

Paul, R., Cohen, D., & Caparulo, B. (1983). A longitudinal study of patients with severe specific developmental language disorders. *Journal of the American Academy of Psychiatry, 22,* 525–534.

Paul, R., & Elwood, T. (1991). Maternal linguistic input to toddlers with slow expressive language development. *Journal of Speech and Hearing Research, 34,* 982–988.

Paul, R., & Fisher, M. (1985). Sentence comprehension strategies in children with autism and developmental language disorders. *Proceedings of the Symposium on Research in Child Language Disorders, 6,* 11–19. Madison: University of Wisconsin, Madison.

Paul, R., Hernandez, R., Herron, L., & Johnson, K. (1995). Narrative development in children with normal, impaired, and late developing language: Early school age. Poster presented at the Convention of the American Speech-Language-Hearing Association, Orlando, FL.

Paul, R., & Jennings, P. (1992). Phonological behavior in toddlers with slow expressive language development. *Journal of Speech and Hearing Research, 35,* 99–107.

Paul, R., & Shiffer, M. (1991). Communicative initiations in normal and late-talking toddlers. *Applied Psycholinguistics, 12,* 419–431.

Paul, R., & Shriberg, L. (1982). Associations between phonology and syntax in speech-delayed children. *Journal of Speech and Hearing Research, 25,* 536–547.

Paul, R., & Smith, R. (1993). Narrative skills in 4-year-olds with normal, impaired, and late-developing language. *Journal of Speech and Hearing Research, 36,* 592–598.

Paul, R., Spangle-Looney, S., & Dahm, P. (1991). Communication and socialization skills at ages 2 and 3 in "late-talking" young children. *Journal of Speech and Hearing Research, 34*, 858–865.

Paul, R., & Unkefer, C. (1995). Familiarity in early language delay. Poster presented at the Convention of the American Speech-Language-Hearing Association, Orlando, FL.

Pembrey, M. (1992). Genetics and language disorder. In P. Fletcher & D. Hall (Eds.), *Specific speech and language disorders in children* (pp. 51–62). London: Whurr.

Perfetti, C. (1985). *Reading ability.* New York: Oxford University Press.

Peters, A. (1985). Language segmentation: Operating principles for the perception and analysis of language. In D. Slobin (Ed.), *The crosslinguistic study of language acquisition, 2, Theoretical issues* (pp. 1029–1067). Hillsdale, NJ: Lawrence Erlbaum.

Peterson, G., & Sherrod, K. (1982). Relationship of maternal language to language development and language delay of children. *American Journal of Mental Deficiency, 86*, 391–398.

Pettit, J., & Helms, S. (1979). Hemispheric language dominance of language-disordered, articulation-disordered, and normal children. *Journal of Learning Disabilities, 12*, 71–76.

Piérart, B., & Harmegnies, B. (1993). Dysphasie simple de l'enfant et langage de la mère. *L'Année Psychologique, 93*, 227–268.

Pierce, A. (1992). *Language acquisition and syntactic theory.* Dordrecht, The Netherlands: Kluwer.

Piggot, G., & Kessler Robb, M. (1994). Prosodic organization in familial language impairment. *McGill Working Papers in Linguistics, 10*, 16–23.

Pinker, S. (1979). Formal models of language learning. *Cognition, 7*, 217–283.

Pinker, S. (1984). *Language learnability and language development.* Cambridge, MA: Harvard University Press.

Pinker, S. (1989). *Learnability and cognition: The acquisition of argument structure.* Cambridge, MA: MIT Press.

Pinker, S. (1991). The rules of language. *Science, 253*, 530–535.

Plante, E. (1991). MRI findings in the parents and siblings of specifically language-impaired boys. *Brain and Language, 41*, 67–80.

Plante, E. (1996). Phenotypic variability in brain-behavior studies of specific language impairment. In M. Rice (Ed.), *Toward a genetics of language* (pp. 317–335). Hillsdale, NJ: Lawrence Erlbaum.

Plante, E., Boliek, C., Binkiewicz, A., & Erly, W. (1996). Elevated androgen, brain development, and language/learning disabilities in children with congenital adrenal hyperplasia. *Developmental Medicine and Child Neurology, 38*, 423–437.

Plante, E., Shenkman, K., & Clark, M. (1996). Classification of adults for family studies of developmental language disorders. *Journal of Speech and Hearing Research, 39*, 661–667.

Plante, E., Swisher, L., Kiernan, B., & Restrepo, M. A. (1993). Language matches: Illuminating or confounding? *Journal of Speech and Hearing Research, 36*, 772–776.

Plante, E., Swisher, L., & Vance, R. (1989). Anatomical correlates of normal and impaired language in a set of dizygotic twins. *Brain and Language, 37*, 643–655.

Plante, E., Swisher, L., Vance, R., & Rapcsak, S. (1991). MRI findings in boys with specific language impairment. *Brain and Language, 41*, 52–66.

Platzack, C. (1990). A grammar without functional categories: A syntactic study of early Swedish child language. *Nordic Journal of Linguistics, 13*, 107–126.

Plaza, M., & Le Normand, M. T. (1996). Singular personal pronoun use: A comparative study of children with specific language impairment and normal French-speaking children. *Clinical Linguistics and Phonetics, 10*, 299–310.

Pollock, K. (1983). Individual preference: Case study of a phonologically delayed child. *Topics in Language Disorders, 3*, 10–23.

Pollock, K., & Keiser, N. (1990). An examination of vowel errors in phonologically disordered children. *Clinical Linguistics and Phonetics, 4*, 161–178.

Poppen, R., Stark, J., Eisenson, J., Forrest, T., & Wertheim, G. (1969). Visual sequencing performance of aphasic children. *Journal of Speech and Hearing Research, 12*, 288–300.

Powell, R., & Bishop, D. (1992). Clumsiness and perceptual problems in children with specific language impairment. *Developmental Medicine and Child Neurology, 34*, 755–765.

Prelock, P., Messick, C., Schwartz, R., & Terrell, B. (1981). Mother–child discourse during the one-word stage. Paper presented at the Syposium on Research in Child Language Disorders, University of Wisconsin, Madison.

Prelock, P., & Panagos, J. (1989). The influence of processing mode on the sentence productions of language-disordered and normal children. *Clinical Linguistics and Phonetics, 3*, 251–263.

Prinz, P. (1982). An investigation of the comprehension and production of requests in normal and language-disordered children. *Journal of Communication Disorders, 15*, 75–93.

Prinz, P., & Ferrier, L. (1983). "Can you give me that one?": The comprehension, production and judgment of directives in language-impaired children. *Journal of Speech and Hearing Disorders, 48*, 44–54.

Purcell, S., & Liles, B. (1992). Cohesion repairs in the narratives of normal language and language disordered school-age children. *Journal of Speech and Hearing Research, 35*, 354–362.

Radford, A. (1988). Small children's small clauses. *Transactions of the Philological Society, 86*, 1–46.

Radford, A. (1990). *Syntactic theory and the acquisition of English syntax.* Oxford: Blackwell.

Ramos, E., & Roeper, T. (1995). Pronoun case assignment by a specifically language impaired child. Poster presented at the Convention of the American Speech-Language-Hearing Association, Orlando, FL.

Rapin, I., & Allen, D. (1983). Developmental language disorders: Nosologic considerations. In U. Kirk (Ed.), *Neuropsychology of Language, Reading and Spelling* (pp. 155–184). New York: Academic Press.

Rapin, I., & Allen, D. (1987). Developmental dysphasia and autism in preschool children: Characteristics and subtypes. In *Proceedings of the First International Symposium on Specific Speech and Language Disorders in Children* (pp. 20–35). Brentford, UK: Association for All Speech Impaired Children.

Rapin, I., & Allen, D. (1988). Syndromes in developmental dysphasia and adult aphasia. In F. Plum (Ed.), *Language, communication, and the brain* (pp. 57–75). New York: Raven Press.

Rapin, I., & Wilson, B. (1978). Children with developmental language disability: Neurological aspects and assessment. In M. Wyke (Ed.), *Developmental dysphasia* (pp. 13–41). London: Academic Press.

Read, C., & Schreiber, P. (1982). Why short subjects are harder to find than long ones. In E. Wanner & L. Gleitman (Eds.), *Language acquisition: The state of the art* (pp. 78–101). Cambridge: Cambridge University Press.

Records, N., & Tomblin, J. B. (1994). Clinical decision making: Describing the decision rules of practicing speech-language pathologists. *Journal of Speech and Hearing Research, 37*, 144–156.

Records, N., Tomblin, J. B., & Freese, P. (1992). The quality of life of young adults with histories of specific language impairment. *American Journal of Speech-Language Pathology, 1*, 44–53.

Rees, N. (1973). Auditory processing factors in language disorders: The view from Procrustes' bed. *Journal of Speech and Hearing Disorders, 38*, 304–315.

Rees, N. (1981). Saying more than we know: Is auditory processing disorder a meaningful concept? In R. Keith (Ed.), *Central auditory and language disorders in children* (pp. 94–120). San Diego: College-Hill Press.

Rescorla, L. (1989). The Language Development Survey: A screening tool for delayed language in toddlers. *Journal of Speech and Hearing Disorders, 54*, 587–599.

Rescorla, L., & Bernstein Ratner, N. (1996). Phonetic profiles of toddlers with specific expressive language impairment (SLI-E). *Journal of Speech and Hearing Research, 39*, 153–165.

Rescorla, L., & Fechnay, T. (1996). Mother–child synchrony and communicative reciprocity in late-talking toddlers. *Journal of Speech and Hearing Research, 39*, 200–208.

Rescorla, L., & Goosens, M. (1992). Symbolic play development in toddlers with expressive specific language impairment. *Journal of Speech and Hearing Research, 35*, 1290–1302.

Rescorla, L., & Schwartz, E. (1990). Outcomes of specific expressive language delay (SELD). *Applied Psycholinguistics, 11*, 393–408.

Restrepo, M. A. (1995). Identifiers of Spanish-speaking children with language impairment who are learning English as a second language. Doctoral dissertation, University of Arizona.

Restrepo, M. A., Swisher, L., Plante, E., & Vance, R. (1992). Relations among verbal and nonverbal cognitive skills in normal language and specifically language-impaired children. *Journal of Communication Disorders, 25*, 205–219.

Reynell, J., & Gruber, C. (1990). *Reynell developmental language scales—U.S. edition.* Los Angeles: Western Psychological Services.

Rice, M. (1991). Children with specific language impairment: Toward a model of teachability. In N. Krasnegor, D. Rumbaugh, R. Schiefelbusch, & M. Studdert-Kennedy (Eds.), *Biological and behavioral determinants of language development* (pp. 447–480). Hillsdale, NJ: Lawrence Erlbaum.

Rice, M., Alexander, A., & Hadley, P. (1993). Social biases toward children with speech and language impairments: A correlative causal model of language limitation. *Applied Psycholinguistics, 14*, 473–488.

Rice, M., & Bode, J. (1993). GAPS in the lexicon of children with specific language impairment. *First Language, 13*, 113–131.

Rice, M., Buhr, J., & Nemeth, M. (1990). Fast mapping word-learning abilities of language-delayed pre-schoolers. *Journal of Speech and Hearing Disorders, 55,* 33–42.

Rice, M., Buhr, J., & Oetting, J. (1992). Specific-language-impaired children's quick incidental learning of words: The effects of a pause. *Journal of Speech and Hearing Research, 35,* 1040–1048.

Rice, M., & Hadley, P. (1995). Language outcomes of the language-focused curriculum. In M. Rice & K. Wilcox (Eds.), *Building a language-focused curriculum for the preschool classroom, 1, A foundation for lifelong communication* (pp. 155–169). Baltimore: Paul H. Brookes.

Rice, M., & Oetting, J. (1993). Morphological deficits in children with SLI: Evaluation of number marking and agreement. *Journal of Speech and Hearing Research, 36,* 1249–1257.

Rice, M., Oetting, J., Marquis, J., Bode, J., & Pae, S. (1994). Frequency of input effects on word comprehension of children with specific language impairment. *Journal of Speech and Hearing Research, 37,* 106–122.

Rice, M., Sell, M., & Hadley, P. (1990). The social interactive coding system (SICS): An on-line, clinically relevant descriptive tool. *Language, Speech, and Hearing Services in Schools, 21,* 2–14.

Rice, M., Sell, M., & Hadley, P. (1991). Social interactions of speech- and language-impaired children. *Journal of Speech and Hearing Research, 34,* 1299–1307.

Rice, M., & Wexler, K. (1995a). Extended optional infinitive (EOI) account of specific language impairment. In D. MacLaughlin & S. McEwen (Eds.), *Proceedings of the 19th Annual Boston University Conference on Language Development, 2,* 451–462. Somerville, MA: Cascadilla Press.

Rice, M., & Wexler, K. (1995b). Tense over time: The persistence of optional infinitives in English in children with SLI. Paper presented at the Boston University Conference on Language Development, Boston.

Rice, M., Wexler, K., & Cleave, P. (1995). Specific language impairment as a period of extended optional infinitive. *Journal of Speech and Hearing Research, 38,* 850–863.

Richman, N., Stevenson, J., & Graham, P. (1982). *Preschool to school: A behavioral study.* London: Academic Press.

Riddle, L. (1992). The attentional capacity of children with specific language impairment. Doctoral dissertation, Indiana University.

Ripley, K. (1986). The Moor House School remedial programme: An evaluation. *Child Language Teaching and Therapy, 2,* 281–300.

Rispoli, M. (1994). Pronoun case overextensions and paradigm building. *Journal of Child Language, 21,* 157–172.

Rissman, M., Curtiss, S., & Tallal, P. (1990). School placement outcomes of young language impaired children. *Journal of Speech-Language Pathology and Audiology, 14,* 49–58.

Robbins, J., & Klee, T. (1987). Clinical assessment of oropharyngeal motor development in young children. *Journal of Speech and Hearing Disorders, 52,* 271–277.

Roberts, J., Rescorla, L., & Borneman, A. (1994). Morphosyntactic characteristics of early language errors: An examination of specific expressive language delay. Poster presented at the Symposium on Research in Child Language Disorders, University of Wisconsin, Madison.

Roberts, S. (1995). Functional categories in the grammars of English- and German-speaking children with specific language impairment. Master's thesis, Purdue University.

Robin, D., Tomblin, J. B., Kearney, A., & Hug, L. (1989). Auditory temporal pattern learning in children with severe speech and language impairments. *Brain and Language, 36,* 604–613.

Robinson, M. (1977). Mothers' questions to three year old children with different language abilities. Master's thesis, Memphis State University.

Robinson, R. (1987). The causes of language disorder: Introduction and overview. *Proceedings of the First International Symposium on Specific Speech and Language Disorders in Children* (pp. 1–19). London: Association for All Speech Impaired Children.

Roediger, H. (1980). Memory metaphors in cognitive psychology. *Memory and Cognition, 8,* 231–246.

Rollins, P. (1995). MLU as a matching variable: Understanding its limitations. Poster presented at the Symposium on Research in Child Language Disorders, University of Wisconsin, Madison.

Rollins, P., Snow, C., & Willett, J. (1996). Predictors of MLU: Semantic and morphological development. *First Language, 16,* 243–259.

Rom, A., & Bliss, L. (1981). Comparison of nonverbal and verbal communicative skills of language impaired and normal speaking children. *Journal of Communication Disorders, 14,* 133–140.

Rom, A., & Bliss, L. (1983). The use of nonverbal pragmatic behaviors by language-impaired and normal-speaking children. *Journal of Communication Disorders, 14,* 251–256.

Rom, A., & Leonard, L. (1990). Interpreting deficits in grammatical morphology in specifically language-impaired children: Preliminary evidence from Hebrew. *Clinical Linguistics and Phonetics, 4,* 93–105.

Roseberry, C., & Connell, P. (1991). Use of an invented language rule in the differentiation of normal and specific language-impaired Spanish-speaking children. *Journal of Speech and Hearing Research, 34,* 596–603.

Rosenthal, W. (1972). Auditory and linguistic interaction in developmental aphasia: Evidence from two studies of auditory processing. *Papers and Reports on Child Language Development, 4,* 19–34.

Roth, F., & Clark, D. (1987). Symbolic play and social participation abilities of language-impaired and normally-developing children. *Journal of Speech and Hearing Disorders, 52,* 17–29.

Rowan, L., Leonard, L., Chapman, K., & Weiss, A. (1983). Performative and presuppositional skills in language-disordered and normal children. *Journal of Speech and Hearing Research, 26,* 97–106.

Rubin, H., & Liberman, I. (1983). Exploring the oral and written language errors made by language disabled children. *Annals of Dyslexia, 33,* 110–120.

Ruscello, D., St. Louis, K., & Mason, N. (1991). School-aged children with phonologic disorders: Co-existence with other speech/language disorders. *Journal of Speech and Hearing Research, 34,* 236–242.

Rutter, M., & Yule, W. (1975). The concept of specific reading retardation. *Journal of Child Psychology and Psychiatry, 16,* 181–197.

Sabbadini, L., Volterra, V., Leonard, L., & Campagnoli, M. G. (1987). Bambini con disturbo specifico del linguaggio: Aspetti morfologici. *Giornale di Neuropsichiatria in Età Evolutiva, 7,* 213–232.

Sahlén, B., & Nettelbladt, U. (1991). Patterns of vulnerability of language in children with severe developmental language disorders. In B. Sahlén (Ed.): *From depth to surface: A case study approach to severe developmental language disorders* (pp. 149–163). Lund: Lund University.

Salthouse, T. (1985). *A theory of cognitive aging.* Amsterdam: North Holland.

Samples, J., & Lane, V. (1985). Genetic possibilities in six siblings with specific language learning disorders, *Asha, 27* (12), 27–32.

Savich, P. (1984). Anticipatory imagery in normal and language disabled children. *Journal of Speech and Hearing Research, 27,* 494–501.

Scarborough, H. (1990). Very early language deficits in dyslexic children. *Child Development, 61,* 1728–1743.

Scarborough, H. (1991). Early syntactic development of dyslexic children. *Annals of Dyslexia, 41,* 207–220.

Scarborough, H., & Dobrich, W. (1990). Development of children with early language delay. *Journal of Speech and Hearing Research, 33,* 70–83.

Scarborough, H., Rescorla, L., Tager-Flusberg, H., Fowler, A., & Sudhalter, V. (1991). The relation of utterance length to grammatical complexity in normal and language-disordered groups. *Applied Psycholinguistics, 12,* 23–45.

Scarborough, H., Wyckoff, J., & Davidson, R. (1986). A reconsideration of the relation between age and mean utterance length. *Journal of Speech and Hearing Research, 29,* 394–399.

Schelletter, C. (1990). Pronoun use in normal and language-impaired children. *Clinical Linguistics and Phonetics, 4,* 63–75.

Schmauch, V., Panagos, J., & Klich, R. (1978). Syntax influences the accuracy of consonant production in language-disordered children. *Journal of Communication Disorders, 11,* 315–323.

Schodorf, J., & Edwards, H. (1983). Comparative analysis of parent–child interactions with language-disordered and linguistically normal children. *Journal of Communication Disorders, 16,* 71–83.

Schöler, H. (1985). Überlegungen zum Erwerb morphologischer Strukturformen bei dysgrammatisch sprechenden Kindern am Beispiel des Pluralmorphems. In I. Füssenich & B. Gläß (Eds.), *Dysgrammatismus—Theoretische und praktische Probleme bei der interdisziplinären Beschreibung gestörter Kindersprache* (pp. 165–179). Heidelberg: Schindele.

Schöler, H., & Kürsten, F. (1995). Specific language impairment: Theoretical approaches and some empirical facts. In M. Kovačević (Ed.), *Language and language communication barriers: Research and theoretical perspectives in three European languages* (pp. 7–44). Zagreb: Croatian University Press.

Schöler, H., & Moerschel, D. (1984). Anmerkungen zu einer Theorie der Sprachbehinderung im Zusammenhang mit der Entwicklung morphologischer Strukturen bei Dysgrammatikern. *Die Sprachheilarbeit, 29,* 95–109.

Schuele, C. M., Rice, M., & Wilcox, K. (1995). Redirects: A strategy to increase peer initiations. *Journal of Speech and Hearing Research, 38,* 1319–1333.

Schütze, C., & Wexler, K. (1995). English root infinitives do not license nominative subjects. Paper presented at the Boston University Conference on Language Development, Boston.

Schwartz, E., & Solot, C. (1980). Response patterns characteristic of verbal expressive disorders. *Language, Speech, and Hearing Services in Schools, 11,* 139–144.

Schwartz, R. (1988). Early action word acquisition in normal and language-impaired children. *Applied Psycholinguistics, 9,* 111–122.

Schwartz, R., Chapman, K., Terrell, B., Prelock, P., & Rowan, L. (1985). Facilitating word combination in language-impaired children through discourse structure. *Journal of Speech and Hearing Disorders, 50,* 31–39.

Schwartz, R., Leonard, L., Folger, M. K., & Wilcox, M. J. (1980). Early phonological behavior in normal-speaking and language disordered children: Evidence for a synergistic view of linguistic disorders. *Journal of Speech and Hearing Disorders, 45,* 357–377.

Schwartz, R., Leonard, L., Messick, C., & Chapman, K. (1987). The acquisition of object names in children with specific language impairment: Action context and word extension. *Applied Psycholinguistics, 8,* 233–244.

Serra-Raventós, M., & Bosch-Galceran, L. (1992). Cognitive and linguistic errors in SLI children: A new perspective from language production models. *Scandinavian Journal of Logopedics and Phoniatrics, 17,* 59–68.

Shatz, M., Bernstein, D., & Shulman, M. (1980). The responsiveness of language disordered children to indirect directives in varying contexts. *Applied Psycholinguistics, 1,* 295–306.

Sheppard, A. (1980). Monologue and dialogue speech of language-impaired children in clinic and home settings. Master's thesis, University of Western Ontario.

Sheridan, M., & Peckham, C. (1975). Follow-up at 11 years of children who had marked speech defects at 7 years. *Child: Care, Health and Development, 1,* 157–166.

Shriberg, L., Gruber, F., & Kwiatkowski, J. (1994). Developmental phonological disorders III: Long-term speech-sound normalization. *Journal of Speech and Hearing Research, 37,* 1151–1177.

Shriberg, L., & Kwiatkowski, J. (1994). Developmental phonological disorders I: A clinical profile. *Journal of Speech and Hearing Research, 37,* 1100–1126.

Shriberg, L., Kwiatkowski, J., Best, S., Hengst, J., & Terselic-Weber, B. (1986). Characteristics of children with phonologic disorders of unknown origin. *Journal of Speech and Hearing Disorders, 51,* 140–161.

Shriner, T. (1967). A comparison of selected measures with psychological scale values of language development. *Journal of Speech and Hearing Research, 10,* 828–835.

Shub, J., Simon, J., & Braccio, M. (1982). The development of symbolic play in language-delayed children. Paper presented at the Boston University Conference on Language Development, Boston.

Siegel, G. (1962). Inter-examiner reliabilitiy for mean length of response. *Journal of Speech and Hearing Research, 5,* 91–95.

Siegel, L., Cunningham, C., & van der Spuy, H. (1979). Interactions of language delayed and normal pre-school children with their mothers. Paper presented at the Meeting of the Society for Research in Child Development, San Francisco.

Siegel, L., Lees, A., Allan, L., & Bolton, B. (1981). Nonverbal assessment of Piagetian concepts in preschool children with impaired language development. *Educational Psychology, 1,* 153–158.

Silva, P. (1987). Epidemiology, longitudinal course, and some associated factors: An update. In W. Yule & M. Rutter (Eds.), *Language development and disorders* (pp. 1–15). London: Mac Keith Press.

Silva, P., McGee, R., & Williams, S. (1985). Some characteristics of 9–year old boys with general back-wardness or specific reading retardation. *Journal of Child Psychology and Psychiatry, 26,* 407–421.

Silverman, L., & Newhoff, M. (1979). Fathers' speech to normal and language delayed children: A comparison. Paper presented at the Convention of the American Speech-Language-Hearing Association, Atlanta.

Sininger, Y., Klatsky, R., & Kirchner, D. (1989). Memory-scanning speed in language-disordered children. *Journal of Speech and Hearing Research, 32,* 289–297.

Skarakis, E. (1982). The development of symbolic play and language in language disordered children. Doctoral dissertation, University of California, Santa Barbara.

Skarakis, E., & Greenfield, P. (1982). The role of old and new information in the verbal expression of language-disordered children. *Journal of Speech and Hearing Research, 25,* 462–467.

Skarakis-Doyle, E., & Prutting, C. (1988). Characteristics of symbolic play in language disordered children. *Human Communication, 12,* 7–17.

Skarakis-Doyle, E., & Woodall, S. (1988). The effects of modeling upon the verbal elaboration of a language disordered child's pretend play. *Human Communication, 12,* 29–35.

Sleight, C., & Prinz, P. (1985). Use of abstracts, orientations, and codas in narration by language- disordered and nondisordered children. *Journal of Speech and Hearing Disorders, 50,* 361–371.

Slobin, D. (1973). Cognitive prerequisites for the development of grammar. In C. Ferguson & D. Slobin (Eds.), *Studies of child language development* (pp. 175–208). New York: Holt, Rinehart and Winston.

Slobin, D. (1985). Crosslinguistic evidence for the language-making capacity. In D. Slobin (Ed.), *The crosslinguistic study of language acquisition, 2, Theoretical issues* (pp. 1157–1249). Hillsdale, NJ: Lawrence Erlbaum.

Slobin, D., & Welch, C. (1971). Elicited imitation as a research tool in developmental psycholinguistics. In C. Lavatelli (Ed.), *Language training in early childhood education* (pp. 170–185). Urbana: University of Illinois Press.

Smit, A., & Bernthal, J. (1983). Voicing contrasts and their phonological implications in the speech of articulation-disordered children. *Journal of Speech and Hearing Research, 26,* 486–500.

Smith, K. (1992). The acquisition of long-distance *wh-* questions in normal and specifically language-impaired children. Paper presented at the Annual Meeting of the Linguistic Society of America, Philadelphia.

Smith-Lock, K. (1995). Morphological usage and awareness in children with and without specific language impairment. *Annals of Dyslexia, 45,* 163–185.

Snyder, L. (1975). Pragmatics in language disabled children: Their prelinguistic and early verbal performatives and presuppositions. Doctoral dissertation, University of Colorado, Boulder.

Snyder, L. (1978). Communicative and cognitive abilities and disabilities in the sensorimotor period. *Merrill-Palmer Quarterly, 24,* 161–180.

Snyder, L. (1982). Defining language disordered children: Disordered or just "low verbal" normal? *Proceedings from the Symposium on Research in Child Language Disorders, 3,* 197–209. Madison: University of Wisconsin, Madison.

Snyder, L., & Downey, D. (1991). The language-reading relationship in normal and reading-disabled children. *Journal of Speech and Hearing Research, 34,* 129–140.

Sommers, R., Kozarevich, M., & Michaels, C. (1994). Word skills of children normal and impaired in communication skills and measures of language and speech development. *Journal of Communication Disorders, 27,* 223–240.

Sommers, R., Logsdon, B., & Wright, J. (1992). A review and critical analysis of treatment research related to articulation and phonological disorders. *Journal of Communication Disorders, 25,* 3–22.

Sommers, R., & Taylor, M. (1972). Cerebral speech dominance in language-disordered and normal children. *Cortex, 8,* 224–232.

Sonksen, P. (1979). The neuro-developmental and paediatric findings associated with significant disabilities of language development in pre-school children. M.D. thesis, University of London.

Spencer, A. (1984). A nonlinear analysis of phonological disability. *Journal of Communication Disorders, 17,* 325–348.

Stampe, D. (1969). The acquisition of phonetic representation. Paper presented at the meeting of the Chicago Linguistic Society, Chicago.

Stampe, D. (1973). A dissertation on natural phonology. Doctoral dissertation, University of Chicago.

Stanovich, K. (1988). The right and wrong places to look for the cognitive locus of reading disability. *Annals of Dyslexia, 38,* 154–177.

Stark, J. (1967). A comparison of the performance of aphasic children on three sequencing tests. *Journal of Communication Disorders, 1,* 31–34.

Stark, J., Poppen, R., & May, M. (1967). Effects of alterations of prosodic features on the sequencing performance of aphasic children. *Journal of Speech and Hearing Research, 10,* 849–855.

Stark, R., Bernstein, L., Condino, R., Bender, M., Tallal, P., & Catts, H. (1984). Four-year follow-up study of language impaired children. *Annals of Dyslexia, 34,* 49–68.

Stark, R., & Heinz, J. (1996). Vowel perception in language impaired and language normal children. *Journal of Speech and Hearing Research, 39,* 860–869.

Stark, R., & Montgomery, J. (1995). Sentence processing in language-impaired children under conditions of filtering and time compression. *Applied Psycholinguistics, 16,* 137–154.

Stark, R., & Tallal, P. (1979). Analysis of stop consonant production errors in developmentally dysphasic children. *Journal of the Acoustical Society of America, 66,* 1703–1712.

Stark, R., & Tallal, P. (1981). Selection of children with specific language deficits. *Journal of Speech and Hearing Disorders, 46,* 114–122.

Stark, R., & Tallal, P. (1988). *Language, speech, and reading disorders in children: Neuropsychological studies.* Boston: Little, Brown.

Steckol, K., & Leonard, L. (1979). The use of grammatical morphemes by normal and language impaired children. *Journal of Communication Disorders, 12,* 291–302.

Stein, A. (1976). A comparison of mothers' and fathers' speech to normal and language-deficient children. Paper presented at the Boston University Conference on Language Development, Boston.

Stemberger, J. (1993). Vowel dominance in overregularizations. *Journal of Child Language, 20,* 503–521.

Stevens, L., & Bliss, L. (1995). Conflict resolution abilities of children with specific language impairment and children with normal language. *Journal of Speech and Hearing Research, 38,* 599–611.

Stevenson, J., & Richman, N. (1976). The prevalence of language delay in a population of three year old children and its association with general retardation. *Developmental Medicine and Child Neurology, 18,* 431–441.

Stockman, I. (1992). Another look at semantic relational categories and language impairment. *Journal of Communication Disorders, 12,* 7–17.

Stoel-Gammon, C. (1989). Prespeech and early speech development of two late talkers. *First Language, 9,* 207–224.

Stoel-Gammon, C., & Herrington, P. (1990). Vowel systems of normally developing and phonologically disordered children. *Clinical Linguistics and Phonetics, 4,* 145–160.

Strominger, A., & Bashir, A. (1977). Longitudinal study of language delayed children. Paper presented at the Convention of the American Speech-Language-Hearing Association, Chicago.

Strong, C., & Shaver, J. (1991). Stability of cohesion in the spoken narratives of language-impaired and normally developing school-aged children. *Journal of Speech and Hearing Research, 34,* 95–111.

Studdert-Kennedy, M., & Mody, M. (1995). Auditory temporal perception deficits in the reading-impaired: A critical review. *Psychonomic Bulletin and Review, 2,* 508–514.

Sturn, A., & Johnston, J. (1993). Thinking out loud: The problem solving language of preschoolers with and without language impairment. Paper presented at the Symposium on Research on Child Language Disorders, University of Wisconsin, Madison.

Sussman, J. (1993). Perception of formant transition cues to place of articulation in children with language impairments. *Journal of Speech and Hearing Research, 36,* 1286–1299.

Swanson, L., & Leonard, L. (1994). Duration of function-word vowels in mothers' speech to young children. *Journal of Speech and Hearing Research, 37,* 1394–1405.

Swanson, L., Leonard, L., & Gandour, J. (1992). Vowel duration in mothers' speech to young children. *Journal of Speech and Hearing Research, 35,* 617–625.

Swisher, L., Plante, E., & Lowell, S. (1994). Nonlinguistic deficits of children with language disorders complicate the interpretation of their nonverbal IQ scores. *Language, Speech, and Hearing Services in Schools, 25,* 235–240.

Swisher, L., Restrepo, M. A., Plante, E., & Lowell, S. (1995). Effect of implicit and explicit "rule" presentation on bound-morpheme generalization in specific language impairment. *Journal of Speech and Hearing Research, 38,* 168–173.

Swisher, L., & Snow, D. (1994). Learning and generalization components of morphological acquisition by children with specific language impairment: Is there a functional relation? *Journal of Speech and Hearing Research, 37,* 1406–1413.

Tallal, P. (1975). Perceptual and linguistic factors in the language impairment of developmental dysphasics: An experimental investigation with the Token Test. *Cortex, 11,* 196–205.

Tallal, P. (1976). Rapid auditory processing in normal and disordered language development. *Journal of Speech and Hearing Research, 19,* 561–571.

Tallal, P. (1980). Auditory temporal perception, phonics, and reading abilities in children. *Brain and Language, 9,* 182–198.

Tallal, P. (1989). Unexpected sex-ratios in families of language/learning-impaired children. *Neuropsychologia, 27,* 987–998.

Tallal, P. (1991). Hormone influences in developmental learning disabilities. *Psychoneuroendocrinology, 16,* 203–211.

Tallal, P., Curtiss, S., & Kaplan, R. (1988). The San Diego Longitudinal Study: Evaluating the outcomes of preschool impairments in language development. In S. Gerber & G. Mencher (Eds.), *International perspectives on communication disorders* (pp. 86–126). Washington, DC: Gallaudet University Press.

Tallal, P., Dukette, K., & Curtiss, S. (1989). Behavioral/emotional profiles of preschool language-impaired children. *Development and Psychopathology, 1,* 51–67.

Tallal, P., Miller, S., Bedi, G., Byma, G., Wang, X., Nagarajan, S., Schreiner, C., Jenkins, W., & Merzenich, M. (1996). Language comprehension in language-learning impaired children improved with acoustically modified speech. *Science, 271*, 81–84.

Tallal, P., & Piercy, M. (1973a). Defects of non-verbal auditory perception in children with developmental aphasia. *Nature, 241*, 468–469.

Tallal, P., & Piercy, M. (1973b). Developmental aphasia: Impaired rate of non-verbal processing as a function of sensory modality. *Neuropsychologia, 11*, 389–398.

Tallal, P., & Piercy, M. (1974). Developmental aphasia: Rate of auditory processing and selective impairment of consonant perception. *Neuropsychologia, 12*, 83–93.

Tallal, P., & Piercy, M. (1975). Developmental aphasia: The perception of brief vowels and extended stop consonants. *Neuropsychologia, 13*, 69–74.

Tallal, P., Ross, R., & Curtiss, S. (1989a). Familial aggregation in specific language impairment. *Journal of Speech and Hearing Disorders, 54*, 167–173.

Tallal, P., Ross, R., & Curtiss, S. (1989b). Unexpected sex-ratios in families of language/learning-impaired children. *Neuropsychologia, 27*, 987–998.

Tallal, P., & Stark, R. (1981). Speech acoustic cue discrimination abilities of normally developing and language impaired children. *Journal of the Acoustical Society of America, 69*, 568–574.

Tallal, P., Stark, R., & Curtiss, B. (1976). Relation between speech perception and speech production impairment in children with developmental dysphasia. *Brain and Language, 3*, 305–317.

Tallal, P., Stark, R., Kallman, C., & Mellits, D. (1980a). Perceptual constancy for phonemic categories: A developmental study with normal and language impaired children. *Applied Psycholinguistics, 1*, 49–64.

Tallal, P., Stark, R., Kallman, C., & Mellits, D. (1980b). Developmental aphasia: The relation between acoustic processing deficits and verbal processing. *Neuropsychologia, 18*, 273–284.

Tallal, P., Stark, R., Kallman, C., & Mellits, D. (1981). A reexamination of some nonverbal perceptual abilities of language-impaired and normal children as a function of age and sensory modality. *Journal of Speech and Hearing Research, 24*, 351–357.

Tallal, P., Stark, R., & Mellits, D. (1985a). Identification of language impaired children on the basis of rapid perception and production skills. *Brain and Language, 25*, 314–322.

Tallal, P., Stark, R., & Mellits, D. (1985b). The relationship between auditory temporal analysis and receptive language development: Evidence from studies of developmental language disorders. *Neuropsychologia, 23*, 527–534.

Tallal, P., Townsend, J., Curtiss, S., & Wulfeck, B. (1991). Phenotypic profiles of language-impaired children based on genetic/family history. *Brain and Language, 41*, 81–95.

Taylor, C. (1995). The not so generalised slowing hypothesis. *University of Reading Working Papers in Linguistics, 2*, 87–108.

Templin, M. (1957). *Certain language skills in children*. Minneapolis: University of Minnesota Press.

Terman, L., & Merrill, M. (1960). *Stanford-Binet intelligence scale*. Boston: Houghton Mifflin.

Terrell, B., & Schwartz, R. (1988). Object transformations in the play of language-impaired children. *Journal of Speech and Hearing Disorders, 53*, 459–466.

Terrell, B., Schwartz, R., Prelock, P., & Messick, C. (1984). Symbolic play in normal and language-impaired children. *Journal of Speech and Hearing Research, 27*, 424–429.

Thal, D., & Barone, P. (1983). Auditory processing and language impairment in children: Stimulus considerations for intervention. *Journal of Speech and Hearing Disorders, 48*, 18–24.

Thal, D., & Bates, E. (1988). Language and gesture in late talkers. *Journal of Speech and Hearing Research, 31*, 115–123.

Thal, D., & Goldenberg, T. (1981). Programming diversity of response: A method for teaching flexibility of language use. *Journal of Childhood Communication Disorders, 5*, 54–65.

Thal, D., Oroz, M., & McCaw, V. (1995). Phonological and lexical development in normal and late-talking toddlers. *Applied Psycholinguistics, 16*, 407–424.

Thal, D., & Tobias, S. (1992). Communicative gestures in children with delayed onset of oral expressive vocabulary. *Journal of Speech and Hearing Research, 35*, 1281–1289.

Thal, D., & Tobias, S. (1994). Relationships between language and gesture in normally developing and late-talking toddlers. *Journal of Speech and Hearing Research, 37*, 157–170.

Thal, D., Tobias, S., & Morrison, D. (1991). Language and gesture in late talkers: A one-year follow-up. *Journal of Speech and Hearing Research, 34*, 604–612.

Thibodeau, L., & Sussman, H. (1979). Performance on a test of categorical perception of speech in normal and communicatively disordered children. *Journal of Phonetics, 7*, 375–391.

Tomblin, J. B. (1983). An examination of the concept of disorder in the study of language variation. *Proceedings from the Symposium on Research in Child Language Disorders*, 4, 81–109. Madison: University of Wisconsin, Madison.

Tomblin, J. B. (1989). Familial concentration of developmental language impairment. *Journal of Speech and Hearing Disorders*, 54, 287–295.

Tomblin, J. B. (1991). Examining the cause of specific language impairment. *Language, Speech and Hearing Services in Schools*, 22, 69–74.

Tomblin, J. B. (1996a). Genetic and environmental contributions to the risk for specific language impairment. In M. Rice (Ed.), *Toward a genetics of language* (pp. 191–210). Hillsdale, NJ: Lawrence Erlbaum.

Tomblin, J. B. (1996b). The big picture of SLI: Results of an epidemiologic study of SLI among kindergarten children. Paper presented at the Symposium on Research in Child Language Disorders, University of Wisconsin, Madison.

Tomblin, J. B., Abbas, P., Records, N., & Brenneman, L. (1995). Auditory evoked responses to frequency-modulated tones in children with specific language impairment. *Journal of Speech and Hearing Research*, 38, 387–393.

Tomblin, J. B., & Buckwalter, P. (1994). Studies of genetics of specific language impairment. In R. Watkins & M. Rice (Eds.), *Specific language impairments in children* (pp. 17–35). Baltimore: Paul H. Brookes.

Tomblin, J. B., Freese, P., & Records, N. (1992). Diagnosing specific language impairment in adults for the purpose of pedigree analysis. *Journal of Speech and Hearing Research*, 35, 832–843.

Tomblin, J. B., & Quinn, M. (1983). The contribution of perceptual learning to performance on the repetition task. *Journal of Speech and Hearing Research*, 26, 369–372.

Tower, D. (1979). Forward. In C. Ludlow & M. Doran-Quine (Eds.), *The neurological bases of language disorders in children: Methods and directions for research* (pp. vii–viii). Bethesda, MD: National Institutes of Health.

Town, C. (1911). Congenital aphasia. *Psychological Clinic*, 5, 167.

Townsend, J., Wulfeck, B., Nichols, S., & Koch, L. (1995). Attentional deficits in children with developmental language disorder. Technical Report CND-9503. Center for Research in Language, University of California at San Diego.

Trantham, C., & Pedersen, J. (1976). *Normal language development*. Baltimore: Williams & Wilkins.

Trauner, D., Wulfeck, B., Tallal, P., & Hesselink, J. (1995). Neurologic and MRI profiles of language impaired children. Technical Report CND-9513, Center for Research in Language, University of California at San Diego.

Treitel, L. (1893). Über Aphasie im Kindesalter. *Sammlung Klinischer Vorträge*, 64, 629–654.

Tyler, A. (1995). Durational analysis of stridency errors in children with phonological impairment. *Clinical Linguistics and Phonetics*, 9, 211–228.

Tyler, A., Edwards, M. L., & Saxman, J. (1990). Acoustic validation of phonological knowledge and its relationship to treatment. *Journal of Speech and Hearing Disorders*, 55, 251–261.

Uchermann, V. (1891). Drei Fälle von Stummheit (Aphasie). *Zeitschrift für Ohrenheilkunde*, 21, 313–322.

Udwin, O., & Yule, W. (1983). Imaginative play in language disordered children. *British Journal of Disorders of Communication*, 18, 197–205.

Ullman, M., & Gopnik, M. (1994). Past tense production: Regular, irregular and nonsense verbs. *McGill Working Papers in Linguistics*, 10, 81–118.

United States Department of Health and Human Services (1995). *The international classification of diseases, ninth revision: Clinical modification*. DHHS Publicaton no. (PHS) 80–1260. Washington, DC: United States Government Printing Office.

Väisse, L. (1866). Des sourds-muets et de certains cas d'aphasie congénitale. *Bulletin de la Société d'Anthropologie de Paris*, 1, 146–150.

Valian, V. (1986). Syntactic categories in the speech of young children. *Developmental Psychology*, 22, 562–579.

van der Lely, H. (1994). Canonical linking rules: Forward versus reverse linking in normally developing and specifically language-impaired children. *Cognition*, 51, 29–72.

van der Lely, H. (1996). Specifically language impaired and normally developing children: Verbal passive vs. adjectival passive interpretation. *Lingua*, 98, 243–272.

van der Lely, H., & Harris, M. (1990). Comprehension of reversible sentences in specifically language impaired children. *Journal of Speech and Hearing Disorders*, 55, 101–117.

van der Lely, H., & Howard, D. (1993). Children with specific language impairment: Linguistic impairment or short-term memory deficit? *Journal of Speech and Hearing Research*, 36, 1193–1207.

van der Lely, H., & Stollwerck, L. (1996). A grammatical specific language impairment in children: An autosomal dominant inheritance? *Brain and Language, 52*, 484–504.

van der Lely, H., & Ullman, M. (1995). The computation and representation of past-tense morphology in specifically language impaired and normally developing children. Paper presented at the Boston University Conference on Language Development, Boston.

Van Gelder, D., Kennedy, L., & Lagauite, J. (1952). Congenital and infantile aphasia: Review of literature and report of a case. *Pediatrics, 9*, 48–54.

Van Hout, A. (1989). Aspects du diagnostic des dysphasies. *ANAE, 1*, 11–15.

Van Kleeck, A., & Carpenter, R. (1978). Effects of children's language comprehension level on language addressed to them. Paper presented at the Convention of the American Speech-Language-Hearing Association, San Francisco.

Van Kleeck, A., & Frankel, T. (1981). Discourse devices used by language disordered children: A preliminary investigation. *Journal of Speech and Hearing Disorders, 46*, 250–257.

Vargha-Khadem, F., Watkins, K., Alcock, K., Fletcher, P., & Passingham, R. (1995). Praxic and nonverbal cognitive deficits in a large family with a genetically transmitted speech and language disorder. *Proceedings of the National Academy of Sciences, 92*, 930–933.

Veit, S. (1986). Das Verständnis von Plural- und Komparativformen bei (entwicklungs)dysgrammatischen Kindern im Vorschulalter. In G. Kegel (Ed.), *Sprechwissenschaft und Psycholinguistik* (pp. 217–286). Opladen: Verlag.

Vellutino, F. (1979). *Dyslexia: Theory and research*. Cambridge, MA: MIT Press.

Vinkler, Z., & Pléh, C. (1995). A case of a specific language impaired child in Hungarian. In M. Kovačević (Ed.), *Language and language communication barriers: Research and theoretical perspectives in three European languages* (pp. 131–158). Zagreb: Croatian University Press.

Waldenburg, L. (1873). Ein Fall von angeborener Aphasie. *Berliner Klinische Wochenschrift, 10*, 8–9.

Warren, S., & Kaiser, S. (1986). Generalization of treatment effects by young language-delayed children: A longitudinal analysis. *Journal of Speech and Hearing Disorders, 51*, 239–251.

Warren, S., McQuarter, R., & Rogers-Warren, A. (1984). The effects of mands and models on the speech of unresponsive language-delayed preschool children. *Journal of Speech and Hearing Disorders, 49*, 43–52.

Watkins, R., & Rice, M. (1991). Verb particle and preposition acquisition in language-impaired preschoolers. *Journal of Speech and Hearing Research, 34*, 1130–1141.

Watkins, R., Rice, M., & Molz, C. (1993). Verb use by language-impaired and normally developing children. *First Language, 37*, 133–143.

Watson, L. (1977). Conversational participation by language deficient and normal children. Paper presented at the Convention of the American Speech-Language-Hearing Association, Chicago.

Webster, B., & Ingram, D. (1972). The comprehension and production of the anaphoric pronouns "he, she, him, her" in normal and linguistically deviant children. *Papers and Reports on Child Language Development, 4*, 55–79.

Wechsler adult intelligence scale—revised (1981). New York: The Psychological Corporation.

Wechsler intelligence scale for children—revised (1974). New York: The Psychological Corporation.

Wechsler preschool and primary scale of intelligence—revised (1989). New York: The Psychological Corporation.

Weeks, T. (1974). *The slow speech development of a bright child*. Lexington, MA: D. C. Heath.

Weeks, T. (1975). The use of nonverbal communication by a slow speech developer. *Word, 27*, 460–472.

Weiner, F. (1981). Systematic sound preference as a characteristic of phonological disability. *Journal of Speech and Hearing Disorders, 46*, 281–286.

Weiner, P. (1969). The perceptual level functioning of dysphasic children. *Cortex, 5*, 440–457.

Weiner, P. (1972). The perceptual level functioning of dysphasic children: A follow-up study. *Journal of Speech and Hearing Research, 15*, 423–438.

Weiner, P. (1974). A language-delayed child at adolescence. *Journal of Speech and Hearing Disorders, 39*, 202–212.

Weiner, P. (1986). The study of childhood language disorders: Nineteenth century perspectives. *Journal of Communication Disorders, 19*, 1–47.

Weinert, S. (1992). Deficits in acquiring language structure: The importance of using prosodic cues. *Applied Cognitive Psychology, 6*, 545–571.

Weinert, S., Grimm, H., Delille, G., & Scholten-Zitzewitz, R. (1989). Was macht sprachgestörten Kindern das Textverstehen so schwer? *Heilpädagogische Forschung, Themenheft: Sprachentwicklungsprobleme/Leseprobleme, 15*, 25–37.

Weismer, G. (1984). Acoustic analysis strategies for the refinement of phonological analysis. *ASHA Monographs*, no. 22, 30–52.

Weismer, G., Dinnsen, D., & Elbert, M. (1981). A study of the voicing distinction associated with omitted, word-final stops. *Journal of Speech and Hearing Disorders, 46*, 320–327.

Weiss, A., Leonard, L., Rowan, L., & Chapman, K. (1983). Linguistic and non-linguistic features of style in normal and language-impaired children. *Journal of Speech and Hearing Disorders, 48*, 154–164.

Wellen, C., & Broen, P. (1982). The interruption of young children's responses by older siblings. *Journal of Speech and Hearing Disorders, 47*, 204–210.

Weller, C. (1979). Training approaches on syntactic skills of language-deviant children. *Journal of Learning Disabilities, 12*, 46–55.

Werner, L. (1945). Treatment of a child with delayed speech. *Journal of Speech Disorders, 10*, 329–334.

Wexler, K. (1994). Optional infinitives. In D. Lightfoot & N. Hornstein (Eds.), *Verb movement* (pp. 305–350). New York: Cambridge University Press.

Whitehurst, G., Arnold, D., Smith, M., Fischel, J., Lonigan, C., & Valdez-Menchaca, M. (1991). Family history in developmental expressive language delay. *Journal of Speech and Hearing Research, 34*, 1150–1157.

Whitehurst, G., Fischel, J., Arnold, D., & Lonigan, C. (1992). Evaluating outcomes with children with expressive language delay. In S. Warren & J. Reichle (Eds.), *Causes and effects in communication and language intervention* (pp. 277–313). Baltimore: Paul H. Brookes.

Whitehurst, G., Fischel, J., Caulfield, M., DeBaryshe, B., & Valdez-Menchaca, M. (1989). Assessment and treatment of early expressive language delay. In P. Aelazo & R. Barr (Eds.), *Challenges to developmental paradigms: Implications for assessment and treatment* (pp. 113–135). Hillsdale, NJ: Lawrence Erlbaum.

Whitehurst, G., Fischel, J., Lonigan, C., Valdez-Menchaca, M., Arnold, D., & Smith, M. (1991). Treatment of early expressive language delay: If, when, and how. *Topics in Language Disorders, 11* (4), 55–68.

Whitehurst, G., Fischel, J., Lonigan, C., Valdez-Menchaca, M., DeBaryshe, B., & Caulfield, M. (1988). Verbal interaction in families of normal and expressive-language-delayed children. *Developmental Psychology, 24*, 690–699.

Whitehurst, G., Novak, G., & Zorn, G. (1972). Delayed speech studied in the home. *Developmental Psychology, 7*, 169–177.

Whitehurst, G., Smith, M., Fischel, J., Arnold, D., & Lonigan, C. (1991). The continuity of babble and speech in children with expressive language delay. *Journal of Speech and Hearing Research, 34*, 1121–1135.

Wiig, E., Semel, E., & Nystrom, L. (1982). Comparison of rapid naming abilities in language-learning-disabled and academically achieving eight-year-olds. *Language, Speech, and Hearing Services in Schools, 13*, 11–23.

Wilcox, M. J., & Leonard, L. (1978). Experimental acquisition of wh- questions in language-disordered children. *Journal of Speech and Hearing Research, 21*, 220–239.

Wilde, W. (1853). *Practical observations on aural surgery and the nature of treatment of diseases of the ear.* Philadelphia: Blanchard & Lea.

Willbrand, M., & Kleinschmidt, M. (1978). Substitution patterns and word constraints. *Language, Speech, and Hearing Services in Schools, 9*, 155–161.

Williams, R. (1978). Play behavior of language-handicapped and normal-speaking preschool children. Paper presented at the Convention of the American Speech-Language-Hearing Association, San Francisco.

Wilson, B., & Risucci, D. (1986). A model for clinical-quantitative classification. Generation I: Application to language-disordered preschool children. *Brain and Language, 27*, 281–309.

Wilson, B., & Risucci, D. (1988). The early identification of developmental language disorders and the prediction of the acquisition of reading skills. In R. Masland & M. Masland (Eds.), *Preschool prevention of reading failure* (pp. 187–203). Parkton, MD: York Press.

Wing, C. (1990). A preliminary investigation of generalization to untrained words following two treatments of children's word-finding problems. *Language, Speech, and Hearing Services in Schools, 21*, 151–156.

Witelson, S., & Rabinovich, R. (1972). Hemispheric speech lateralization in children with auditory linguistic deficits. *Cortex, 8*, 412–426.

Wolf, M. (1982). The word-retrieval process and reading in children and aphasics. In K. E. Nelson (Ed.), *Children's Language, 3*, 437–493. Hillsdale, NJ: Lawrence Erlbaum.

Wolff, P., Michel, G., & Ovrut, M. (1990). Rate variables and automatized naming in developmental dyslexia. *Brain and Language, 39*, 556–575.

Wolfus, B., Moscovitch, M., & Kinsbourne, M. (1980). Subgroups of developmental language impairment. *Brain and Language, 10*, 152–171.

Wood, P. (1980). Appreciating the consequences of disease: The classification of impairments, disabilities, and handicaps. *World Health Organization Chronicle, 34*, 376–380.

Worster-Drought, C., & Allen, I. (1929). Congenital auditory imperception (congenital word deafness): With report of a case. *Journal of Neurology and Psychopathology, 9*, 193–208.

Wren, C. (1980). Identifying patterns of syntactic disorder in six-year-old children. *Proceedings from the Symposium on Research in Child Language Disorders, 1*, 113–123. Madison: University of Wisconsin, Madison.

Wright, S. (1993). Teaching word-finding strategies to severely language-impaired children. *European Journal of Disorders of Communication, 28*, 165–175.

Wulbert, M., Inglis, S., Kriegsmann, E., & Mills, B. (1975). Language delay and associated mother–child interactions. *Developmental Psychology, 11*, 61–70.

Wulfeck, B., & Bates, E. (1995). Grammatical sensitivity in children with language impairment. Technical Report CND-9512. Center for Research in Language, University of California at San Diego.

Wyke, M. (Ed.) (1978). *Developmental dysphasia*. London: Academic Press.

Wyke, M., & Asso, D. (1979). Perception and memory for spatial relations in children with developmental dysphasia. *Neuropsychologia, 17*, 231–239.

Wyllie, J. (1894). *The disorders of speech*. Edinburgh: Oliver & Boyd.

Yoder, P. (1989). Maternal question use predicts later language development in specific language disordered children. *Journal of Speech and Hearing Disorders, 54*, 347–355.

Yoder, P., Kaiser, A., & Alpert, C. (1991). An exploratory study of the interaction between language teaching methods and child characteristics. *Journal of Speech and Hearing Research, 34*, 155–167.

Yule, W., & Rutter, M. (1976). Epidemiology and social implications of specific reading retardation. In R. Knight & D. Bakker (Eds.), *The neuropsychology of learning disorders* (pp. 25–39). Baltimore: University Park Press.

Zardini, G., Battaini, S., Vender, C., & D'Angelo, A. (1985). Un test di ripetizione di frasi. Analisi delle performances di un gruppo di bambini con disfasia di evoluzione. *Giornale di Neuropsichiatria in Età Evolutiva, 5*, 235–244.

Zwitman, D., & Sonderman, J. (1979). A syntax program designed to present base linguistic structures to language-disordered children. *Journal of Communication Disorders, 12*, 323–337.

Name Index

Subject Index